INTERNAL COMBUSTION PROCESSES OF LIQUID ROCKET ENGINES

INTERNAL COMBUSTION PROCESSES OF LIQUID ROCKET ENGINES

MODELING AND NUMERICAL SIMULATIONS

Zhen-Guo Wang

National University of Defense Technology, Changsha, China

National Defense Industry Press

Contents

Preface   **x**

1 Introduction   **1**
  1.1   Basic Configuration of Liquid Rocket Engines   2
       1.1.1   Propellant Feed System   2
       1.1.2   Thrust Chamber   6
  1.2   Internal Combustion Processes of Liquid Rocket Engines   13
       1.2.1   Start and Shutdown   13
       1.2.2   Combustion Process   15
       1.2.3   Performance Parameters in Working Process   18
  1.3   Characteristics and Development History of Numerical Simulation of the Combustion Process in Liquid Rocket Engines   19
       1.3.1   Benefits of Numerical Simulation of the Combustion Process in Liquid Rocket Engines   19
       1.3.2   Main Contents of Numerical Simulations of Liquid Rocket Engine Operating Process   19
       1.3.3   Development of Numerical Simulations of Combustion Process in Liquid Rocket Engines   21
  1.4   Governing Equations of Chemical Fluid Dynamics   22
  1.5   Outline of this Book   24
  References   25

2 Physical Mechanism and Numerical Modeling of Liquid Propellant Atomization   **26**
  2.1   Types and Functions of Injectors in a Liquid Rocket Engine   27
  2.2   Atomization Mechanism of Liquid Propellant   28
       2.2.1   Formation of Static Liquid Droplet   28
       2.2.2   Breakup of Cylindrical Liquid Jet   29
       2.2.3   Liquid Sheet Breakup   36

      2.2.4   *Droplet Secondary Breakup*   43
  2.3 Characteristics of Atomization in Liquid Rocket Engines   48
      2.3.1   *Distribution Function of the Droplet Size*   51
      2.3.2   *Mean Diameter and Characteristic Diameter*   53
      2.3.3   *Measurement of Spray Size Distribution*   55
  2.4 Atomization Modeling for Liquid Rocket Engine Atomizers   59
      2.4.1   *Straight-flow Injector*   60
      2.4.2   *Centrifugal Injector*   60
      2.4.3   *Impinging-stream Injectors*   64
      2.4.4   *Coaxial Shear Injector*   70
      2.4.5   *Coaxial Centrifugal Injectors*   70
  2.5 Numerical Simulation of Liquid Propellant Atomization   75
      2.5.1   *Theoretical Models of Liquid Propellant Atomization*   75
      2.5.2   *Quasi-fluid Models*   80
      2.5.3   *Particle Trajectory Models*   81
      2.5.4   *Simulation of Liquid Jet Atomization Using Interface Tracking Method*   85
      2.5.5   *Liquid Jet Structure – Varying Flow Conditions*   91
  References   94

3 Modeling of Droplet Evaporation and Combustion   **97**
  3.1 Theory for Quasi-Steady Evaporation and Combustion of a Single Droplet at
      Atmospheric Pressure   97
      3.1.1   *Quasi-Steady Evaporation Theory for Single Droplet in the Static*
             Gas without Combustion   98
      3.1.2   *Quasi-Steady Evaporation Theory for Droplet in a Static Gas with*
             Combustion   103
      3.1.3   *Non-Combustion Evaporation Theory for a Droplet in a*
             Convective Flow   107
      3.1.4   *Evaporation Theory for a Droplet in a Convective Medium with*
             Combustion   108
  3.2 Evaporation Model for a Single Droplet under High Pressure   109
      3.2.1   *ZKS Droplet High Pressure Evaporation Theory*   110
      3.2.2   *Application of the Liquid Activity Coefficient to Calculate the*
             Gas–Liquid Equilibrium at a High Pressure   115
  3.3 Subcritical Evaporation Response Characteristics of Propellant Droplet in
      Oscillatory Environments   117
      3.3.1   *Physical Model*   118
      3.3.2   *Examples and the Analysis of Results*   120
  3.4 Multicomponent Fuel Droplet Evaporation Model   123
      3.4.1   *Simple Multicomponent Droplet Evaporation Model*   124
      3.4.2   *Continuous Thermodynamics Model of Complex Multicomponent*
             Mixture Droplet Evaporation   135
  3.5 Droplet Group Evaporation   145
      3.5.1   *Definition of Group Combustion Number*   146
      3.5.2   *Droplet Group Combustion Model*   146
  References   149

4   Modeling of Turbulence   **151**
   4.1   Turbulence Modeling in RANS   152
      4.1.1   Algebraic Model   153
      4.1.2   One-Equation Model   154
      4.1.3   Two-Equation Models   156
      4.1.4   Turbulence Model Modification   161
      4.1.5   Nonlinear Eddy Viscosity Model   165
      4.1.6   Reynolds-Stress Model   170
      4.1.7   Comments on the Models   173
   4.2   Theories and Equations of Large Eddy Simulation   174
      4.2.1   Philosophy behind LES   174
      4.2.2   LES Governing Equations   175
      4.2.3   Subgrid-Scale Model   176
      4.2.4   Hybrid RANS/LES Methods   182
   4.3   Two-Phase Turbulence Model   187
      4.3.1   Hinze–Tchen Algebraic Model for Particle Turbulence   187
      4.3.2   Two-Phase Turbulence Model $k\text{-}\varepsilon\text{-}k_p$ *and* $k\text{-}\varepsilon\text{-}A_p$ 188
   References   189

5   Turbulent Combustion Model   **192**
   5.1   Average of Chemical Reaction Term   192
   5.2   Presumed PDF—Fast Chemistry Model for Diffusion Flame   194
      5.2.1   Concepts and Assumptions   195
      5.2.2 $\kappa-\varepsilon-Z-g$ Equations   197
      5.2.3   Probability Density Distribution Function   197
      5.2.4   Presumed PDF   198
      5.2.5   Truncated Gaussian PDF   200
   5.3   Finite Rate EBU—Arrhenius Model for Premixed Flames   201
   5.4   Moment-Equation Model   202
      5.4.1   Time-Averaged Chemical Reaction Rate   203
      5.4.2   Closure for the Moments   203
   5.5   Flamelet Model for Turbulent Combustion   204
      5.5.1   Diffusion Flamelet Model   205
      5.5.2   Premixed Flamelet Model   206
   5.6   Transported PDF Method for Turbulent Combustion   208
      5.6.1   Transport Equations of the Probability Density Function   208
      5.6.2   The Closure Problem of Turbulence PDF Equation   211
      5.6.3   Transport Equation for the Single-Point Joint PDF with
          Density-Weighted Average   212
      5.6.4   Solution Algorithm for the Transport Equation of Probability
          Density Function   212
   5.7   Large Eddy Simulation of Turbulent Combustion   214
      5.7.1   Governing Equations of Large Eddy Simulation for
          Turbulent Combustion   214
      5.7.2   Sub-Grid Scale Combustion Models   218
   References   226

6   Heat Transfer Modeling and Simulation   **228**
  6.1   Convective Heat Transfer Model of Combustor Wall   228
     6.1.1   *Model of Gas Convection Heat*   229
     6.1.2   *Convection Cooling Model*   232
  6.2   Heat Conduction Model of Combustor Wall   235
     6.2.1   *Fourier Heat Conduction Law*   235
     6.2.2   *1D Steady Heat Conduction*   235
     6.2.3   *2D Steady Heat Conduction*   237
     6.2.4   *Unsteady Heat Conduction*   237
  6.3   Radiation Heat Transfer Model   238
     6.3.1   *Basic Law of Radiation*   238
     6.3.2   *Empirical Model of Radiation Heat Flux Density Calculation*   245
     6.3.3   *Numerical Simulation of Combustion Heat Radiation*   246
  References   254

7   The Model of Combustion Instability   **255**
  7.1   Overview   255
     7.1.1   *Behavior of Combustion Instability*   256
     7.1.2   *Classification of Combustion Instability*   257
     7.1.3   *Characteristics of Combustion Instability*   259
  7.2   Acoustic Basis of Combustion Instability   260
     7.2.1   *Rayleigh Criterion for Acoustic Oscillations Arising from*
            Heat or Mass Supply   260
     7.2.2   *Acoustic and Acoustic Oscillations*   261
     7.2.3   *Acoustic Modes in the Combustion Chamber*   263
     7.2.4   *Self-Excited Oscillations in Rocket Engines*   267
  7.3   Response Characteristics of Combustion Process in Liquid Rocket Engines   269
     7.3.1   *Response Characteristics of the Propellant Supply System*   269
     7.3.2   *Response Characteristics of Spray Atomization Process* 271
     7.3.3   *Response Characteristics of Droplet Evaporation Process*   272
  7.4   Sensitive Time Delay Model $n-\tau$ 272
     7.4.1   *Combustion Time Delay*   272
     7.4.2   *Sensitive Time Delay Model*   273
  7.5   Nonlinear Theory for Combustion Stability in Liquid Rocket Engines   283
     7.5.1   *Nonlinear Field Oscillator Model*   286
     7.5.2   *Continuous Stirred Tank Reactor Acoustic Model*   287
     7.5.3   *Spatio-Temporal Interaction Dynamic Model*   291
     7.5.4   *General Thermodynamic Analysis of Combustion Instability*   293
  7.6   Control of Unstable Combustion   295
     7.6.1   *Passive Control*   295
     7.6.2   *Active Control*   297
     7.6.3   *A Third Control Method*   298
  References   300

8   Numerical Method and Simulations of Liquid Rocket Engine
      Combustion Process   **302**
      8.1   Governing Equations of Two-Phase Multicomponent Reaction Flows   302
            8.1.1   Gas Phase Governing Equation   303
            8.1.2   Liquid Particle Trajectory Model   305
            8.1.3   Turbulence Model   308
            8.1.4   Droplets Atomizing Model   309
            8.1.5   Droplet Evaporation Model   311
            8.1.6   Chemical Reaction Kinetics Model   313
      8.2   Numerical Methodology   314
            8.2.1   Overview   314
            8.2.2   The Commonly-Used Discretization Scheme   315
            8.2.3   Discrete Equations   320
            8.2.4   Discretization of the Momentum Equation Based on the
                    Staggered Grid   323
            8.2.5   The SIMPLE Algorithm of Flow Field Computing   326
            8.2.6   PISO Algorithm   329
      8.3   Grid Generation Techniques   334
            8.3.1   Structured Grid Generation Technology   334
            8.3.2   Unstructured Mesh Generation Techniques   338
      8.4   Simulations of Combustion in Liquid Rocket Engines and Results Analysis   340
            8.4.1   Numerical Analysis of Dual-States Hydrogen Engine Combustion
                    and Heat Transfer Processes   340
            8.4.2   Numerical Heat Transfer Simulation of a Three-Component
                    Thrust Chamber   349
            8.4.3   Numerical Simulation of Liquid Rocket Engine Combustion
                    Stability   356
      References   376

Index   **377**

Preface

Liquid rocket engines are the main propulsion system for a spacecraft. The widespread applications of liquid rocket engines in the future demands further studies of combustion mechanisms in liquid rocket engines to improve their performance. Numerical modeling of the combustion process can improve our understanding of the incorporated physical mechanism and help in the design of liquid rocket engines. Since the 1970s, numerical simulations of combustion in liquid rocket engines have developed into a new interdisciplinary subject involving computational fluid dynamics, computational heat transfer, computational combustion, software design, and flow visualization. Owing to its significance in engine design, this new subject has attracted many researchers. With the rapid development of computer techniques and numerical methods, numerical modeling and simulations of atomization and combustion in liquid rocket engines will become an ever important research area.

The author has dedicated himself to the area of Aeronautical and Astronautical Science and technology since the 1980s. The present book is based on the teaching and supervision of undergraduate and postgraduate students in the past 30 years. The book highlights the advanced research work in the field of combustion modeling in liquid rocket engines, such as liquid propellant atomization, evaporation of liquid droplets, turbulent flows, turbulent combustion, heat transfer, and combustion instability. All these will contribute to our understanding of the combustion mechanism and to the improvement of combustion modeling, facilitating numerical simulations of combustion process in liquid fuelled engines.

The book consists of eight chapters. Chapter 1 describes the configuration and fundamentals of liquid rocket engines, and presents an overview of numerical simulations of combustion in liquid rocket engines. Chapters 2–7 detail the modeling of combustion sub processes in liquid rocket engines, i.e., atomization modeling, evaporation modeling, turbulence modeling, combustion modeling, heat transfer modeling, and combustion instability modeling. Chapter 8 presents a full description of numerical models for combustion, numerical methodology for governing equation solution, and grid generation. Finally, three applications are run to demonstrate the capability of the numerical models to predict the combustion process in liquid rocket engines.

1

Introduction

A liquid rocket engine, which is also called a liquid propellant rocket engine, is a chemical rocket engine using liquid chemicals (liquid propellant) as the energy source and the working fluid. Liquid rocket engine technology has drawn researchers' attention and been quite a hot topic in aerospace and aeronautic research during the last 70 years. In the short long history of human aviation, i.e., from the A-4 engine of the German V2 missile, to the F-1 engine of the U.S. lunar landing rocket "Saturn 5" and further to reusable space shuttle main engines, every milestone event is closely linked with the progress made in liquid rocket engine technology. Because liquid rocket engines have the characteristics of high specific impulse, repeatable starting, arbitrary working hours setting, multiple usage, adjustable thrust, etc., they are bound to occupy the dominant position in the area of aerospace propulsion long into the future.

The liquid rocket engine uses liquid fuels as the propellant. In a liquid rocket engine, the liquid chemical propellants combust in the combustion chamber and produce very high pressure gas. The gas is accelerated when it flows downstream through the nozzle and produces impulse, i.e., thrust, for the aircraft. There are several types of liquid propellants. The scheme, structure, ignition and thermal protection, etc. of the liquid rocket engine have a close relationship with the characteristics of the propellants used by the engine system.

The expansion of liquid rocket application requires more in-depth studies on the basic theory and design method of the liquid rocket engine. Numerical simulation of the combustion process in a liquid rocket engine is also an important research direction. This chapter introduces the basic configuration and working process of liquid rocket engines, and then discusses the main objective and research method of the numerical simulation of the combustion process in a liquid rocket engine.

Internal Combustion Processes of Liquid Rocket Engines: Modeling and Numerical Simulations,
First Edition. Zhen-Guo Wang.
© 2016 National Defense Industry Press. Published 2016 by John Wiley & Sons Singapore Pte Ltd.

1.1   Basic Configuration of Liquid Rocket Engines

A liquid rocket engine consists of a thrust chamber (which consists of an injector, a combustor, and a nozzle), a propellant feed system, propellant tanks and various automatic regulators, etc. This section mainly introduces the propellant feed system and the thrust chambers, which are closely associated with the combustion process.

1.1.1   Propellant Feed System

The propellant feed system is employed to deliver the propellants from the containing tanks to the thrust chamber and can be divided into two categories according to the working mode, namely, the pressure feed system and the turbo-pump feed system.

1.1.1.1   Pressure Feed System

The pressure feed system pushes the propellants to the thrust chamber or the propellant gas generator by the high pressure gas in the tanks of the propellants. The high pressure gas, i.e., the pressed gas, can be pre-stored in cylinders as the storage gas and can also be generated by a liquid or solid gas generator during the working process of the liquid rocket engine. The main requirements for the pressed gas are as follows: (i) high density while under the pressed state, (ii) low relative molecular mass under the pressed state, (iii) minor solubility with propellant, (iv) no or minor chemical reaction with the propellants, and (v) no solid and liquid impurities.

The pressure feed system can employ inert gases as the pressed gas. This kind of pressure feed system has two type of working mode, namely, the regulated pressure mode and the blow-down mode. The former employs a pressure regulator to maintain the pressure in the propellant tank, and also maintains the thrust at a constant value. The latter stores the propellant and the pressed gas in one tank. The pressure drops during the adiabatic expansion of gas, fewer propellants are injected into the combustor and therefore the pressure in the combustion chamber also drops. Typical pressure feed systems are (i) those with high-pressure gas cylinders and (ii) those with gas generators. The former can employ air, nitrogen, helium, and some other inert gas as the pressed gas. The main drawback of air is that the contained oxygen has a relatively high boiling point, and therefore it cannot be used to press cryogenic propellants. Helium can be used to press all existing liquid propellants. Although such a pressure feed system has a relatively large size and heavy mass, it has the characteristics of a simple structure and high reliability. It is also simple to employ and ensures repeatable starting of the engine.

In pressure feed systems with a gas generator, a single-component liquid fuel gas generator using a monopropellant as the source of the driven pressure and the propellant decomposition can be realized by catalysis or heating according to the kind of propellant. In dual-component liquid fuel gas generators, the high pressure gas can be obtained from the two propellant components by burning under oxygen-rich or fuel-rich conditions. The temperature of the gas is determined by the propellant component mixed ratio in the gas generator.

The structure of the pressure feed system is simple and reliable. However, as the propellant tanks must withstand high internal pressure, the pressure feed system is relatively bulky and it is often employed by spacecraft-attitude-control engines. Sometimes, to ensure the reliability of manned flight, although the engine thrust is large, a pressure feed system is also employed, such as the service module engine, drop class, and upgraded engines of Apollo spacecraft.

1.1.1.2   Turbo-Pump Feed System

A turbo-pump feed system employs pumps to deliver propellants, and the pump acquires the driven force from a turbo. In the turbo-pump feed system, a turbo-pump assembly is necessary. The basic requirements for a liquid rocket engine turbo-pump are as follows:

1. If the mass flow rate of a given propellant is given, we need to ensure the pressure at the engine outlet matches the requirement of the engine system.
2. The turbo-pump should be as small and light as possible.
3. The turbo-pump is to have as high an efficiency as possible.
4. The turbo-pump should ensure stable operation at all engine operating conditions and the pressure pulsation and mechanical vibration must be minor.
5. The turbo-pump is to be compatible with corrosive liquid and cryogenic liquids. Friction is not allowed between the components of the oxidizer pump because the heat created by the friction may produce a local high temperature, even an explosion.
6. The turbo-pump is to be capable of sucking propellants that contain a small amount of gas or steam.

There are three common types of cycle program for the turbo-pump feed system, namely, gas generator cycle, expansion cycle, and staged combustion cycle. The gas generator cycle and the staged combustion cycle can employ most of the commonly used liquid propellants. The expansion cycle engine is commonly used in an engine that employs liquid hydrogen as thrust chamber coolant, because liquid hydrogen is a good absorbing-heat medium and it does not decompose.

In the gas generator cycle, the turbine inlet gas is from an independent gas generator, turbine exhaust gas by passing through a small area ratio turbine nozzle, or by injecting in the main stream of the engine through the opening of a nozzle divergence cone. The gas generator propellant can be monopropellant or bipropellant, both of which are from the main propellant feed system. Figure 1.1 shows the bipropellant gas generator cycle turbo-pump feed system; the fuel in the pump portion is injected into the bipropellant gas generator and combust, producing working fluid to drive the turbine. To make sure that the temperature of the combustion products in the gas generator is suitable for the requirements of the turbine, we need to control the propellant mixing ratio in the gas generator, and the gas temperature should be in the range 700–900 °C. Since a bipropellant gas generator system does not require an auxiliary propellant or another tank, its structure is certainly simple, and it is widely employed in liquid rocket engines.

The gas generator cycle is relatively simple. The pressure in the fluid pipes and pumps is relatively low. Therefore, it is the most commonly used turbo-pump cycle. For an engine using the gas generator cycle, the specific impulse of the thrust chamber is slightly higher than that of the engine. However, the thrust of the thrust chamber is always slightly lower than that of the engine.

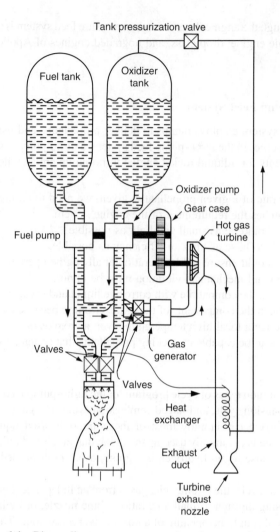

Figure 1.1   Bipropellant gas generator cycle turbo-pump feed system.

The expander cycle is typically used in an engine that employs liquid hydrogen as the fuel. An expander cycle turbo-pump feed system is shown in Figure 1.2. After absorbing heat from the cooling jacket, the liquid hydrogen becomes heated hydrogen gas. Before entering the main thrust chamber, hydrogen first drives the turbine, and then all the hydrogen gas is injected into the engine combustion chamber, mixed with the oxidant, and combusted. Further, the combustion gas is efficiently expanded in the nozzle and then exhausted. The expansion cycle has high specific impulse. Such an engine is also simple and relatively lightweight. However, the cooling jacket of the liquid hydrogen limits the amount of the absorbed heat, so that the turbine work capability is limited, thereby limiting improvement of the combustion chamber pressure. The pressure in the chamber is generally 7–8 MPa. If the chamber pressure is higher than 8 MPa, this cycle mode is not recommended.

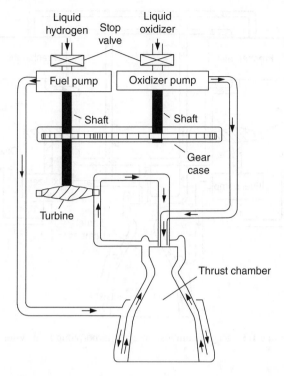

Figure 1.2 Expander cycle turbo-pump feed system.

Figure 1.3 shows a turbo-pump feed system with staged combustion cycle. In this system, the coolant and fuel firstly flow into a thrust chamber cooling jacket and then are injected into a high pressure pre-combustion chamber where the fuel combusts with part of oxygen. The combustion in the pre-combustion chamber provides the high energy gas for the turbine. After driving the turbine, the gas flows into the main combustion chamber, fully combusts with the oxygen, is exhausted, and is ejected through nozzle.

In a staged combustion cycle, the high pressure pre-combustion can be a monopropellant gas generator or bipropellant gas generator; we can adopt oxygen-rich pre-combustion, such as with the Russian RD120 engine (using liquid oxygen/kerosene propellant) and Russian RD253 engine (using nitrogen tetroxidizer/unsymmetrical dimethylhydrazine propellant). We can also employ fuel-rich pre-combustion such as used in the Space Shuttle Main Engine (using liquid hydrogen/liquid oxygen propellants). Because one of the propellant components goes entirely into the pre-combustion, the flow rate of the turbine working fluid is quite large, so that the turbine output power is greatly improved, and thus a high combustion chamber pressure is allowed to obtain high performance and reduce the size of the thrust chamber. The staged combustion engine has the highest specific impulse; however, the engine is the heaviest and most complex.

In the turbo-pump feed system, the turbine exhaust gas contains energy, and therefore it is possible to improve the specific impulse of the liquid rocket engine through using this energy. If the turbine exhaust gas flows into the liquid rocket engine combustion chamber, where it is

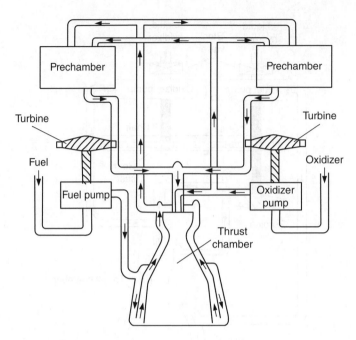

Figure 1.3   Staged combustion cycle turbo-pump feed system.

combusted with other propellants, this type of pump circulation is called a closed loop. If the turbine gas exhaust goes directly to the surrounding environment or goes into the main flow through the engine nozzle expansion opening section, then this pump circulation is called open cycle. Clearly, the gas generator cycle is an open-cycle, and the expansion cycle and the staged combustion cycle are closed loops. In contrast, an open-cycle engine is relatively simple, operates at low pressure, and the research and development costs are low; however, the closed cycle engine can achieve a higher specific impulse.

1.1.2   Thrust Chamber

A thrust chamber is a device in which chemical energy is turned into mechanical energy. Typically, a device in which chemical energy is converted into heat energy is called a combustion chamber while a device in which heat energy is converted into kinetic energy is called a nozzle. In addition to the combustion chamber and nozzle, a liquid rocket engine thrust chamber also has a unique component—the injector, which is located in the combustion chamber head. Propellant components are injected into the combustion chamber from the injector head, and then they atomize, evaporate, mix with each other, and combust in the combustion chamber. The chemical energy of the propellants is thus converted into heat, producing high-temperature, high-pressure gas. Then, the combustion gas eject from the nozzle at high speed after expansion and acceleration, producing thrust.

Since the thrust chamber works under harsh conditions of high temperature, high pressure, and high flow scour, its structure should satisfy the requirements of high efficiency (combustion efficiency and nozzle efficiency), stable working conditions (reliable ignition start, stable combustion), reliable cooling measures, and good economy (simple structure, light weight, good technology, and low cost), etc.

1.1.2.1 Constituents of a Thrust Chamber

Injectors

Injectors are usually located in the front of the combustion chamber. The function of the injector is to inject the propellants into the combustion chamber at a fixed flow rate, and let the propellant atomize and mix in a certain ratio to form a homogeneous mixture of fuel and oxidant to facilitate gasification and burning. The injector provides a cooling protective film to prevent the thrust chamber wall from being overheated. In addition, injectors also bear and transfer thrust.

The commonly used injectors can be classified into orifice injectors, swirl injectors, and coaxial tube injectors. Figure 1.4 shows the injector classification.

As shown in Figure 1.5, for an impinging stream pattern, propellant is ejected from numerous independent small holes. The fuel and oxidizer jets then collide with each other, after which a thin liquid fan is produced, which helps the liquid atomize into droplets and helps to produce an even distribution. For a self-impinging stream pattern, oxidizer jets collide with nearby oxidizer jets. Similarly, the fuel jets collide with nearby fuel jets. For a triplet impinging stream pattern, a jet of one component is used to collide with two jets of the other component. When the volume flow rates of the oxidant and fuel are not the same, the triplet impinging stream pattern is more effective.

A shower head stream pattern injector usually employs propellant jet that does not hit and eject from the surface perpendicularly. Mixing is achieved through turbulence and diffusion processes. The engine used in the V-2 missile adopted this kind of injector. A splashed pattern injector helps propellant liquids mix; it applies the principle of a propellant jet impacting with

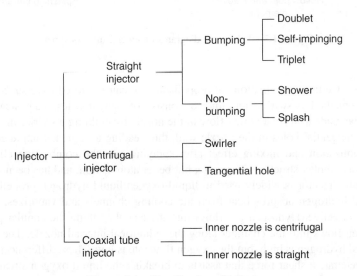

Figure 1.4 Classification of injectors.

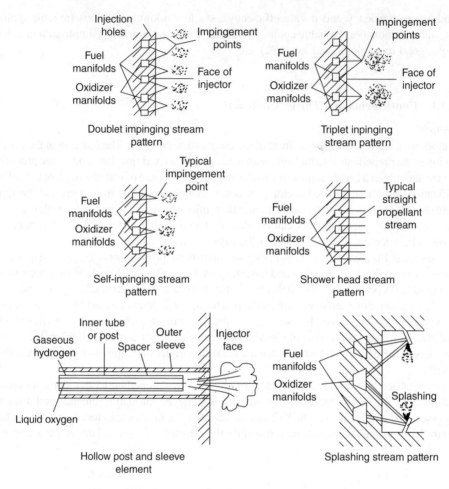

Figure 1.5   Schematic diagrams of several injector types.

solid surfaces. Certain combinations of propellants that can be stored have successfully used this spraying method. A swirl pattern injector consists of many nozzles as a basic unit; the propellant can be made to form a vortex flow in the nozzle by building a swirler in the nozzle or with drilled tangential holes in the nozzle wall, thus leading to a greater angle cone spray to improve atomization and mixing effect after spraying into the combustion chamber. This injector has a complex structure, large size, but better atomization, and has been widely used. A coaxial tube injector is widely used in liquid oxygen/liquid hydrogen propellant engines. When liquid hydrogen absorbs heat from the cooling channels and vaporizes, this injector is good and effective. Hydrogen gas flows into the chamber along the annular passage, and liquid oxygen flows into the chamber along the cylindrical internal nozzle. The flow rate of the vaporized hydrogen is high, but the oxygen flow rate is low. These differences in flow rate and speed generate a shear force that assists in breaking the liquid oxygen stream into small droplets.

An injector consists of a number of nozzles. There are two basic types of nozzles: swirl injector and orifice injector. Injector performance depends on the structure and properties of the nozzle. It has a great impact on the complete and stable combustion processes of the propellants in the combustion chamber.

Combustion Chamber

The combustion chamber is a volume where the atomization, mixture, and combustion processes of the propellants take place. The combustion chamber withstands a high temperature combustion gas pressure, its head is equipped with an injector assembly, and its exit is united with a nozzle. The shape and size of the combustion chamber volume has an important effect on the propellant combustion efficiency. The combustion structure generally is spherical, circular, or cylindrical in shape. A spherical chamber was widely used by liquid rocket engines prior to the 1950s. Although a combustion chamber with this shape has a good bearing capacity, combustion stability, light structural weight, and a small heated area in the same volume, etc., it is rarely employed nowadays because of its structural complexity and difficulties in head nozzle arrangement and processing. The cross-sectional area of the combustion chamber is annular, which was developed to adapt to the so-called plug nozzle and expansion bias nozzle; it has few practical applications. Now the most popular combustion chamber is the cylindrical combustion chamber. This is because it is simple and easy to manufacture, and it is also not expensive to build.

Nozzle

As the high-temperature gas in the combustion chamber flows downstream it expands and accelerates in the nozzle, turning the thermal energy into kinetic energy and producing a high-speed jet. The nozzle employed in the rocket engine usually consists of three parts, namely, nozzle convergence region, throat section, and divergent section. Various nozzle cross-sectional configurations are round in the divergent section. Nozzles can be classified as cone-shaped, bell-shaped, plug, and expansion deflection, according to their different longitudinal section. The nozzle should ensure minor total pressure loss of the flow, and the flow at the outlet should be in parallel with the engine axis.

1.1.2.2   Cooling of Thrust Chamber

The primary objective of cooling is to prevent the chamber and nozzle walls from being overheated. Once overheated, they will no longer be able to withstand the imposed loads or stresses, thus causing the chamber or nozzle to fail. Basically, there are two cooling methods in common use today, namely, the steady state method and the unsteady state method. For the steady state method, the heat transfer rate and the temperatures of the chambers achieve thermal equilibrium. This includes regenerative cooling and radiation cooling. For the unsteady state method, the thrust chamber does not reach thermal equilibrium, and the temperature of the combustion chamber and nozzle continues to increase in the period of operation. The heat sink cooling and ablative cooling methods are examples of the unsteady cooling method. Film cooling and special insulation are supplementary techniques that are used occasionally with both steady and unsteady cooling methods so as to locally augment their cooling capability. In the following section, we will describe these cooling methods in detail.

Regenerative Cooling

In the regenerative cooling method, before the propellant (normally fuel) is injected into the combustion chamber, the propellant flows into the cooling passage that surrounds the combustion chamber for heat exchange through forced convection. The term "regenerative" means that the heat is regenerative. The combustion gas transfers heat to the chamber wall and the wall transfers the heat to the coolant (one propellant component). The heated coolant is injected into the combustion chamber, and this achieves a reuse of the heat energy. This cooling technique is used primarily with bipropellant chambers from medium to large thrust. It has been effective in applications with high chamber pressure and high heat transfer rates. The applications show that the energy loss of this cooling method is trivial. In addition, this cooling method has little impact on the external environment. However, the drawback is its complexity, such as the complex structure of the cooling passage of the thrust chamber. Furthermore, the flow in the cooling passage brings hydraulic loss.

For a thrust chamber using regenerative cooling, the structural strength of the whole thrust chamber, the cooling reliability, and the structure mass, etc., are closely related to the structure of the cooling passages. Popular thrust chamber cooling passage structures of regenerative cooling systems are mainly as follows:

1. A smooth slit pattern cooling passage formed between the inner and outer walls:

    The structure of the smooth slit pattern cooling passage formed between the inner and outer walls is simple (Figure 1.6). However, in the case of a small flow of cooling liquid, to ensure the desired flow rate, the channel gap size must be small (<0.4–0.5 mm), which is difficult to achieve in the manufacturing process. Further, when the pressure in the cooling passage is high, the thin inner wall can easily deform as it is not rigid enough.
2. A cooling passage of inner and outer walls connected to each other:

    There are three main forms of this type of passage. One form is produced by welding the inner and outer walls together in the special punching holes (Figure 1.7). The holes are

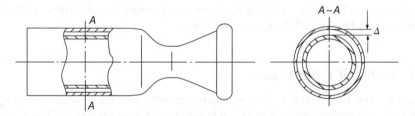

Figure 1.6 Schematic of a smooth slit pattern cooling passage.

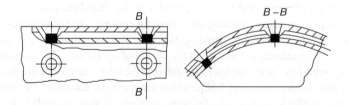

Figure 1.7 Schematic of connected cooling passage at the indentation.

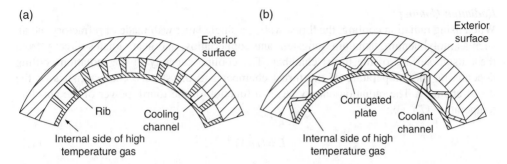

Figure 1.8 Schematic of brazed inner and outer wall of a cooling passage: (a) solder along the outer edge of the ribs; (b) solder along the inner and outer walls of the corrugated boards.

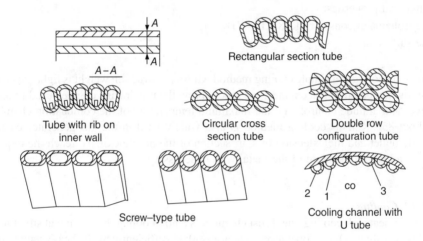

Figure 1.9 Tube bundle cooling passage.

located in the outer wall, and their shape can be circular or elliptical. Another form is produced by soldering the outer wall with the ribs milled on the inner wall (Figure 1.8a); the third form is obtained by inserting a corrugated plate and solder along the inner and outer walls of the corrugated plate (Figure 1.8b).

3. Tube bundle cooling passage:

  Engines manufactured in the United States of America widely adopt the thrust chamber tube bundle structure shown in Figure 1.9. This kind of thrust chamber body part is combined by special tubules (0.3–0.4 mm) with certain types of surface. Tubules are made of materials with good heat transfer performance (generally nickel alloy), and are soldered together. To ensure the strength of the tube bundle type thrust chamber we should use special enhancement techniques, for example, installment of staged reinforcing ring and use of entire bearing coat.

Radiation Cooling

When using radiation cooling, the thrust wall is a single layer wall made of refractory metal, such as molybdenum, tantalum, tungsten, and copper alloy, and heat radiates directly from the outside surface of the thrust chamber. The cooling capacity of the radiation cooling depends on the temperature of the thrust chamber and the surface characteristics of the thrust chamber. The radiation energy (E) is a function of the fourth power of the absolute temperature (T), i.e.:

$$E = f\varepsilon\sigma A T^4 \tag{1.1}$$

where:

f is a geometric factor that is determined by the relative position and shape of near objects;
ε is the blackness, which is a dimensionless coefficient determined by the surface condition and
   the material properties;
σ is the Boltzmann constant $(1.38 \times 10^{-23} \mathrm{JK^{-1}})$;
A is the surface area.

Radiation cooling is a simple cooling method whose construction weight is light; it is widely used in low temperature gas engines, e.g., monopropellant hydrazine engines used in aircraft maneuver and attitude control, and the maximum temperature of the combustion chamber is only about 850 K. This cooling method is also widely used in the gas generator chamber, the nozzle outlet and the extension or skirt section of the nozzle exit. To reach the requested heat flux, a high temperature of the metal wall is necessary.

Heat Sink Cooling

When using heat sink cooling, the thrust chamber is a non-cooling, heavy, metal structure and its wall is very thick. During operation, heat is absorbed sufficiently by the heavy wall before it reaches a temperature that would cause damage. Therefore, the heat-absorbing capacity of the thrust chamber wall material determines the longest operation period of the thrust chamber. This method is mainly used in the case of a low pressure chamber and low heat transfer rate, such as with the heavier experimental engine.

Cooling methods also include ablative cooling, film cooling, and special insulation cooling. When using ablative cooling, the thrust wall is made of ablative materials (typically resin materials). At temperatures of 650–800 K, ablative materials absorb heat and decompose into a porous carbonaceous layer and pyrolysis gas, and the gas forms a fuel-rich protective boundary layer on the carbon surface. Film cooling makes use of various measures (such as a dedicated head cooling jet orifice, holes and slots forming the films on the walls, a wall made up of porous material) to inject the liquid propellant component or cool air. A liquid layer and its vapor film form on the inner wall of the chamber to isolate the burned gas, reducing the wall temperature. The special insulation method adopts refractory metal and refractory material as the coating to improve the allowed upper temperature limit of the wall and reduce the heat flux to the chamber walls.

1.2 Internal Combustion Processes of Liquid Rocket Engines

Liquid rocket engine operation characteristics include performance, stability, and compatibility. Actual specific impulse is used to evaluate performance. Stability refers to an engine combustion process that does not produce any instability, it is measured using dynamic stability indicators. Compatibility refers to the ability of the thrust chamber wall to withstand high temperature and high pressure gas; it is described by the ability of the thrust chamber wall and propellant to work compatibly. Performance, stability, and compatibility are three key issues to be solved in the development of liquid rocket engine. Since these tissues are correlated, the possibility of finding a comprehensive solution to meet all three requirements should be explored to improve the combustion process. This section briefly describes the working process of a liquid rocket engine and then introduces the combustion process.

1.2.1 Start and Shutdown

The process involved from sending a starting instruction to establishment of the primary working condition is called the start. During the start, the engine goes through a series of procedures and related processes to ensure the transition from starting preparation state to primary working condition state. During the startup, there are unstable processes in the combustion chamber and engine equipment; the flow conditions of these processes will determine the reliability and performance of a liquid rocket engine. For example, so-called water hammer occurring at startup can destroy a liquid rocket engine's rated working condition, and even cause engine damage. Therefore, it is important to ensure a reliable engine start, as most liquid rocket engine failures occur during the start.

Usually, liquid rocket engines need to complete the following processes at startup: the pressure in the propellant tank first increases to the specific pressure; for liquid rocket engines using cryogenic propellant, it is necessary to cool the propellant lines; the propellant feed system (e.g. turbo-pump) goes into the specified working state; for liquid rocket engines using non-hypergolic propellant, the initial ignition flame should be produced in the thrust chamber and gas generator to ensure ignition of the propellant injected into the thrust chamber and gas generator; the propellant valve should be opened to ensure the propellants eject into the combustion chamber and gas generator.

The operations listed above are typical, but some of them are not necessary for certain types of liquid rocket engines and some operations are not needed according to special requirements. In turbopump-fed liquid propellant rocket engines, the schemes of turbine start can be classified into self-start and external-energy start according to the way the energy needed is provided. Self-start does not require additional boot devices. External energy start schemes include gunpowder start (use gunpowder-generated gas to drive turbine) and cylinder start (use of bottled pressed gas to drive the turbine).

The transition process from sending shutdown instructions to the thrust dropping to zero is called the liquid rocket engine shutdown. Engine shutdown is necessary. For example, we need to shut down engines when the rocket reaches the desired speed, the spacecraft complete the necessary maneuvers, and experiments are completed or fail on the test bench. Shutdown methods and various operations depend on the requirements of a liquid rocket engine, and these requirements depend on working conditions and aircraft functions.

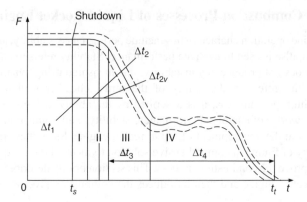

Figure 1.10   Typical thrust attenuation characteristics at shutdown.

According to different requirements proposed for a rocket system, the shutdown method can be divided into the following modes: shutdown after propellant consumption, shutdown guaranteeing minimum aftereffect impulse, fault shutdown, and multiple-shutdown. A working liquid rocket engine on an aircraft in active flight usually adopts the first shutdown method, e.g., shutdown of the liquid rocket engine on ballistic missiles.

The difference between shutdown guaranteeing minimum aftereffect impulse and the first shutdown is that for a given case the former makes an additional requirement that a minimum aftereffect impulse should be ensured. The so-called impulse is thrust impulse produced in the period from sending the shutdown instruction to the thrust dropping to zero. Figure 1.10 shows typical thrust attenuation characteristics during the shutdown. First, from sending the shutdown instruction to the main valve being turned off, the propellant injected into the combustion chamber has to be delayed for a certain time to change into combustion gas, namely, there is a combustion delay Δt_1, so the whole process moves to the right along the time axis, and the thrust remains unchanged. Secondly, after sending the shutdown instruction, the valve needs the time Δt_2 to operate due to the inertia of the control circuit, when the propellant flow and thrust remain unchanged. After Δt_2, the valve begins to close, and when the flow cross-sectional area begins to change, the thrust and flow also change. Because of the inertia of the valve it needs the time Δt_{2v} to shut down completely. After the valve is closed completely, the gas in the combustion chamber disappears quickly, the pressure in the combustion chamber declines sharply; the time taken for this process is Δt_3. The evaporation and combustion of residual propellants corresponding to Δt_4 in Figure 1.10 is determined by the unstable process by which the remaining propellant from the main valve to the injector head cavity flows into the combustion chamber. When the chamber pressure drops below the propellant component saturation vapor pressure, the propellant in the head cavity is injected into the combustion chamber under the saturated vapor pressure effect. Due to the difference between the two component saturated vapor pressures, the first entering the combustion chamber evaporates first and discharges, leading to partial thrust; When both components have entered the combustion chamber, combustion occurs. At this stage, the mixed ratio in the combustion chamber is changing, often deviating from the optimum value, and the combustion is unstable. Shutdown failure happens easily. Since the process is in a non-controlled state, it directly affects the aftereffect impulse deviation. The aftereffect impulse can be reduced by adopting the staged shutdown method which reduces the thrust during shutdown, or by shorting the pipeline, reducing the valve actuation time and reducing the residual volume of the residual propellant. We can also force to empty the residual propellant to mitigate the aftereffect impulse.

Multiple shutdown and startup of liquid rocket engines are used in orbiter aircraft and aircraft that need to start periodically. To achieve multiple shutdowns, liquid rocket engines must be able to automatically transit to the state that startup preparation is completed; consequently, most multiple-shutdown liquid rocket engines use the same equipment to complete the startup and shutdown.

1.2.2 Combustion Process

The combustion process of liquid propellant is a very complex process from the injection of liquid propellants to formation of the combustion products. The internal combustion process in a liquid rocket engine might be the most complicated combustion and flow phenomena. This is mainly because (i) the combustion process in a liquid rocket engine actually contains many physical/chemical sub-processes of several types, multiple characteristics, and different temporal and spatial scales. Whether these sub-processes can be simulated using a specific model is highly dependent on the propellants and rocket engines. (ii) Generally, many sub-processes occur simultaneously and are strongly coupled. It is usually very difficult to decouple these into independent processes. (iii) The propellant mass flow rate in the combustion chamber can reach hundreds or even thousands of kilos per second while the residence time of propellants in the combustion chamber is very short at the order of several milliseconds. But, the combustion efficiency requirement is very demanding, up to 0.98–0.99. (iv) In the combustion chamber, the propellant concentration gradient, temperature gradient, and pressure gradient are significantly high in the vicinity of injectors. Therefore, the flowfield is very complicated. Furthermore, the pressure, velocity, and temperature in the combustion chamber of rocket engines are much higher than in other engines (e.g., jet engines), which induces more difficulties for combustion. (v) Combustion instability of different frequencies is prone to occur, which may cause a performance reduction and damage to the facility.

Figure 1.11 presents the individual sub-processes of spray combustion in a liquid rocket engine. These sub-processes occur in the two-phase flow in the combustion chamber.

Based on available experimental results, the flowfield can be qualitatively divided into a series of discrete zones (Figure 1.12):

1. Injection–atomization zone:

    The injected propellant mainly atomizes and forms liquid drops in this zone. Since the liquid fuel and oxidizer are sprayed independently through different injectors and inject into the combustion chamber via the orifices with a certain distribution and orientation, the mass flux, mixing ratio, atomization performance, and the properties of the gases varies significantly in different directions, which consequently leads to a mixing process. The majority of the gas in the injection–atomization zone is the gaseous propellant component or the recirculated combustion gas from the downstream. The recirculation of combustion gas is mainly caused by the shear stress between the propellant jet and the surrounding gas. This shear stress also twists the jet surface and causes jet breakup, which is helpful for atomization. Gas–liquid injectors are designed according to the atomization mode caused by shear stress

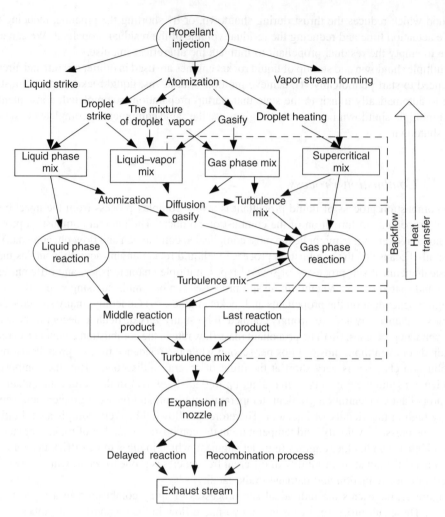

Figure 1.11   Schematic of internal combustion process in a liquid rocket engine.

breakup. However, for most of the liquid–liquid injectors, atomization is usually achieved using liquid jet impingement or swirl injectors. The accomplishment of atomization needs a certain distance, usually 1–5 cm. The formation and distribution of liquid drops proceeds simultaneously. Once liquid drops form, they are surrounded by the gases. The temperature of the surrounding gas is higher than that of the liquid drops and therefore the liquid drops are heated. Since the initial temperature of the liquid drops is significantly lower than the propellant saturation temperature under the thrust chamber pressure, the vaporization of propellants in this zone can be neglected. As the temperature of the liquid drops and surrounding gas elevates, the vaporization rate keeps increasing, and chemical reactions between the fuel vapor and oxidizer vapor start, with transfer to the next zone.

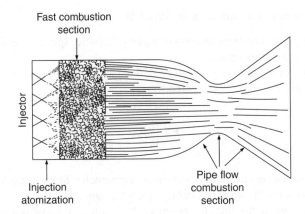

Figure 1.12   Schematic of zone division of combustion process in a liquid rocket engine.

2. Rapid combustion zone:

   The characteristic of this zone is that the propellant jet has completely become a liquid mist, which vaporizes quickly, mixes, and reacts to produce combustion products. Although the spray diffusion and flow recirculation reduces the transverse gradient of flow parameters in the injection–atomization zone, it still cannot be neglected for most liquid rocket engines. Therefore, the mass, heat, and momentum exchange in radial and circumferential directions, leading to the acceleration of mixing and evaporation processes. As a result, the combustion products are quickly produced, which leads to gas acceleration dominantly in axial direction and also causes transverse flow from the high combustion rate zone to the low combustion zone and recirculation of combustion gas to the injection–atomization zone.

3. Stream tube combustion zone:

   The transverse movement of atomization and combustion gas almost disappears. The evaporation and combustion occur in stream tubes parallel to the engine axis. In addition, the mixing between stream tubes relies on turbulent oscillation. Since the velocity of a combustible gas is high, the residence time of propellant in the combustion chamber is about 3–5 ms. Given that the turbulent oscillation frequency is about 1000–2000 Hz, turbulent fluctuation occurs in the combustion chamber no more than ten times and, thus, turbulent exchange is not significant. High-speed photography shows that the flow in this zone is close to laminar flow. As the distance from injection plate increases, the local residence time decreases and the relative velocity between droplet and gas flow decreases. The stream tube combustion zone then ends at the sonic section of the nozzle.

4. Supersonic expansion zone:

   As the expansion of combustion products continues in the nozzle, the vaporization and combustion can be assumed to stop as status parameters such as the pressure and temperature decrease and the residence time minimizes. The loss of energy in the supersonic zone in the nozzle should be linked to the two-dimensional nozzle flow, boundary layer loss, and the chemical hysteresis of the combustion products decomposition.

1.2.3 Performance Parameters in Working Process

The rocket engine performance parameters (e.g., total impulse, specific impulse, and the mixture ratio) are defined in this section.

The total impulse of a rocket engine can be defined as the integration of thrust (P, which is time-dependent) over the entire combustion time, t:

$$I_t = \int_0^t P \, dt \tag{1.2}$$

The total impulse is a very important performance parameter. Since it contains the thrust and its duration, it represents the level of engine's working capability.

The specific impulse is the impulse produced by burning 1 kg of propellant in a rocket engine:

$$I_s = \frac{I_t}{m_p} \tag{1.3}$$

where m_p is the total effective propellant mass and I_s is the average specific impulse in the operation process.

For liquid rocket engines, the specific impulse is the thrust generated by consuming 1 kg propellant per second:

$$I_s = \frac{P}{\dot{m}} \tag{1.4}$$

The specific impulse is an important performance parameter that has a significant influence on the performance of space launchers and spacecrafts.

The density specific impulse is defined as the thrust generated by a unity of propellant mass flowrate:

$$I_{s,p} = \frac{P}{\dot{V}} = \frac{P}{\dot{m}/\rho_T} = I_s \rho_T \tag{1.5}$$

where ρ_T is the density of propellant.

The mixture ratio of propellant is defined as the ratio of oxidizer mass flow rate to the fuel mass flow rate:

$$MR = \frac{\dot{m}_o}{\dot{m}_f} \tag{1.6}$$

MR has a significant effect on the specific impulse of a rocket engine. It also influences the propellant density (ρ_T) and thus has a strong effect on the structure, mass, and dimensions of rockets and spacecrafts and further on the performance of the launchers.

1.3   Characteristics and Development History of Numerical Simulation of the Combustion Process in Liquid Rocket Engines

1.3.1   Benefits of Numerical Simulation of the Combustion Process in Liquid Rocket Engines

1. Shortening of the development period and reduction of research cost:

    The complexity of the operation process of a liquid rocket engine means that the calculations show significant discrepancy in comparison with practical testing results. For a long period, the design and development of a rocket engine relied on experiments. It has to go through many specific and full-scale engine tests, resulting in huge costs and long research periods. Numerical calculation of the combustion process in a liquid rocket engine can reproduce the practical combustion process on a computer. The unreasonable testing schemes can be abandoned based on simulation results. This can significantly reduce the number of tests and increase the testing success rate. The overall research period can be shortened and research costs can be reduced. Additionally, numerical simulations of liquid rocket engine combustion can help researchers perform quick design optimization and complex calculations.

2. Detection of potential problems and prediction of engine performance:

    During the operation process in liquid rocket engines, the high-temperature and high-pressure combustion gases pose a demanding environment for most testing instruments. The experimental data measured from one test is limited and is not enough to analyze experimental phenomenon. However, numerical simulations can provide a large amount of combustion process information, which can be used to help identify the cause of experimental phenomenon and provide clues for solving the problem. As the numerical simulation technique develops, it can precisely predict the complicated working process and performance in engines. The carrying out of numerical simulations will allow researchers to estimate the performance before hot-fire tests and will be beneficial in results analysis and performance evaluation after hot-fire tests. The numerical simulation results are completely repeatable, which can allow researchers to observe the work process at any stage and conduct quantitative studies. From this viewpoint, numerical simulations of the combustion process in liquid rocket engines are a kind of hot-fire tests on computers using a numerical method.

1.3.2   Main Contents of Numerical Simulations of Liquid Rocket Engine Operating Process

As mentioned above, the combustion process in a liquid rocket engine is a combustion and flow process with complex boundary conditions. The main parameters of a liquid rocket engine are determined by the flow and combustion process. However, combustion instability may cause structure failure of engines, which may also be related to the combustible gas flow process. Therefore, it can said that the numerical simulation of combustible gas flow and combustion process is the key problem in the numerical simulation of the combustion process:

1. Modeling of combustion process:

    The internal flow in liquid rocket engines can be described by three-dimensional unsteady compressible Navier–Stokes equations. The effect of two-phase flow, chemical reaction, heat convection, heat radiation should be taken into account. Generally, the combustible gas flow in a combustion chamber has three-dimensional characteristics. In addition, this three-dimensional effect will affect the combustion gas flow in the nozzle.

Numerical models of the combustion process in liquid rocket engines contain an atomization model, liquid droplet evaporation model, flow turbulence model, combustion model, and heat transfer model. Since the flow, combustion, and heat transfer processes are coupled, these models are inter-related. The spray and combustion of liquid propellant is a complicated process containing heat transfer, fluid movement, mass and concentration diffusion, and other chemical dynamic processes. Therefore, it is very difficult to set up a comprehensive theoretical model. In liquid rocket engines, the flow of combustion gas is turbulent. Turbulence simulation usually applies Reynold average equations with turbulence models to enclose the equations. However, there is no turbulence model that can work at all conditions. Numerical simulations of internal flow in engines have to properly deal with the droplet distribution, droplet velocity and thermal properties of further formed two-phase mixture, particles dynamics and energy characteristics, physical properties of particle state, and size distribution. Liquid droplet evaporation is influenced by high pressure, multiple species, and pressure oscillation. A proper model describing this process should be developed. Heat transfer in liquid rocket engines occurs in three fundamental ways, namely, heat conduction, heat convection, and heat radiation

Generally, the main ways heat transfer takes place in liquid rocket engines are heat convection and heat radiation with even higher heat convection. The energy conservation equation with heat radiation is an integral–differential equation, which is very complicated to solve especially in the calculation of spatial angle coefficients. The physical properties of radiation heat transfer are not easy to determine in liquid rocket engines when considering temperature and pressure variations. Combustion instability is an unsteady phenomenon occurring in liquid rocket engine combustion. The most common form is periodic combustion oscillation.

Combustion oscillation can occur spontaneously, and often appears at a specific characteristic time in the combustion process. This characteristic time is determined not only by the combustor geometry but is also related to gas flow variation. When variations go beyond the system stable limit, the flow turbulence is amplified by the interaction of combustion and average flow, which can lead to severe damage. Modeling of combustion instability is very difficult but some progress has been made.

2. Solution of Numerical Model

Numerical simulation of the combustion process in liquid rocket engines relies on solving governing equations including combustion and heat transfer models with complex boundary conditions. Consequently, numerical solution of simulation models becomes a key point where the computation of heat transfer and combustion is also required. Computational fluid dynamics is a new subject, arising with the development of computer science and technology. The discretization method has three schemes: finite difference, finite element, and finite volume. Numerical schemes of temporal discretization can be divided into three categories: explicit, implicit, and explicit–implicit. Every scheme contains first- and second-order resolution schemes and other high-order schemes developed in recent years. For numerical simulation of flow field, the selection of discretization methods and numerical schemes depends on the governing equations, solution scales, requirements of numerical solution, computation efficiency and accuracy, and the available computer resources.

The numerical solution process in computational fluid dynamics can be generally divided into three procedures:

1. pre-processing, including model setup, division of simulation domain, and grid generation;
2. numerical solution, containing scheme selection, equation discretization, initialization, boundary condition specification, coding and debugging, and computation;
3. post-processing, involving visualization of flow properties using numerical methods and flowfield analysis.

3. Analysis of simulation results:

After completion of numerical simulations, the analysis and study of the operation process of rocket engines can be carried out using various methods. With the help of image flow visualization techniques, the internal flow structures, flow processes, combustion process, and their interaction can be investigated in detail to predict engine performance.

1.3.3 Development of Numerical Simulations of Combustion Process in Liquid Rocket Engines

To maintain high reliability and reduce the test cost of liquid rocket engine design, usually several modifications are made to the current engine design based on engineering experience. Then these design modifications are evaluated using a large number of tests. However, the engine tests are always expensive and time-consuming. A single engine test may cost tens of thousands, hundreds of thousands, and even several million RMB Yuan. The measurement and diagnostics are also very challenging because of the high temperature and pressure in the combustion chamber and their rapid variation. It is very difficult to obtain enough reliable measurement data, accurate performance improvement, and the cause of failure. Therefore, the design of combustion chambers, especially injectors, is the most time-consuming period in the development of a liquid rocket engine.

To avoid the reliance of liquid rocket engine design on pure experience, the Chemical Rocket Propulsion Cooperation Bureau and Performance Standardize Team was founded for this work in the USA in 1965. It published an assessment guidance for liquid rocket engines. The guidance presents all the performance losses except the energy release (combustion process) loss. As the combustion process was believed to be too complicated to describe using analytical methods, it was supposed that propellants burn completely and the propellant enthalpy is reduced to account for this effect in the calculation of heat equilibrium. These approaches all show difficulties in numerical simulations of the combustion process. In the 1970s, as combustion computation developed, the computation techniques were applied to liquid rocket engine design. Many combustion computation models were developed for specific newly designed engine prototypes in the 1980s, such as the combustion computation model for the combustion chamber design of auto-ignition liquid rocket engines F-20 and F-5, the ARICC model for the injector/combustion chamber of space shuttles and advanced rockets, PHEDRE model for Ariane rocket engines, REFLAN3D-SPRAY model for liquid oxygen/hydrocarbon rocket engines, and CAFILRE code for liquid hydrogen/liquid oxygen rocket engines. In the 1980–1990s, most books in China on liquid rocket engines still focused on the introduction of thermal calculation, and aerodynamic and heat transfer calculation. For example, Professor

Fengchen Zhuang published the book *The Theories, Models and Applications of Liquid-Propellant Rocket Engine Spray Combustion* [1].

Another issue to solve in the combustion simulation of liquid rocket engines is the turbulent two phase flow and combustion process using full Navier–Stokes equations. To further improve the numerical accuracy of models and understand more details (temperature distribution, concentration distribution, velocity distribution, and pressure distribution) of the entire flow field in the combustion process from start to shutdown, a more complicated model covering all combustion sub-processes should be developed. These models and programs have improved applications of liquid rocket engine atomization models. The author of this book has conducted much work in liquid rocket engine research and development [2–7]. This book is a summary of work conducted in recent years.

Numerical modeling of combustion process in liquid rocket engines has been a multidisciplinary subject since the 1970s. It combines computational fluid dynamics, computational heat transfer, computational combustion, computer software design, and flow visualization. Therefore, it is one of the most active research subjects in the liquid rocket engine research field. The numerical simulation of liquid rocket engine working processes has been applied in every aspect of liquid rocket engine research. It is an efficient research method, and also can assist the design and experimental study. Therefore, further development of this research subject can significantly improve understanding of fundamental physical phenomena and the engine design.

1.4   Governing Equations of Chemical Fluid Dynamics

Combustion is a flow process containing chemical reactions. No matter how complex the combustion processes are they all obey the fundamental laws of nature, namely, the conservation of mass, species, momentum, and energy. The mathematical equations representing these laws are from chemical fluid mechanics, also referred to as the fundamental governing equations of a combustion process. These equations are the fundamentals of numerical simulations of the combustion process. They will be presented in the following and the physical meaning of these equations is briefly described. For the mathematical expression of multidimensional variables, the tensor notation is applied. Subscriptions i, j, k indicate the coordinate direction. The variables with only one subscription are vectors, e.g., u_i. Variables with two subscriptions are tensors, such as viscous stress τ_{ij}. The same subscription appearing twice for a variable indicates Einstein's summation convention of this variable from index 1 to 3.

In Cartesian coordinates, the governing equations contain the following four equations:

1. Continuity equation:

$$\frac{\partial \rho}{\partial t} + \frac{\partial \rho u_i}{\partial x_i} = 0 \tag{1.7}$$

2. Momentum equation:

$$\frac{\partial \rho u_i}{\partial t} + \frac{\partial \rho u_i u_j}{\partial x_j} = \frac{\partial p}{\partial x_i} + \frac{\partial \tau_{ij}}{\partial x_j} + g_i - f_i \tag{1.8}$$

where ρ is the density of the fluid mixture; p is the pressure; u_i is the velocity in the i direction; g_i and f_i are the components of gravity and other drag in the i direction, respectively; τ_{ij} is the viscous stress tensor, which can be related to strain rate tensor S_{ij} using generalized Newton's law:

$$\tau_{ij} = 2\mu S_{ij} - \frac{2}{3}\mu S_{kk}\delta_{ij} \tag{1.9}$$

$$S_{ij} = \frac{1}{2}\left(\frac{\partial u_i}{\partial x_j} + \frac{\partial u_j}{\partial x_i}\right) \tag{1.10}$$

where S_{kk} is the fluid volumetric dilatation $div\mathbf{u}$, which represents the volumetric expansion and compression; μ denotes the fluid dynamic viscosity coefficient; δ_{ij} is the second order unit sensor, if $i = j$, $\delta_{ij} = 1$; otherwise $i \neq j$, $\delta_{ij} = 0$.

3. Energy equation:

$$\frac{\partial \rho h_0}{\partial t} + \frac{\partial \rho u h_0}{\partial x_j} = \frac{\partial}{\partial x_j}\left(u_i\tau_{ij}\right) + \frac{\partial}{\partial x_j}\left(\lambda\frac{\partial T}{\partial x_j}\right) + \rho q_R + \frac{\partial}{\partial x_j}\left[\sum_l (\Gamma_l - \Gamma_h)\frac{\partial m_l}{\partial x_j}\right] \tag{1.11}$$

where h_0 is the stagnation enthalpy, also the total enthalpy, $h_0 = h + u_i u_l/2$, $h = \sum_l m_l h_l$, m_l and h_l are the mass fraction and specific enthalpy of specie l in the fluid mixture; Γ_l and Γ_h are the transportation coefficient of specie l and exchange coefficient of enthalpy; q_R is the radiant heat flux. If using the definition of total enthalpy, T in the diffusion term on the right-hand side of Equation 1.11 can be replaced by h_0. Therefore, we can get an alternative form of the energy equation:

$$\frac{\partial \rho h_0}{\partial t} + \frac{\partial \rho u_j h_0}{\partial x_j} = \frac{\partial}{\partial x_j}\left(\Gamma_h\frac{\partial h_0}{\partial x_j}\right) + S_k \tag{1.12}$$

where the source term is:

$$S_k = \frac{\partial p}{\partial t} + \frac{\partial}{\partial x_j}\left(u_i\tau_{ij}\right) + \rho q_R + \frac{\partial}{\partial x_j}\left[\left(\lambda - \sum_l m_l c_{pl}\Gamma_k\right)\frac{\partial T}{\partial x_j} + \sum_l (\Gamma_l - \Gamma_h)h_l\frac{\partial m_l}{\partial x_j} - \Gamma_h\frac{\partial}{\partial x_j}\left(\frac{u_i u_i}{2}\right)\right] \tag{1.13}$$

4. Species transport equations:

$$\frac{\partial \rho m_l}{\partial t} + \frac{\partial \rho u_j m_l}{\partial x_j} = \frac{\partial}{\partial x_j}\left(\Gamma_l\frac{\partial m_l}{\partial x_j}\right) + R_l \tag{1.14}$$

where R_l is the production rate of specie l caused by chemical reaction.

The above equations constitute the fundamental governing equations of chemical fluid mechanics. Obviously, these equations are exactly the same in formulation. They all contains four basic terms, which are an unsteady terms representing temporal variation, a convective term caused by macroscopic movement, a diffusive term related to the molecule movement, and source terms that do not belong to the above three terms. If φ can represent general dependent variables (u_i, h_0, m_l), the fundamental equations can be written as a uniform expression of:

$$\frac{\partial \rho q}{\partial t} + \frac{\partial \rho u_j \varphi}{\partial x_j} = \frac{\partial}{\partial x_j}\left(\Gamma_\phi \frac{\partial \phi}{\partial x_j}\right) + S_\phi \tag{1.15}$$

where Γ_ϕ and S_φ are the exchange coefficient and source terms depending on variable φ. They also can be referred to as transportation equations since the equations actually describe the diffusion and transportation processes of various physical variables. The variables following the transportation equations can be named as transportable variables. The transportation equations can be expressed in a uniform form. This fact reveals that the transportation process of variables has similar physical and mathematical characteristics. This can also benefit the numerical simulation as numerical methods and program coding can be expressed in a uniform form. All the equations can be solved using the corresponding Γ_ϕ and S_φ.

The above equations and gas mixture status equations can form closed equations. Theoretically, if the source term can be calculated using the knowledge from other subjects and boundary conditions are determined, the numerical solution of the combustion processes describing engines or other systems can be obtained. In fact, it would not be so easy since the flow and combustion in science and engineering are almost always turbulent process. Therefore, relevant modeling of turbulent flow and turbulent combustion is also necessary.

1.5   Outline of this Book

This book introduces numerical modeling of combustion processes in liquid rocket engines and its applications. The introduction of sub-processes such as atomization, liquid droplet evaporation, turbulent mixing, heat transfer, and combustion instability is first given as well as modeling and analysis methodology. Then, numerical tests and several applications are presented.

There are eight chapters in this book. The main content of each chapter is as follows:

- Chapter 1 is an introduction. The fundamental structures and working process of liquid rocket engines are first described, followed by the characteristics and history of numerical simulations of combustion process in liquid rocket engines. The governing equations of the chemical fluid mechanics are provided as well.
- Chapter 2 presents the atomization theory and research techniques, and atomization models for various types of injectors.
- Chapter 3 introduces the droplet evaporation combustion model. The droplet evaporation models under the normal and high pressure conditions are detailed. The response characteristics of a droplet in the pressure oscillation environment, multi-components droplet evaporation, and droplet group evaporation are discussed as well.
- Chapter 4 modelizes the turbulent flow and introduces the turbulence models for RANS simulation and sub-grid models for large-eddy simulation.

- Chapter 5 provides modeling of interaction of combustion and turbulence. The turbulent combustion model for RANS and sub-grid model for large-eddy simulation is given in this chapter.
- Chapter 6 describes models for heat transfer, heat convection, and heat radiation.
- Chapter 7 presents characteristics and theoretical models of combustion instability. Methods of controlling combustion instability and evaluation of these methods are detailed.
- Chapter 8 presents numerical models for atomization combustion in liquid rocket engines. Methods of grid generation and numerical methods of equations and models solution are provided as well as applications.

References

[1] Zhuang F C. *The Theory and Modeling of Spray Combustion in Liquid Rocket Engines with Applications* (in Chinese), National University of Defense Technology Press, Changsha, 1995.

[2] Wang Z G. *Internal work process of liquid-propellant rocket engine combustion numerical simulation*, National University of Defense Technology, PhD dissertation, 1993 (in Chinese).

[3] Wang Z G, Wu X X, Zhuang F C. The application of computation fluid dynamics in the liquid-propellant rocket engine, *Journal National University Defense Technology*, 1994, **16** (4), 1–7 (in Chinese).

[4] Wang Z G, Zhou J, Yan X Q, Zhuang F C. The numerical analysis of hydrogen and oxygen gas jet combustion process in prechamber, *Journal National University of Defense Technology*, 1996, **18** (3), 5–10 (in Chinese).

[5] Wang Z G, Zhou J, Yan X Q. Hot flow field calculation of hydrogen and oxygen gas coaxial single injector combustor, *Propulsion Technology*, 1996, **17** (4), 42–47 (in Chinese).

[6] Liu W D, Wang Z G, Zhou J. The three-dimensional numerical simulation of hydrogen and oxygen liquid (l) rocket engine atomizing combustion, *Propulsion Technology*, 1999, **20** (1), 19–28 (in Chinese).

[7] Huang Y H, Wang Z G, Zhou J. Numerical simulation of liquid-propellant rocket engine combustion stability, *Chinese Science B*, 2002, **32** (4), 377–384 (in Chinese).

2

Physical Mechanism and Numerical Modeling of Liquid Propellant Atomization

In the combustor of a liquid rocket engine, the liquid propellant should first disintegrate into small droplets through the atomization process so that the fuel and oxidizer can evaporate quickly and mix well before combustion. Atomization plays a significant role in combustion stability and efficiency. It is important in engine design to investigate the atomization mechanism and atomization characteristics of injectors.

The atomization mechanism is essentially the same for all kinds of injectors used in liquid rocket engines. The liquid propellant first has to be expanded into a thin liquid film or jet [1]. When the jet speed is not high enough for the flow to develop into turbulence, perturbation waves on the liquid sheet or jet flow will grow rapidly under the action of the surface tension and aerodynamic force, resulting in the breakup of the liquid sheet/jet. The initial perturbations mainly arise from velocity fluctuations, nozzle vibration, and nozzle burr. When the jet develops into turbulent flow, turbulent eddies inside the jet determine the atomization process.

The main factors influencing the atomization performance of an injector:

1. internal flow characteristics determined by nozzle structures and operating conditions;
2. ambient gas parameters;
3. physical properties of liquid.

These factors should be included in the theoretical modeling of atomization. A large amount of work has been carried out on the breakup mechanism of liquid jet and liquid sheet and the resulting spray characteristics. However, there is still no one atomization model that can predict atomization characteristics for all kinds of injectors due to the complexity of the atomization process.

This chapter will describe the atomization mechanism of liquid propellant, evaluation parameters of atomization performance, and atomization model for nozzles that are in common

Internal Combustion Processes of Liquid Rocket Engines: Modeling and Numerical Simulations,
First Edition. Zhen-Guo Wang.
© 2016 National Defense Industry Press. Published 2016 by John Wiley & Sons Singapore Pte Ltd.

use in liquid rocket engines. Finally, the progress made in numerical simulations of atomization using interface tracking methods in recent years will be detailed.

2.1   Types and Functions of Injectors in a Liquid Rocket Engine

In a liquid rocket engine, propellant atomization is accomplished through the nozzles on the injecting panel under a certain injecting pressure. There are two advantages of atomization for the liquid fuel:

1. Beneficial for evaporation: Since evaporation is a surface process it occurs more quickly with a larger surface area for a certain amount of liquid. As the total surface area of drops with a certain amount of liquid mass is inversely proportional to their diameter, droplets with a smaller diameter will exhibit larger surface, which can accelerate the evaporation process. For example, while a spherical water drop of 1 kg only has a total surface area of 0.0483 m^2, the total surface area can increase to 60 m^2 after this water drop is atomized into droplets with a diameter of 50 μm. In a liquid rocket engine, the diameter of propellant droplets after atomization is in the range 25–500 μm, and thus the total surface of 1 cm^3 liquid propellant increases by tens of thousands of times through the atomization, which then accelerates the evaporation process.
2. Beneficial for mixing and combustion: while a kerosene droplet of 1 mm diameter needs at least 1 s to burn out in air, it can burn out in 0.025 s after being broken up into droplets of 50 μm diameter.

Depending on the pattern of injection holes on the face of an injector, injectors can be categorized into several types. The commonly-used injecting units include impinging injectors, coaxial straight-flow injectors, and coaxial swirl injectors. Impinging injectors can be classified into unlike impinging injectors and self-impinging injectors. The unlike doublet impinging injectors are often used in rocket engines with storable propellants, such as low-thrust control rocket engines, apogee rocket engines, and ascent engines of Apollo lunar modules. Self-impinging injectors are applied in rocket engines with spontaneously ignitable propellants. Coaxial injectors are commonly used in the LOX/H$_2$ cryogenic engines. Coaxial straight-flow injectors are commonly used in cryogenic propellant rocket engines in America, e.g., the main engines of space shuttles and the RL-10 engines of Atlas Centaur. In comparison with the coaxial straight-flow injectors, coaxial swirl injectors are less sensitive to unsteady combustion, and the injector number required in an engine with coaxial swirl injectors is less than that with coaxial straight-flow injectors. As part of the RL-10 series [2, 3], RL-10A-3, which used coaxial swirl injectors to improve the performance, was successfully developed. The American Space Transportation Main Engine (STME) also employed coaxial swirl injectors, and its stability was validated in experiments. In China, gas liquid coaxial swirl injectors have been employed in the development of cryogenic propellant rocket engines, with YF-75 as a representative example. A 50t LOX/H$_2$ rocket engine that can be used as the core stage of a CZ5 launcher has been successfully developed, and will undertake the mission of carrying a heavier payload together with 120t liquid oxygen/kerosene engines.

2.2   Atomization Mechanism of Liquid Propellant

In a liquid rocket engine, liquid propellants are fed into the combustion chamber in the form of a cylindrical jet or conical liquid sheet depending on the injector configuration. A significant part of the design of liquid rocket engines is to investigate the atomization mechanism of the cylindrical jet and conical liquid sheet. In the early research of atomization theory, the size of droplets resulting from fragmentation of the cylindrical jet and conical liquid sheet was calculated through theoretical models. However, the error of the calculated results is considerable in comparison with the experimental data due to the lack of influential factors in the model. As more of these factors are included in further developments, the theoretical model can produce better results. In this section, the size of liquid droplets dribbled from a pipe is analyzed, and the important role of the viscosity and surface tension in the formation of liquid droplets is also discussed. Then breakup models of liquid jet/sheet and droplet are introduced.

2.2.1   Formation of Static Liquid Droplet

The formation of static liquid droplets is the typical form of atomization. Droplets dripping from a pipe is a classic example. When the gravitational pull on the liquid exceeds the surface tension acting on the pipe exit, the suspension state of the liquid is breached, and the liquid falls in the form of a droplet. The mass of the resulting droplet is determined by the gravity and the surface tension. The gravitational force on the droplet is equal to the surface tension acting on the pipe exit, resulting in the following relation:

$$m_d g = \pi d \sigma \tag{2.1}$$

where:

d is the diameter of the pipe,
m_d is the mass of the droplet,
g is the gravitational acceleration,
σ is the surface tension coefficient.

The droplet diameter is:

$$D = \left(\frac{6 d \sigma}{\rho_l g} \right)^{1/3} \tag{2.2}$$

Here D is the diameter of the formed droplet, and ρ_l is the density of the droplet.

For water, $\rho_l = 1000 \text{ kg m}^{-3}$, $\sigma = 0.07237 \text{ N m}^{-1}$; and as for kerosene, $\rho_l = 800 \text{ k m}^{-3}$, $\sigma = 0.023 \text{ N m}^{-1}$.

When the pipe diameter (d) is 1 mm, the diameter of the generated water drop (D) is 3.6 mm and the kerosene droplet diameter (D) is 2.6 mm.

When $d = 10 \text{ μm}$, the generated water and kerosene and drops are 765 and 560 μm, respectively, in diameter.

For a droplet dripping from a wet plate, the theoretical equation [4] for calculating the drop diameter is:

$$D = 3.3 \left(\frac{\sigma}{\rho_1 g} \right)^{1/2} \tag{2.3}$$

This equation shows the diameter of the droplet formed from a liquid film under the action of gravity (droplet diameter is 9 mm for water and 5.6 mm for kerosene). It is observed that the diameter of the droplet generated from the sluggishly flowing liquid under the action of gravity is usually large. A static suspending droplet is rare in the practical atomization of liquid propellant, and is practically significant only in the case of leakage owning to the nozzle abrading. However, the formation and the breakup mechanism of a static drop is the foundation of atomization study.

2.2.2   Breakup of Cylindrical Liquid Jet

Research on the breakup mechanism of a cylindrical liquid jet was first performed by Rayleigh [5]. In 1878, he deduced the relation between the liquid jet diameter and the unsteady perturbation wavelength and the expression for the diameter of the formed droplet. Based on the work of Rayleigh, Weber [6] proposed, in 1931, a more extensive jet breakup theory that accounted for the influence of liquid viscosity. Weber presented the expression for liquid jet breakup length, breakup time, and the size of the droplet. In 1933, Tyler [7] correlated the perturbation wavelength and the frequency of droplet shedding, producing similar results to the work of Rayleigh. The experiments carried out by Haenlein [8] in 1932 provided strong evidence for theoretical analysis, and the process of liquid jet atomization was divided into four classic regimes. In 1936, Ohnesorge [9] carried out dimensionless analysis based on photograph experiments and classified liquid jet breakup into three regimes based on a dimensionless parameter Oh (Ohnesorge number). Oh is the product of the Re (Reynolds number) and We (Weber number):

$$Oh = Re^{0.5} We = \mu_1 / \sqrt{\rho_1 \sigma d} \tag{2.4}$$

In 1978, Reitz [10] summarized the work of Haenlein and Ohnesorge, and proposed four regimes of liquid jet breakup depending on Re and Oh.

2.2.2.1   Regimes of Liquid Jet Breakup

1. Liquid jet breakup is classified into four regimes by Haenlein (Figure 2.1):
    1) only consider the action of surface tension and ignore the aerodynamic force;
    2) under the combined action of the surface tension and aerodynamic force;
    3) the instability caused by aerodynamic force;
    4) liquid jet breakup at the outlet of the nozzle.
2. Ohnesorge categorized liquid jet atomization into three regimes based on dimensional analysis and presented a breakup regime map in term of Re and Oh as shown in Figure 2.2.

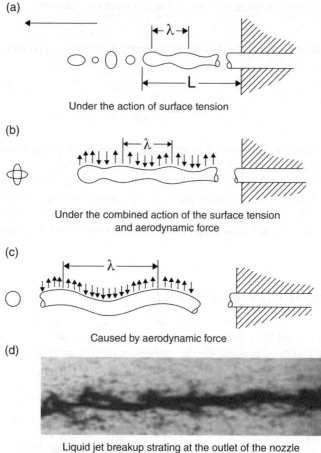

(a) Under the action of surface tension

(b) Under the combined action of the surface tension and aerodynamic force

(c) Caused by aerodynamic force

(d) Liquid jet breakup strating at the outlet of the nozzle

Figure 2.1　Four regimes of liquid jet breakup by Haenlein.

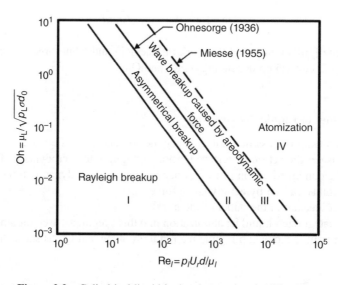

Figure 2.2　Cylindrical liquid jet breakup regime by Ohnesorge.

1) When Re is low, the liquid jet breaks up into big droplets of uniform size. This breakup regime is called Rayleigh breakup.
2) When Re is moderate, the liquid jet vibrates around its own axis, and instability grows under the action of aerodynamic forces, resulting in liquid jet breakup. In this breakup regime the formed droplets show a wider size distribution.
3) At a high Re, breakup of the liquid jet starts at the nozzle outlet. This breakup regime is called atomization.
3. Reitz [10] obtained similar results as Ohnesorge in a study of the diesel jet breakup mechanism (Figure 2.3). Reitz [10] also presented a qualitative correlation between the liquid jet diameter and the droplet diameter for different breakup regimes.

Juan Liu [11] *et al.* applied high-speed photography to observe the liquid column breakup at low Re. Figure 2.4 shows the jet patterns of a 1 mm diameter cylindrical liquid jet within 0–60 mm downstream of the nozzle outlet. When Re is equal to 5630, asymmetric disturbance modes are observed on the liquid jet. At the outlet of the nozzle, waves with a small wavelength are favored. As the liquid jet goes downstream, the amplitude of surface waves increases and waves with a long wavelength dominate the surface instability. This corresponds with the first-order mode in linear stability analysis. When Re increases to 9760, the wavelength of the instability waves on the liquid jet surface decreases, and droplets are observed to shed from the liquid jet. At a Re of 12580, hollow holes where the light intensity is strong are observed inside the liquid jet. As Re increases to 17800, aerodynamic forces grow stronger, resulting in the formation of liquid ligaments on the jet surface. Under the action of the aerodynamic forces these liquid ligaments are stretched in the upstream direction. With increasing injection pressure the liquid jet diameter at the nozzle exit grows; this is caused by interface deformation and even shedding of the liquid filaments and droplets under the action of the aerodynamic forces and nozzle internal friction.

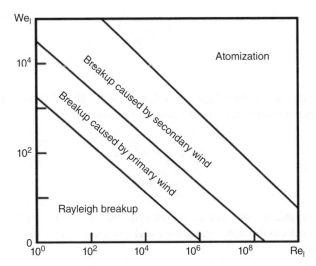

Figure 2.3 Cylindrical liquid jet breakup regime by Reitz.

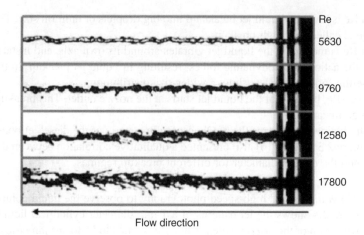

Figure 2.4   Influence of Re on breakup and deformation of a cylindrical jet.

2.2.2.2   Liquid Jet Breakup at Low Re

Rayleigh [5] employed a perturbation method to analyze the breakup process of a liquid jet. The idea is to linearize the disturbance equation deduced from the Euler equations of inviscid and incompressible flow, and then to solve the growth rate of the disturbance wave. Rayleigh suggested that the disturbance wave with the largest growth rate determines the breakup process and the diameter of the resulting droplets. The analysis of Rayleigh is based on the following assumptions: the liquid is inviscid, the aerodynamic force is negligible due to the low jet velocity, and the disturbance is axisymmetric.

Rayleigh deduced the highest disturbance growth rate (q_{max}) and the corresponding disturbance wavelength, λ_{opt}:

$$q_{max} = 0.97 \left(\frac{\sigma}{\rho_1 d^3} \right)^{0.5} \tag{2.5}$$

$$\lambda_{opt} = 4.51d \tag{2.6}$$

The liquid column of one disturbance wavelength peels off and shrinks into a droplet. In addition, the droplet diameter can be calculated as follows:

$$4.51d \frac{\pi}{4} d^2 = \frac{\pi}{6} D^3 \tag{2.7}$$

$$D = 1.89d \tag{2.8}$$

Taylor [12] correlated the disturbance wavelength and the droplet breakup frequency, and obtained similar results as Rayleigh except for a slight difference in the disturbance wavelength and droplet diameter:

$$\lambda_{opt} = 4.69d \tag{2.9}$$

$$D = 1.92d \tag{2.10}$$

Based on the study of Rayleigh, the liquid viscosity and aerodynamic force have been accounted for by Weber [6]. Weber suggested that any disturbance induces a symmetrical oscillation of the liquid jet, and assumed that there exists a minimum disturbance wavelength, λ_{min}. When the initial disturbance wavelength is smaller than λ_{min}, the disturbance will be damped by the surface tension of the liquid jet. When the initial disturbance wavelength is larger than λ_{min}, the disturbance will amplify under the surface tension, resulting in disintegration of the liquid jet. In all the initial disturbances whose wavelength is larger than λ_{min}, the one that grows fastest determines the breakup process of the liquid jet and the diameter of the generated droplet. The minimum disturbance wavelength deduced by Weber is equal to the liquid jet perimeter:

$$\lambda_{min} = \pi d \tag{2.11}$$

1. For an inviscid fluid, the minimum disturbance wavelength that can induce instability is:

$$\lambda_{min} = \pi d \tag{2.12}$$

and the wavelength of the disturbance that grows fastest is:

$$\lambda_{opt} = \sqrt{2}\pi d = 4.44d \tag{2.13}$$

2. For viscid fluid, the minimum disturbance wavelength that can induce instability is:

$$\lambda_{min} = \pi d \tag{2.14}$$

and the wavelength of the fastest-growing disturbance is:

$$\lambda_{opt} = \sqrt{2}\pi d \left(1 + \frac{3\mu_l}{\sqrt{\rho_1 \sigma d}} \right)^{0.5} = \sqrt{2}\pi d (1 + 3Oh)^{0.5} \tag{2.15}$$

3. In the presence of aerodynamic forces, λ_{min} and λ_{opt} show smaller values. For gas–liquid relative velocity $U_r = 15$ m s^{-1}:

$$\lambda_{min} = 2.2d \tag{2.16}$$

$$\lambda_{opt} = 2.8d \tag{2.17}$$

The breakup length and time of the liquid jet was obtained by Weber based on the analysis. In his derivation, the axisymmetric disturbance (δ) increases as an exponential function of the growth rate q_{max}. When δ increases to the radius of the liquid jet, the liquid jet breaks up.

Assuming t_b is the time taken for the liquid jet to move from the nozzle outlet to the breakup point:

$$\frac{d}{2} = \delta_0 \exp(q_{max} t_b) \tag{2.18}$$

$$t_b = \frac{1}{q_{max}} \ln\left(\frac{d}{2\delta_0}\right) \tag{2.19}$$

Taking Equation 2.5 into Equation 2.19 gives:

$$t_b = 1.03\left(\frac{\rho_1 d^3}{\sigma}\right)^{0.5} \ln\left(\frac{d}{2\delta_0}\right) \tag{2.20}$$

The length of the liquid jet is:

$$L_{bu} = U_r t_b = 1.03 d \, \mathrm{We}^{0.5} \ln\left(\frac{d}{2\delta_0}\right) \tag{2.21}$$

where We is Weber's number:

$$\mathrm{We} = \frac{\rho_1 U_r^2 d}{\sigma} \tag{2.22}$$

For a viscid liquid, the breakup length is:

$$L_{bu} = d \, \mathrm{We}^{0.5}(1 + 3\mathrm{Oh})\ln\left(\frac{d}{2\delta_0}\right) \tag{2.23}$$

In the breakup length correlations proposed by Weber, the initial disturbance $d/(2\delta_0)$ depends on the nozzle configuration and the experimental conditions of the flow. The commonly-used function [13] for the initial disturbance is:

$$\ln\left(\frac{d}{2\delta_0}\right) = 7.68 - 2.66\mathrm{Oh} \tag{2.24}$$

Results based on Weber's theory show considerable difference from experimental data. Sterling [14] and Sterling and Sleicher [15] suggested that the lack of a transverse velocity distribution is the reason for the error, and proposed a correlation for the breakup length of the laminar liquid jet:

$$L_{bu} = d \, \mathrm{We}^{0.5}(1 + 3\mathrm{Oh})\ln\left(\frac{d}{2\delta_0}\right)/f(\mathrm{Oh}, \mathrm{We}) \tag{2.25}$$

where function $f(\mathrm{Oh}, \mathrm{We})$ depends on the liquid jet velocity and fluid properties.

2.2.2.3 Liquid Jet Breakup at Moderate Re

Liquid jet breakup at a moderate Re number is a more complicated and meaningful case. In this situation there is a high relative velocity between the liquid jet and the ambient gas with either a low speed liquid jet fed into the high velocity ambient gas or a high speed liquid jet fed into the low velocity ambient gas, making it necessary to take account of aerodynamic forces on the liquid jet surface.

As stated above, Weber analyzed the influence of aerodynamic forces, and concluded that with increasing relative velocity the maximum growth rate of disturbance (q_{max}) increased and the disturbance wavelength decreased. Therefore, the liquid jet breaks up quickly and the size of the resulting droplets is relatively small.

Owing to the complexity of the governing equations when considering aerodynamic forces, it is very difficult to study liquid jet breakup through a theoretical analysis. Previous studies of liquid jet breakup are mainly focused on the following four cases: high viscosity, low viscosity, long wavelength, and short wavelength. Lefebvre [13] has summarized the results and drew the following conclusions:

1. Liquid viscosity decreases the growth rate of the disturbance and increases the breakup time and length of the liquid jet.
2. Breakup of a high-viscosity liquid jet is caused by long-wavelength disturbances and the size of the formed droplets is relatively large.
3. Breakup of a low-viscosity liquid jet is determined by short-wavelength disturbances and the formed droplets are relatively small.
4. When the relative velocity is high, both sinusoidal and transverse disturbances can induce irregular breakup of the liquid jet, and the formed droplet is relatively large.
5. The breakup length increases as the ambient pressure grows.

2.2.2.4 Turbulent Liquid Jet Breakup at High Re

In this case, the liquid jet develops into full turbulence at the nozzle exit. The radial velocity grows in the turbulent liquid jet because of the vigorous momentum exchange, which leads to the rapid disintegration of the liquid jet. Aerodynamic forces are not necessary for the breakup of the liquid jet—even in a vacuum environment the liquid jet can break up under the influence of turbulence. Several mechanisms have been proposed for turbulent liquid jet breakup at high Re:

1. Aerodynamic force disturbance theory: Lin and Kang [17] applied the spatial patterns to study the development of short-wavelength disturbance waves in the case when a high-speed viscous liquid is injected into a high-density gas. They suggested that atomization was caused by the sheet wave resonance due to the air pressure fluctuation on the air–liquid interface. In addition, they also found that the viscosity plays an important role in the atomization process. When gas interacts with the high-speed liquid jet, unstable shear waves develop in the boundary layer under the action of the viscosity, which contributes to the disintegration of the liquid jet.
2. Turbulence disturbance theory: in this theory, the radial velocity arising from the liquid turbulence developed inside nozzle causes atomization.

3. Cavitation disturbance theory: A large-amplitude pressure disturbance due to cavitations inside the nozzle induces the atomization. Whereas theories (2) and (3) can interpret the phenomenon of atomized droplets observed at the nozzle exit, theory (1) cannot.
4. Boundary-condition discontinuity theory: the liquid jet is atomized because of the boundary-condition discontinuity after being injected from a nozzle.
5. Pressure oscillation theory: The pressure oscillation in a fuel feeding system has a certain influence on the atomization process. However, this theory still needs further research.

Bracco and Reitz [18] of Princeton University proposed an improved aerodynamic force disturbance mechanism. When resolving the equations that govern the atomization process, the solution calculated through the linear perturbation method includes an uncertain parameter reflecting the initial amplitude. They suggested that the atomization characteristics of different nozzles can be comprehensively analyzed and predicted by connecting this parameter with the geometrical parameters of the nozzle. The spray cone angle, initial droplet diameter, and the breakup length predicted by this theory are in good agreement with experimental results.

2.2.3   Liquid Sheet Breakup

Studies on the stability of the liquid sheet date back to 1868 when Helmholtz came up with the instability problem of two incompressible fluids with different densities. In 1871, Kelvin analyzed this stability problem and investigated the ripples caused by wind under the combined action of the liquid surface tension and aerodynamic forces. The wave growth studied by Helmholtz and Kelvin is the classic Kelvin–Helmholtz instability, and the subsequent research on the liquid sheet breakup caused by surface waves is based on this.

Fraser [19], Dombrowski and Hopper [16], and Eisenklam [20], from 1953 to 1963, carried out numerous theoretical and experimental studies on the breakup mechanism of the planar liquid sheet. They revealed details of the liquid sheet disintegration process by high-speed photography, and suggested that rim disintegration, perforation disintegration, and wave disintegration are the three basic modes of liquid sheet breakup. In 1953 York et al. [21] also studied the breakup process of the planar liquid sheet theoretically. They thought that the formation and growth of waves on the gas–liquid interface plays an important role in the liquid sheet breakup, and derived a correlation for the growth rate of disturbances in terms of wavelength and We. This theory was then extended to investigate the breakup of a conical liquid sheet from a swirl injector. In 1953 Squire [22] studied the stability of a planar liquid sheet of constant thickness and proposed a mathematic expression with the largest disturbance wavelength using an inviscid potential flow theory. In 1955 Hagerty and Shea studied the stability of a planar liquid sheet. They obtained the correlation function and experimental value of β, which is the disturbance growth factor in York's equation, and deduced the minimum frequency and wavelength of the unstable disturbance that leads to disintegration.

Breakup mechanisms of the planar, fan, and conical liquid sheet are discussed in the following subsections.

2.2.3.1   Planar Liquid Sheet Breakup

Fraser [19] and Dombrowski and Hopper [16], between 1953 and 1963, employed high-speed photography to reveal details of the liquid sheet breakup process. They categorized the breakup process into four modes:

1. Rim breakup: Under the action of the surface tension the liquid sheet contracts into a thick rim at its edge. The rim then begins to disintegrate because of aerodynamic forces and surface tension. This pattern of liquid sheet breakup occurs when the liquid viscosity and surface tension are quite high. The droplets formed in this breakup mode are relatively large.
2. Wave disintegration: The disturbance waves on the liquid sheet increase until liquid sheets of one or half wavelength shed off and form liquid filaments or slices that contract into liquid droplets due to the surface tension. The droplets formed in this disintegration pattern show a wide size distribution.
3. Perforation: Holes appear in the liquid sheet at a distance downstream of the nozzle exit. The size of the holes increases, leading to the formation of liquid ribbons or filaments between two adjacent cavities. The liquid ribbons and liquid filaments then separate, and break up into drops of different sizes. Since the sizes of irregularly formed liquid filaments and ribbons are uniformly distributed, the droplets resulting from such atomization show a relatively uniform size.
4. Turbulence breakup: When the injection speed of the liquid is sufficiently high, the liquid disintegrates into small drops at the nozzle exit. To date, there is no appropriate theory that can describe this complex process.

Research on theoretical modeling of the planar liquid sheet breakup is mainly focused on the simple wave disintegration shown in Figure 2.5. It is difficult to carry out a theoretical analysis of the other three complex breakup modes.

The theoretical analysis of the wave disintegration by York et al. [21], Squire [22], and Hagerty and Shea [23] is detailed in the following subsections.

Analysis of York et al. [21]
The basic assumptions are: (i) the liquid is inviscid and irrotational; (ii) the formation and development of the liquid sheet surface waves determines the liquid sheet breakup; (iii) the initial disturbance increases exponentially over time:

$$\delta = \delta_0 \exp(\beta t) \tag{2.26}$$

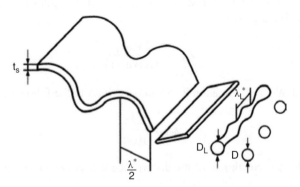

Figure 2.5   Planar liquid sheet breakup.

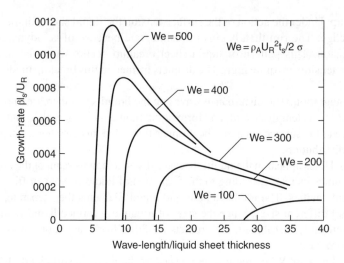

Figure 2.6   Growth rate of disturbance of different wavelengths on a planar liquid sheet (York).

where δ_0 is the amplitude of the initial disturbance, β is the growth rate of the disturbance, and t is time.

The breakup time of a liquid sheet deduced by York is:

$$t = \beta^{-1} \ln\left(\frac{h_0}{2\delta_0}\right) \tag{2.27}$$

where h_0 is the initial liquid sheet thickness.

Equation 2.27 indicates that a thicker liquid sheet requires a longer breakup time in a logarithmic function. York also proposed a correlation between the maximum growth rate and disturbance wavelength at different We (Figure 2.6).

Analysis of Squire [22]

Based on the work of York, Squire suggested that initial disturbances of wavelength beyond a critical value will be amplified by the surface tension while disturbances of wavelength smaller than the critical value will be damped. The minimum wavelength of disturbances that can induce liquid sheet breakup is:

$$\lambda_{min} = \frac{2\pi h_0 \rho_1}{\rho_A (We - 1)} \tag{2.28}$$

Considering that $We \gg 1$, Equation 2.28 can be written in the following form:

$$\lambda_{min} = \frac{2\pi\sigma}{\rho_A u_r^2} \tag{2.29}$$

The wavelength and growth rate of the disturbance that grows most rapidly is:

$$\lambda_{max} = \frac{4\pi\sigma}{\rho_A u_r^2} = 2\lambda_{min} \tag{2.30}$$

$$\beta_{max} = \frac{\rho_A U_r^2}{\sigma(\rho_l h_0)^{0.5}} \tag{2.31}$$

Equation 2.31 indicates that the wavelength of the disturbance with the maximum growth rate is proportional to the surface tension. When liquid sheets of one or half wavelength peel off and break up, a larger surface tension leads to larger droplet size. The gas density and gas–liquid relative velocity have an adverse influence in contrast with the surface tension.

Analysis of Hagerty and Shea [23]

The growth rate of the sinusoidal disturbance was measured in experiments. By comparing the experimental data with the theoretical model, a correlation between the frequency and disturbance growth-rate was derived (Figure 2.7):

$$\beta = \frac{n^2 U_r^2 (\rho_A/\rho_l) - n^3 \sigma/\rho_l}{\tanh(n h_0/2)} \tag{2.32}$$

where $n = \bar{\omega}/u$ is the wave number of the disturbance, U_r is the gas–liquid relative velocity, and tanh is a hyperbolic tangent function.

The minimum frequency of the disturbance causing liquid sheet breakup is:

$$f_{min} = \frac{\rho_A U_r^3}{2\pi\sigma} = \frac{U_r We}{2\pi h_0} \tag{2.33}$$

$$\lambda_{min} = \frac{2\pi\sigma}{\rho_A U_r^2} \tag{2.34}$$

The minimum disturbance wavelength obtained by Hagerty and Shea is identical with that of Squire (Figure 2.7).

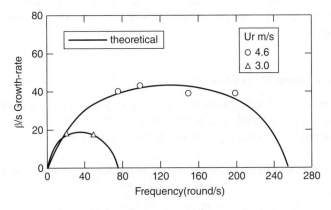

Figure 2.7   Growth rate of disturbances of different frequencies on a planar liquid sheet (Hagerty–Shea).

2.2.3.2 Fan Liquid Sheet Breakup

Fraser [19] extended the theory of Hagerty and Shea and Squire to study the breakup mechanism of a fan liquid sheet with a low viscosity. He found that the fastest-growing disturbance determined the liquid sheet disintegration. In addition, liquid sheets of half wavelength shed off and form filaments that subsequently break up into drops. The size of the droplets can be computed using the method employed to determine the diameter of droplets formed in liquid jet breakup.

The liquid sheet thickness (h_0) can be calculated from the nozzle configuration parameters, surface tension, and gas–liquid relative velocity:

$$h_0 = \left(\frac{1}{2H^2}\right)^{1/3} \left(\frac{k\rho_A u_r^2}{\rho_1 \sigma}\right)^{1/3} \tag{2.35}$$

$$H = \ln(\delta^*/\delta_0) \tag{2.36}$$

Here, δ^* is the disturbance amplitude where the liquid sheet disintegrates, and δ_0 is the initial disturbance amplitude:

$$D_1 = \left(\frac{2}{\pi}\lambda_{max} h_0\right)^{1/2} \tag{2.37}$$

where λ_{max} is the wavelength of the fastest-growing disturbance.

Following Equation 2.8, the droplet diameter can be computed as:

$$d = 1.89 D_1 = 1.89 \times \sqrt{\frac{2}{\pi}\lambda_{max} h_0} \tag{2.38}$$

Substituting Equations 2.30 and 2.35 into Equation 2.38 yields:

$$d \propto \left(\frac{\rho_1}{\rho_A}\right)^{1/6} \left(\frac{k\sigma}{\rho_1 u_r^2}\right)^{1/6} \tag{2.39}$$

Equation 2.39 indicates that the diameter of the drops formed when fan liquid sheets break up is related to the gas/liquid density ratio, gas–liquid relative velocity, surface tension, and nozzle configuration parameters. Though the relation is not quantitative, it is still instructive in engineering applications.

2.2.3.3 Conical Liquid Sheet Breakup

Since conical liquid sheet breakup is more complex than planar liquid sheet breakup, the theoretical analysis of conical liquid sheet breakup is more difficult. Figure 2.8 presents four different liquid patterns with increasing pressure in the swirl injector:

1. At stage 1, liquid is injected from the nozzle in the form of droplets or a distorted thin column.

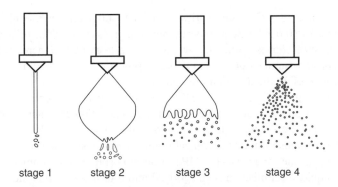

stage 1 stage 2 stage 3 stage 4

Figure 2.8 Four stages of the development of a conical liquid sheet.

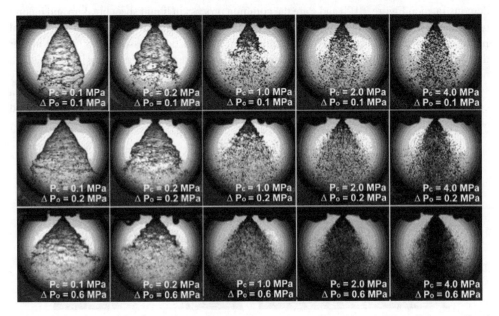

Figure 2.9 Atomization at different injection pressure drop and ambient pressure in a centrifugal injector [24].

2. At stage 2, the conical liquid sheet appears at the nozzle outlet, and contracts in the downstream into a closed liquid bulb under the action of surface tension.
3. At stage 3, the closed liquid sheet bulb expands into a shape of tulip. Large droplets continually shed off at the irregular edges of the liquid sheet.
4. At stage 4, the liquid sheet expands into a conical shape. As the conical liquid sheet expands it becomes thinner. The liquid sheet then becomes unstable and breaks up into filaments and droplets.

Figure 2.9 presents photographs of conical liquid sheet breakup at different injection pressure drop and ambient pressure, taken in the rocket propulsion laboratory of Seoul National

University [24]. The ambient pressure varies from 0.1 to 4.0 MPa, and the injection pressure drop varies from 0.1 to 0.6 MPa. At a constant injection pressure drop, the breakup position of the conical liquid sheet moves upstream towards the nozzle exit when increasing the ambient pressure. At a constant ambient pressure, the breakup position of the liquid sheet also moves upstream when increasing the injection pressure drop.

In a rocket engine, the designed pressure drop of a swirl injector is generally 20–30% of the combustion chamber pressure. As the pressure in the combustion chamber varies from 2.0 to 5.0 MPa, the pressure drop of the swirl nozzle varies from 0.4 to 1.5 MPa. Liu *et al.* [25] used high-speed shadowgraphs to study the conical liquid sheet breakup mechanism with an injection pressure drop in the range 0.4–1.0 MPa in an atmospheric environment. As shown in Figure 2.10 the liquid sheet breakup is categorized into four modes: (a) liquid sheet avulsion; (b) liquid sheet crest breakup; (c) liquid sheet trough breakup; (d) liquid sheet burr breakup. Figure 2.10a presents the breakup mode at a low injection pressure drop. The breakup modes demonstrated in Figure 2.10b and c are the common process in liquid sheet breakup and belong to regular fracture phenomenon. Figure 2.10d shows the breakup morphology of a conical liquid sheet injected from a recessed injector when the injection pressure drop (Δp) is higher than 0.5 MPa. Generally, a liquid sheet cannot break up immediately after being injected from a nozzle for two reasons. First, the amplitude of the surface wave near the nozzle exit is relatively small. Second, the liquid sheet in this region is thick. Only when the fluctuation propagates a certain distance downstream and the amplitude grows beyond a critical value will the liquid sheet disintegrate.

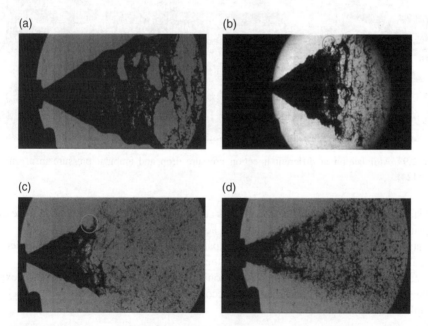

Figure 2.10 Conical liquid sheet breakup mechanism of a centrifugal injector at about the designed working conditions: (a) liquid sheet avulsion; (b) liquid sheet crest breakup; (c) liquid sheet trough breakup; (d) liquid sheet burr breakup.

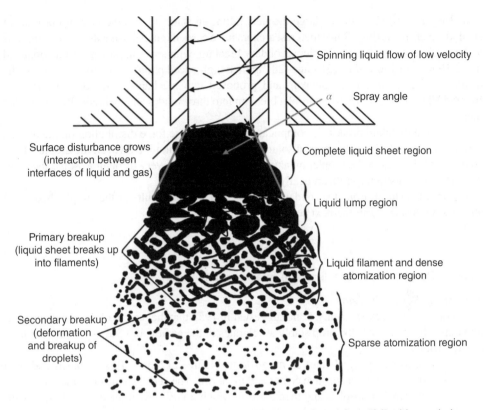

Spinning liquid flow of low velocity

α Spray angle

Surface disturbance grows
(interaction between
interfaces of liquid and gas)

Complete liquid sheet region

Liquid lump region

Primary breakup
(liquid sheet breaks up
into filaments)

Liquid filament and dense
atomization region

Secondary breakup
(deformation
and breakup of
droplets)

Sparse atomization region

Figure 2.11   Atomization of conical liquid sheet breakup. Adapted from [26] with permission.

Experimental observations indicate that the breakup process of the conical liquid sheet can be divided into several regions [26] (Figure 2.11): complete liquid sheet region, liquid lump region, liquid filament region, dense atomization region, and the sparse atomization region. The complete liquid sheet region is the region where the liquid sheet does not break up and the surface disturbance continuously grows. In the liquid lump and dense atomization regions the liquid sheet undergoes primary atomization and breaks up into liquid filaments. In the sparse atomization region the liquid filament breaks up into droplets and secondary atomization occurs.

In the above subsections, several forms of liquid sheet breakup have been analyzed and theoretical analysis results have been presented. The theoretical research on the breakup process of a liquid sheet can act as qualitative guidance, but is not validated for engineering applications.

2.2.4   Droplet Secondary Breakup

If droplets resulting from a liquid jet or sheet breakup enter the airflow with a high relative velocity, the droplets are subject to strong aerodynamic forces. When the aerodynamic forces exceed the restoring surface tension the droplets can disintegrate into smaller ones. This process is called secondary breakup.

In 1904 Lenard [27] studied large droplets falling in static air and the breakup process of small droplets in air flow. Since then researchers have carried out extensive theoretical analysis and experimental studies on droplet secondary breakup, with some important results obtained in the 1950s and 1960s. In 1951 Lane [28] applied high-speed photography to demonstrate the details of droplet secondary breakup and proposed two mechanisms for droplet breakup. In 1955 Hinze [29] classified the droplet breakup into three modes and theoretically interpreted them.

The droplet breakup process is complicated. Aerodynamic force distribution on the droplet surface is required in a mathematical solution for droplet breakup. However, the droplet shape continually changes in the droplet breakup process, making it very difficult to solve the aerodynamic force distribution theoretically.

The mechanism of droplet secondary breakup and several results of the simple theoretical analysis are discussed in the next section.

2.2.4.1 Droplet Secondary Breakup in a Laminar Airflow

Analysis of Results by Lane

Based on numerous experiments, Lane suggested in 1951 that the pattern of droplet disintegration depends on whether the drop stably accelerates in the airflow or suddenly enters the high speed airflow.

When a drop stably accelerates in an airflow (Figure 2.12) it is first flattened and then deforms into a bowl with a concave center and thick rim. The center of the deformed drop first breaks up into smaller droplets and then the rim, which holds 70% of the drop volume, disintegrates into relatively large droplets. For this breakup mode, Lane deduced an empirical

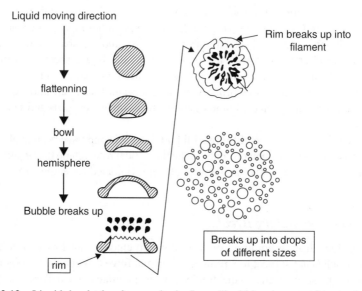

Figure 2.12   Liquid droplet breakup modes by Lane (liquid breakup at stable acceleration).

expression of minimum gas–liquid relative speed needed for secondary breakup in an ambient atmosphere:

$$u_r \propto \left(\frac{\sigma}{D}\right)^{1/2} \tag{2.40}$$

For a water drop the minimum relative velocity required for secondary breakup is:

$$u_r = 784/\sqrt{D} \tag{2.41}$$

When a droplet suddenly enters a high speed airflow the pattern of secondary breakup is significantly different from the previous one. In the airflow direction, a droplet protrudes as a lentoid and its edge is cut into a slice. After that, the slice contracts into filaments that subsequently break up into droplets. While Lane did not deduce any empirical expression for this breakup mode he found that the critical We for secondary breakup is much smaller in this breakup mode than for the previous one.

Analysis Results of Hinze
In 1955, Hinze [29] employed high-speed photography to demonstrate the detailed breakup process and classified the droplet breakup into three basic modes (Figure 2.13):

1. The droplet is first flattened into a lentoid shape under the aerodynamic forces. The lentoid droplet then deforms into a concave lotus-seat shape, which breaks up into small droplets.
2. The droplet is flattened into a cylindroid cigar shape, and then breaks up into filaments.
3. The droplet deforms into an irregular shape and the swelling part sheds off and produces small droplets.

Hinze suggested that the deformation of the droplet mainly depends on three factors: aerodynamic forces, surface tension, and viscous forces. The combined action of these three forces determines the droplet breakup mode. Hinze ignored the liquid viscosity and suggested that the critical condition for droplet breakup is when the aerodynamic forces equal the surface tension:

$$C_D \frac{\pi D^2}{4} 0.5 \rho_A u_r^2 = \pi D \sigma \tag{2.42}$$

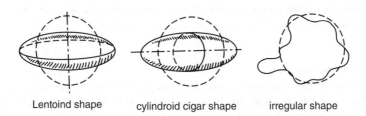

Lentoind shape   cylindroid cigar shape   irregular shape

Figure 2.13   Liquid droplet breakup modes according to Hinze.

Equation 2.42 can be converted into:

$$\frac{\rho_A u_r^2 D}{\sigma} = \frac{8}{C_D}$$

(2.43)

$$We_{crit} = \frac{8}{C_D}$$

(2.44)

The maximum stable diameter of the droplet at a certain relative velocity is:

$$D_{max} = \frac{8\sigma}{C_D \rho_A u_r^2}$$

(2.45)

The maximum relative velocity keeping the droplet stable is:

$$u_{r,max} = \left(\frac{8\sigma}{C_D \rho_A D}\right)^{1/2}$$

(2.46)

Influence of Viscosity
Since liquid viscosity retards the deformation of the droplet, the secondary breakup is more difficult for a drop with a high viscosity. Hinze employed the Oh to show the influence of liquid viscosity on secondary breakup, and proposed an expression for the critical We for droplet breakup versus liquid viscosity:

$$We_{crit} = We_{crit}^0 [1 + f(Oh)]$$

(2.47)

We_{crit}^0 is the critical We when the viscosity is zero and the corresponding Oh is also zero. A droplet with high liquid viscosity corresponds to a large Oh, and the critical We increases considerably as Oh increases.

Hanson et al. [30] deduced the specific correlation:

$$We_{crit} = We_{crit}^0 + 14Oh$$

(2.48)

Hanson pointed out that the error of this correlation is about 20%. When Oh is relatively small, droplet breakup can be categorized into six regimes with increasing We: non-deformation, non-oscillatory deformation, oscillatory deformation, bag breakup, multimode breakup, and shear breakup (Figure 2.14 shows some of these regimes).

Viscosity of the Ambient Gas
When considering the gas viscosity, the aerodynamic pressure is less important because there is a boundary layer velocity that is relatively low around the droplet. The main factors influencing the droplet breakup are the viscous shear force and the surface tension. When the viscous shear force exceeds the surface tension, secondary breakup happens. Under the action of the viscous

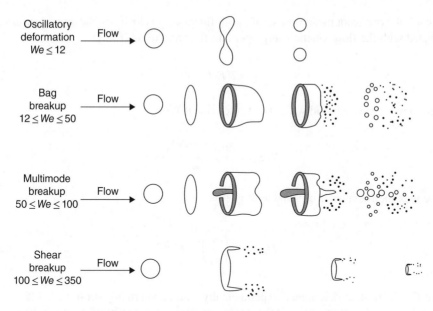

Figure 2.14   Some liquid breakup modes at low Oh.

force the droplet is pulled into a prolate-spheroid shape. The We_{crit} for droplet breakup depends on the ambient airflow and the droplet parameters.

The correlation for the critical We [31] is:

$$We_{crit} = \frac{1 + (\mu_l/\mu_g)}{1 + 1.189(\mu_l/\mu_g)} \qquad (2.49)$$

2.2.4.2   Droplet Secondary Breakup in Turbulent Flows

The gas flows are turbulent in an engine. Because of the gas-phase turbulent vortex, a droplet is continually subject to random disturbing forces. The droplet movement caused by random disturbance is more obvious for smaller droplets. Kolmogorov [32] and Hinze [29] studied the droplet breakup in a turbulent region. They suggested that droplet breakup in a turbulent region was related to the turbulent kinetic energy and that the maximum droplet diameter after atomization was determined by the turbulent aerodynamic forces. The influence of turbulent kinetic energy on atomization grows stronger when the wavelength of surface waves on the injected liquid column or the gas–liquid relative velocity increases. When the wavelength of the surface waves exceeds twice the droplet diameter, the gas–liquid velocity difference generates relatively strong aerodynamic forces and the turbulent kinetic energy causes droplet breakup. A correlation for the critical We in isentropic flows is:

$$We_{crit} = \frac{\rho_g \bar{u}^2 D_{max}}{\sigma} \qquad (2.50)$$

where $\bar{u}$ is the rms (root mean square) of the air fluctuating velocity on the droplet surface; it is correlated with the flow kinetic energy per unit time and mass (E) by:

$$\bar{u}^2 = 2(ED_{max})^{2/3} \tag{2.51}$$

For a low-viscosity liquid with Oh ≤ 1, the critical We is:

$$We_{crit} = \frac{2\rho_g E^{2/3} D^{5/3}_{max}}{\sigma} \tag{2.52}$$

The maximum diameter of a stable droplet is:

$$D_{max} = C\left(\frac{\sigma}{\rho_g}\right)^{3/5} E^{-2/5} \tag{2.53}$$

where C is a constant determined experimentally, and is commonly set to be 0.725.

The above analysis indicates that the maximum diameter of a stable droplet is only related to the surface tension (σ), the ambient gas density (ρ_g), and the kinetic energy per unit time and mass (E).

2.2.4.3 Droplet Breakup in a High Speed Flow

Lee and Ritz [33] analyzed droplet breakup in a high-speed air flow with high-resolution photographs and observed three breakup modes at different air flow speeds and densities: the bag breakup (Figure 2.15), stretching breakup (Figures 2.16 and 2.17), and catastrophe breakup (Figure 2.18). They concluded that the breakup mode is determined by We rather than Re.

Dinh et al. [34] studied the droplet breakup in a liquid-fuel detonation engine (Figure 2.19). Under the condition of supersonic air flows, a droplet breaks up more rapidly. In addition, under the condition of subsonic air flows the droplet surface is smoother.

Tianwen Fang [35] utilized the interface tracking method to study the deformation and breakup process of a single droplet. Figure 2.20 demonstrates the deformation and breakup process of this droplet in an airflow of 432 m s^{-1} ($Ma = 2.0$). Under the action of a supersonic airflow the deformation of the droplet starts on the upstream side. As the droplet deforms into a liquid disk, the small droplets start shedding off at the disc periphery at $t = 170$ μs, and then the main droplet becomes smaller.

2.3   Characteristics of Atomization in Liquid Rocket Engines

In a liquid rocket engine, the size of droplets resulting from atomization is widely distributed. The distribution function of droplet size, droplet mean diameter, and other parameters are commonly used to evaluate the atomization performance.

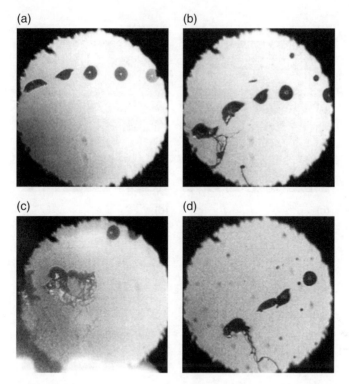

Figure 2.15 The bag breakup (We = 72, density and velocity of flow are respectively (a) 1.2 kg m^{-3}, 82 m s^{-1}; (b) 4.3 kg m^{-3}, 42 m s^{-1}; (c) 7.5 kg m^{-3}, 32 m s^{-1}; (d) 10.6 kg m^{-3}, 27 m s^{-1}) [33].

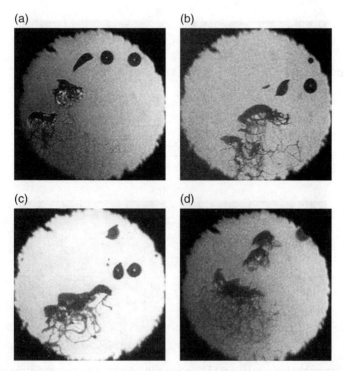

Figure 2.16 Stretching breakup (We = 148, density and velocity of flow are respectively (a) 1.2 kg m^{-3}, 118 m s^{-1}; (b) 4.3 kg m^{-3}, 61 m s^{-1}; (c) 7.5 kg m^{-3}, 46 m s^{-1}; (d) 10.6 kg m^{-3}, 39 m s^{-1}) [33].

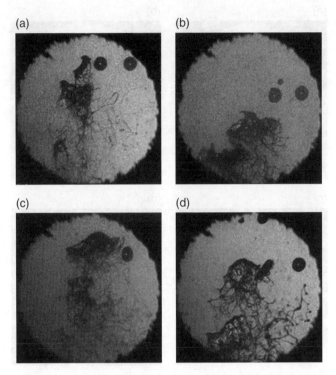

Figure 2.17 Stretching breakup (We = 270, density and velocity of flow are respectively (a) 1.2 kg m^{-3}, 159 m s^{-1}; (b) 4.3 kg m^{-3}, 82 m s^{-1}; (c) 7.5 kg m^{-3}, 62 m s^{-1}; (d) 10.6 kg m^{-3}, 52 m s^{-1}) [33].

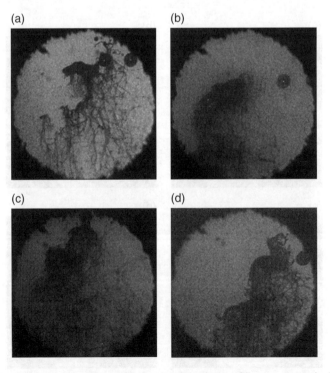

Figure 2.18 Catastrophe breakup (We = 532, density and velocity of flow are respectively (a) 1.2 kg m^{-3}, 223 m s^{-1}; (b) 4.3 kg m^{-3}, 115 m s^{-1}; (c) 7.5 kg m^{-3}, 87 m s^{-1}; (d) 10.6 kg m^{-3}, 73 m s^{-1}) [33].

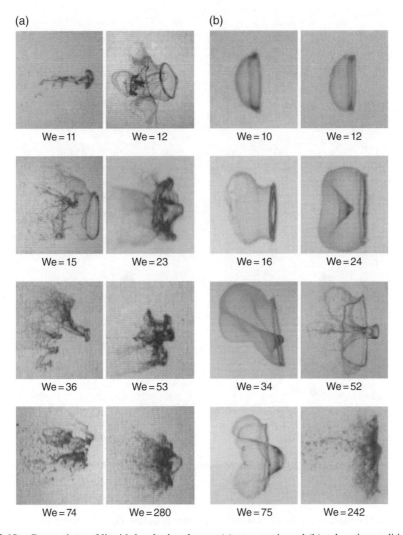

Figure 2.19   Comparison of liquid droplet breakup at (a) supersonic and (b) subsonic conditions [34].

2.3.1   *Distribution Function of the Droplet Size*

The distribution function of the droplet size describes the percentage of droplets of different diameters in a spray in terms of number, volume, or mass. The commonly used size distribution functions are:

1. Normal distribution function:

$$\frac{\mathrm{d}N}{\mathrm{d}D} = \frac{1}{\sqrt{2\pi}s_\mathrm{n}} \exp\left[-\frac{1}{2s_\mathrm{n}^2}(D-\bar{D})^2 \right] \tag{2.54}$$

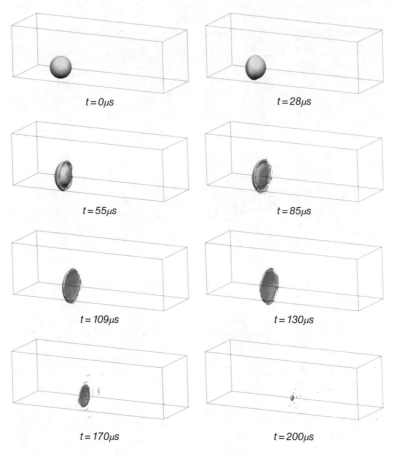

Figure 2.20   Breakup of droplet in an airflow of 432 m s^{-1}.

Here, N is the number of droplets whose diameter is below D, s_n is the standard deviation in normal distribution, and $\bar{D}$ is the algorithmic mean value of droplet diameters.

2. Logarithm–normal distribution function:

$$\frac{\mathrm{d}N}{\mathrm{d}D} = \frac{1}{\sqrt{2\pi}s_g} \exp\left[-\frac{1}{2s_g^2}\left(\ln D - \ln \bar{D}_{ng}\right)^2 \right] \qquad (2.55)$$

where $\bar{D}_{ng}$ is the geometric mean diameter and s_g is the geometric standard deviation.

3. Nukiyama–Tanasawa distribution function:

$$\frac{\mathrm{d}N}{\mathrm{d}D} = aD^p \exp(-bD)^q \qquad (2.56)$$

where a, b, and p are constants. This equation is applied to straight-flow injectors.

4. Rosin–Rammler distribution function:

$$Q = 1 - \exp\left[-\left(\frac{D}{c}\right)^N\right] \tag{2.57}$$

where c and N are constants and Q represents the volume percentage of droplets whose diameter is below D. This equation is applied to centrifugal injectors.

Rizk and Lefebvre [36] studied the sprays of swirl injectors and found that the Rosin–Rammler distribution function was appropriate for most spray cases, but the error is considerable in the range of large particles. Therefore, they modified the equation:

$$Q = 1 - \exp\left[-\left(\frac{\ln D}{\ln c}\right)^N\right] \tag{2.58}$$

5. Upper-limit distribution function:

This equation was proposed by Mugele and Evans [37] based on the analysis of various theories and detailed comparison between the empirical distribution functions and experimental data:

$$\frac{dQ}{dD} = k \exp\left(-\frac{k^2 y^2}{\sqrt{\pi}}\right) \tag{2.59}$$

$$y = \ln\frac{aD}{D_{max} - D},$$

where D_{max} is the maximum diameter; k and a are constants.

2.3.2 Mean Diameter and Characteristic Diameter

The size of droplets in a spray is widely distributed. A mean diameter is used for convenience in the study of atomization. The concept of the mean droplet diameter was proposed by Mugele and Evans: the mean diameter is the diameter of droplets in an assumed spray field of uniform-sized droplets which substitutes the real spray field.

There are several definitions of the mean diameter of spray droplets. Different averaging methods give different means. Assuming that the droplet diameter range is $[D_0, D_m]$, the mean diameter can be calculated by integrating the droplet diameter in this range as follows:

Length mean diameter:

$$D_{10} = \frac{\int_{D_o}^{D_m} D\left(\frac{dN}{dD}\right) dD}{\int_{D_o}^{D_m} \left(\frac{dN}{dD}\right) dD} \tag{2.60}$$

surface-area mean diameter:

$$D_{20} = \sqrt{\frac{\int_{D_o}^{D_m} D^2 \left(\frac{dN}{dD}\right) dD}{\int_{D_o}^{D_m} \left(\frac{dN}{dD}\right) dD}} \tag{2.61}$$

volume mean diameter:

$$D_{30} = \left(\frac{\int_{D_o}^{D_m} D^3 \left(\frac{dN}{dD}\right) dD}{\int_{D_o}^{D_m} \left(\frac{dN}{dD}\right) dD}\right)^{\frac{1}{3}} \tag{2.62}$$

volume to surface area ratio diameter:

$$D_{32} = \frac{\int_{D_o}^{D_m} D^3 \left(\frac{dN}{dD}\right) dD}{\int_{D_o}^{D_m} D^2 \left(\frac{dN}{dD}\right) dD} \tag{2.63}$$

where D_{32} is the Sauter mean diameter (SMD).

Obviously, the total surface area of droplets with a smaller D_{32} is larger for a certain liquid volume. Assuming that N_i is the number of droplets with a diameter of D_i, the total liquid mass in the spray is:

$$G = \sum N_i \frac{\pi}{6} \rho_1 D_i^3 \tag{2.64}$$

If the size of all droplets is assumed to be the SMD and the droplet number is N_s, the total spray mass can be computed by:

$$G = N_s \frac{\pi}{6} \rho_1 SMD^3 \tag{2.65}$$

From Equations 2.64 and 2.65, we can obtain the following equation:

$$\sum N_i D_i^3 = N_s SMD^3 \tag{2.66}$$

Since the total area is assumed to be same, we can obtain:

$$\sum N_i D_i^2 = N_s SMD^2 \tag{2.67}$$

Dividing Equation 2.66 by Equation 2.67, the SMD can be calculated by:

$$\text{SMD} = \frac{\sum N_i D_i^3}{\sum N_i D_i^2} \qquad (2.68)$$

The SMD of droplets reflects the combustion performance of the liquid spray and is a commonly-used characteristic parameter in the combustion field.

In further research on the size distribution of spray droplets, not only the droplet mean diameter and distribution function are presented but also extra characteristic points in the distribution curve have been analyzed. The extra points represent the volume percentage of droplets with a diameter below a certain value. For example, the percentage of the total volume of droplets with a diameter below $D_{0.1}$, $D_{0.5}$, and $D_{0.9}$ is, respectively, 10%, 50%, and 90%. The term $D_{0.5}$ is a mass medial diameter (MMD). Assuming that the droplet diameter in the spray follows the Rosin–Rammler distribution, the following correlation between SMD and MMD is obtained:

$$\text{MMD} = \text{SMD}\,\Gamma\left(1 - \frac{1}{N}\right) 0.693^{1/N} \qquad (2.69)$$

where Γ is the gamma function and N is a distribution constant.

2.3.3 Measurement of Spray Size Distribution

To evaluate the atomization performance qualitatively and quantitatively it is necessary to carry out experimental measurements of the spray. The experimental techniques are generally classified into three categories: mechanical, electronic, and optical measurement methods. Mechanical techniques include droplet solidification, dewaxing, sedimentation, and indentation methods. Electronic techniques include electrode, traverse and hot-wire methods. Optical techniques are the immersion method and direct stroboscopic photography, laser holographic method, high-speed photography, laser interference fringes spectrum method, laser scattered light intensity correlation method, multi-source laser scattering method, laser Doppler method, and diffraction-based Malvern method. At present the commonly-used approaches are the Malvern method based on diffraction and PDA (phase Doppler anemometry) based on scattering.

2.3.3.1   Malvern Principle

The Malvern apparatus is mainly composed of a laser generator, receiving terminal, and laser diffraction particle sizing system that employs a He-Ne laser, expanded beam lens, spatial filter, receiving lens, and light target (multiple loop photoelectric detector). The laser generator includes a He-Ne laser, expanded beam lens, spatial filter and collimation lens. The receiving lens and light target are integrated into the receiving terminal, which is connected to a computer by a data wire from the light target. The optical system is shown in Figure 2.21.

The laser produced by the He-Ne laser generator becomes a beam of parallel light with a diameter of 9 mm after being expanded, filtered, and collimated. The beam of parallel light generates the scattered light passing through the spray field in the test section. The scattered

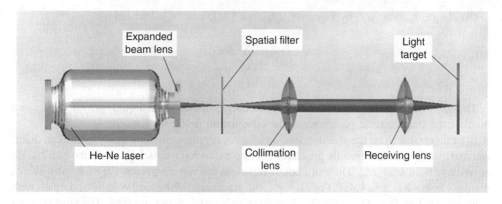

Figure 2.21   Optical system of a laser scattering particle detector.

light, which contains the information on the droplet's size and distribution, is focused on the light target by the receiving lens.

A laser scattering particle detector resolves the size distribution of particles by detecting the distribution of optical energy scattered by particles. Actually, the optical energy distribution of a droplet cluster is the superposition of distributions of the optical energy scattered by a single particle.

For a single particle, if the total energy of the incident light is considered to be 100%, according to the Fraunhofer diffraction theory the scattered optical energy distribution is:

$$L(\rho) = 1 - J_0^2(\rho) - J_1^2(\rho) \tag{2.70}$$

where J_0 and J_1 are, respectively, the zeroth and first order Bessel functions; L is the relative optical energy distribution, $\rho = (2\pi R/\lambda)\theta$, where R is particle radius, and θ is the scattering angle.

Consequently, the relative energy of scattered light cast on the annulus with an inner diameter of R_{in} and an outer diameter of R_{out} is:

$$
\begin{aligned}
L_{\mathrm{in,\,out}} &= L(\rho_{\mathrm{out}}) - L(\rho_{\mathrm{in}}) \\
&= \left[J_0^2(\rho_{\mathrm{in}}) + J_1^2(\rho_{\mathrm{in}}) \right] - \left[J_0^2(\rho_{\mathrm{out}}) + J_1^2(\rho_{\mathrm{out}}) \right]
\end{aligned} \tag{2.71}
$$

$$\rho_{\mathrm{in}} = \frac{2\pi R}{\lambda} \cdot \frac{R_{\mathrm{in}}}{l}, \rho_{\mathrm{out}} = \frac{2\pi R}{\lambda} \cdot \frac{R_{\mathrm{out}}}{l} \tag{2.72}$$

where l is the focal length of the lens. Equation 2.71 can written in the following simplified form:

$$L_{\mathrm{in,out}} = \left(J_0^2 + J_1^2 \right)_{\mathrm{in,\,R}} - \left(J_0^2 + J_1^2 \right)_{\mathrm{out,R}} \tag{2.73}$$

The subscripts "in", "out", and "R" represent, respectively, the annulus inner diameter (R_{in}), annulus outer diameter (R_{out}), and the radius of the detected particle (R).

The equation set of the Malvern scattering optical energy distribution can be obtained by applying the energy distribution of the light scattered by a single particle to a particle cluster.

We assume the inner and outer diameter of the i-th annulus is, respectively, $R_{in,i}$ and $R_{out,i}$, and denote the annulus as $(R_{in,i}, R_{out,i})$. The energy laser beam cast on a unit area per unit time is constant and is denoted as C.

For a spherical particle with a radius R, its projected area is πR^2. Assuming the number of particles with a radius of R is N, the total energy projected on N particles by the incoming parallel light is:

$$E_0 = C\pi R^2 N = C' R^2 N \qquad (2.74)$$

With Equation 2.72 the energy scattered by N particles whose radius is R on the annulus per unit time is:

$$
\begin{aligned}
E_i &= E_0 \left[\left(J_0^2 + J_1^2 \right)_{in,i,R} - \left(J_0^2 + J_1^2 \right)_{out,i,R} \right] \\
&= C' R^2 N \left[\left(J_0^2 + J_1^2 \right)_{in,i,R} - \left(J_0^2 + J_1^2 \right)_{out,i,R} \right]
\end{aligned}
\qquad (2.75)
$$

For a cluster of particles with different diameters, assume that the number of particles with a radius of R_j ($j = 1$–n) is N_j; then, the scattering energy on a certain annulus can be obtained by superposition:

$$E_i = C' \sum_{j=1}^{n} N_j R_j^2 \left[\left(J_0^2 + J_1^2 \right)_{in,i,Rj} - \left(J_0^2 + J_1^2 \right)_{out,i,Rj} \right] \qquad (2.76)$$

If the particles are grouped based upon diameters with the group number equal to the annulus number of light target (the arithmetic mean diameter of each group is used in calculations), the equation set is closed and has a unique solution, and thus the size distribution of the particles can be calculated.

Since the multivariate linear equation set is complex, the following method can be employed to simplify the solution process. It is assumed that the diameter distribution accords with a function such as the Rosin–Rammler distribution ($Q = \exp(-(D/\bar{D})^n)$). First, take a set of n and D into the equations, and then compare the calculated and measured optical energy of each annulus, and repeat until the convergence of the iteration.

2.3.3.2 Phase Doppler Anemometry PDA Method

The idea behind a phase Doppler particle analyzer is based on the analysis by Hulst [38] of scattering when light passes through transparent spherical particles with geometrical optics. In 1984, Bachalo and Houser [39, 40] developed the PDA technique. The technology is significant in spray measurement because it is capable of measuring the size, speed, number distribution and volumetric flow of the particles, induces no interference when measuring the spray, and possesses the features of high spatial resolution and multi-direction velocity components measurement [41]. The principle of PDA is demonstrated in Figures 2.22 and 2.23.

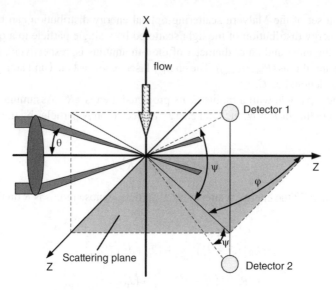

Figure 2.22   Optical principle of phase Doppler measurement.

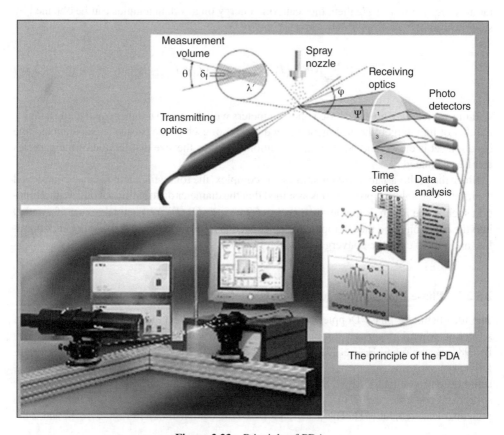

Figure 2.23   Principle of PDA.

The frequency of scattered light produced by moving particles is different from the frequency of the incoming light cast on the particles. The frequency difference is proportional to the particle velocity, and thus the speed of the moving particle can be calculated from the frequency difference. When multiple detectors are used, every detector receives the same Doppler frequency differences, but the scattered light received by different detectors shows a different phase. Since the phase difference is proportional to particle diameter, the particle diameter can be computed from the phase difference.

The particle speed (U) is calculated from the Doppler frequency (f_D) measured by any one of the detectors:

$$U = \frac{\lambda}{2\sin(\theta/2)}f_D \tag{2.77}$$

where θ is the angle between two incoming light beams.

The particle diameter (D) is calculated from the phase difference (ϕ) between the signals of two detectors.

When the reflected light is dominant in scattering, the correlation between ϕ and D is as follows:

$$\phi = \frac{2\pi D}{\lambda} \times \frac{\sin(\theta/2)\sin\psi}{\sqrt{2(1-\cos\theta/2\times\cos\psi\cos\varphi)}} \tag{2.78}$$

When refracted light is dominant in scattering, the correlation is:

$$\phi = \frac{-2\pi D}{\lambda} \times \frac{n_{rel}\sin(\theta/2)\sin\psi}{\sqrt{2(1+\cos\theta/2\times\cos\psi\cos\varphi)}\left(1+n_{rel}^2-n_{rel}\right)\sqrt{2(1+\cos\theta/2\times\cos\psi\cos\varphi)}} \tag{2.79}$$

where φ and ϕ are, respectively, the scattering angle and the azimuthal angle, and n_{rel} is the refractive index of the droplet.

The receiving system is required to receive the same pattern of scattered light (scattered light includes three patterns, namely, reflection, refraction, and secondary refraction) to guarantee the linear correlation between the particle diameter and the phase difference. Generally, the scattering angle of refraction is 30–70°, that of reflection is 80–110°, and that of secondary reflection is 135–150°.

2.4 Atomization Modeling for Liquid Rocket Engine Atomizers

The injector configurations in a liquid rocket engine are mainly impinging and coaxial. Efficient atomization by utilizing the flow of propellant itself guarantees improvement of the evaporation, mixing, and combustion inside the engine and enhancement of the engine performance. Since the injector has a significant influence on liquid rocket engine performance and stability margin, many researchers have carried out an extensive study on several commonly-used injector configurations and developed the corresponding atomization models.

2.4.1 Straight-flow Injector

Owing to its simple construction and easy machinability, the straight-flow injector is the main injector configuration applied in vehicles, boilers, and aviation engines. The flow inside a straight-flow injector satisfies the Bernoulli equation:

$$p_1 + \frac{1}{2}\rho u_1^2 = p_2 + \frac{1}{2}\rho u_2^2 \tag{2.80}$$

where p_1 and p_2 are, respectively, the pressure at the inlet and outlet, and u_1 and u_2 are, respectively, the velocity at the inlet and outlet.

Since u_1 is considerably smaller than u_2 and can be ignored, the outlet speed of the injector can be calculated by:

$$u_2 = \sqrt{2\Delta p/\rho} \tag{2.81}$$

The theoretical mass flow of an injector is:

$$\dot{m}_{\text{theo}} = \rho u_2 A = A\sqrt{2\rho\Delta p} \tag{2.82}$$

The flow coefficient of the injector is:

$$C_d = \frac{\dot{m}}{A\sqrt{2\rho\Delta p}} \tag{2.83}$$

Here, $\dot{m}$ is the practical mass flow of the injector.

2.4.2 Centrifugal Injector

The centrifugal injector is a simple mechanical pressure atomization injector (Figure 2.24). The injector is divided into four parts according to its function in the formation of ligaments at the

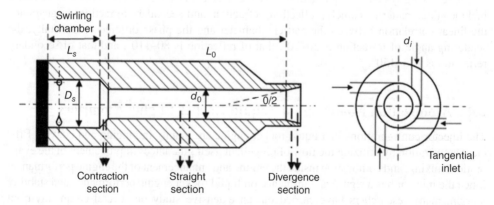

Figure 2.24 Configuration of the centrifugal injector.

injector outlet. The four parts are the swirling chamber, contraction section, straight section, and divergence section. The configuration of a swirling injector is simple. Since the swirling injector is energy-saving in atomization and reliable in operation, it is commonly applied in gas turbines, aviation engines, and rocket engines.

The theoretical methods used to determine the flow inside a centrifugal injector can be divided into two categories: one is the Abramovic theory based on the maximum flow principle, and the other is the method based on the momentum equation. The Abramovic theory is simple, convenient, and widely applied. The calculated results of the method based on the momentum equation agree well with experimental data, but this method is complicated. The model and equations of the flow coefficient, atomization angle, and droplet diameter distribution of centrifugal injector are summarized in the following sections.

Flow Coefficient

Currently, the most widely used flow coefficient equation is derived from the Abramovic theory, which is based on the maximum flow principle. The derivation process is presented below.

If we assume that the radius of a tangential hole is r_n and the number of holes is N, the total flow area is:

$$A_p = N\pi r_n^2 \tag{2.84}$$

The radius of the swirling chamber is R_s; the radius and area at the outlet of the swirling chamber are, respectively, r_0 and $A_o = \pi r_0^2$. According to the principle of momentum conservation, the following correlation is derived:

$$V_a r_o = V_n \cdot R_s \tag{2.85}$$

Here V_n is the liquid velocity at the outlet of the tangential hole, and can be calculated from:

$$V_n = \frac{\dot{m}}{\rho_1 A_p} \Rightarrow V_a = \frac{\dot{m} R_s}{\rho_1 A_p r_o} \tag{2.86}$$

According to the Bernoulli equation:

$$P = p + \frac{1}{2}\rho_1\left(u_a^2 + V_a^2\right) \tag{2.87}$$

Furthermore, if we assume that u is constant in the whole process and that the environment pressure equals the static pressure (P) at any axial position, the following correlation is obtained:

$$\Delta P = \frac{1}{2}\rho_1\left(u_a^2 + V_a^2\right) \tag{2.88}$$

The mass flow at the injector outlet can be calculated by:

$$\dot{m} = \rho_1(A_o - A_a)u_a \tag{2.89}$$

and thus:

$$u_a = \frac{\dot{m}}{\rho_1(A_o - A_a)} \tag{2.90}$$

Substituting Equations 2.86 and 2.90 into Equation 2.88 results in:

$$\Delta P = \frac{1}{2}\rho_1 \left[\left(\frac{\dot{m}R_s}{\rho_1 A_p r_o}\right)^2 + \left(\frac{\dot{m}}{\rho_1(A_o - A_a)}\right)^2 \right] \tag{2.91}$$

If we assume that the expression of the practical mass flow is:

$$\dot{m} = C_d \cdot A_o \sqrt{2\rho_1 \Delta p} \tag{2.92}$$

we can substitute Equation 2.90 into Equation 2.91 to obtain:

$$\frac{1}{C_d^2} = \frac{1}{u^2 X} + \frac{1}{(1-K)^2} \tag{2.93}$$

$$X = \frac{A_a}{A_o}; \quad K = \frac{A_p}{\pi r_o R_s}$$

Though the correlation between the flow coefficient (C_d) and the injector configuration/operation parameters has been obtained, $X = A_a/A_0$ is still unknown because the radius of air core at the injector outlet (r_a) is not available. For any centrifugal injector, the maximum flow coefficient cannot be achieved when the air core radius is too large or small. When the air core radius is large, the mass flow is small owing to the small effective sectional area. In the other case, when the air core radius is small, a considerable portion of fuel pressure energy is transferred into tangential kinetic energy, and thus the mass flow reduces owing to the decrease of axial velocity component.

Assuming that the injector is always at the operating condition of the maximum mass flow:

$$\frac{dC_d}{dX} = 0 \tag{2.94}$$

$$2K^2 X^2 = (1-X)^3 \tag{2.95}$$

the flow coefficient can be obtained from Equations 2.93 and 2.95:

$$C_d = \sqrt{\frac{(1-X)^3}{1+X}} \tag{2.96}$$

After the variable X ($X = (d_o - 2t)^2/d_o^2$) is obtained, it is easy to calculate the thickness of the liquid film at the injector outlet.

Since X is a function of configuration parameters (the geometric parameter K), the thickness of the liquid film at the injector outlet is a function of K, which contradicts the practical situation. For example, with increasing injection pressure drop the flow inevitably increases and thus the liquid film thickness grows.

In the above derivation, there are some assumptions that relate the flow coefficient (C_d) of the centrifugal injector with the geometric parameter, which is the combination of the injector diameter, swirling chamber diameter, and tangential hole diameter. It is irrational to consider the C_d of the injector as related only to the geometric parameter. In practice, the injector flow coefficient is related to many other construction parameters of the injector, including the length of swirling chamber and straight section and the angle of the contraction section.

Rizk and Lefebvre [42] deduced the equation for C_d based on the theoretical derivation and experimental results. The equation generally matches the experimental data:

$$C_d = 0.35 \left(\frac{A_P}{D_s d_0} \right)^{0.5} \left(\frac{D_s}{d_0} \right)^{0.25} \tag{2.97}$$

Jones [43] carried out tests on 159 different injector configurations and obtained a more accurate correlation between the flow coefficient and the configuration parameters, liquid properties, and injection conditions:

$$C_d = 0.45 \left(\frac{d_0 \rho_L U}{\mu_L} \right)^{-0.02} \left(\frac{l_0}{d_0} \right)^{-0.03} \left(\frac{L_s}{D_s} \right)^{0.05} \left(\frac{A_P}{D_s d_0} \right)^{0.52} \left(\frac{D_s}{d_0} \right)^{0.23} \tag{2.98}$$

Atomization Taper Angle

The atomization taper angle, which reflects the spatial distribution of the spray, is an important performance parameter. Rizk and Lefebvre [42] thought that the droplet disintegrates from the spray cone in the tangential direction in the plane which is tangential to the spray. The equation of the half-cone angle can be deduced by mathematical analysis:

$$\cos^2 \alpha = \frac{1-X}{1+X} \tag{2.99}$$

Prediction of Droplet Size

The droplet mean diameter is an important parameter in evaluating the atomization performance. Owing to the complexity of the injector configuration and variety of influencing factors, equipment measuring the diameter of atomized particles, and the measuring positions, the mean diameter of spray atomized by the injector cannot be predicted by a uniform equation. Lefebvre [44] systematically studied the mean spray diameter of centrifugal injectors and obtained the following correlation between SMD and liquid properties/injector working conditions:

$$\text{SMD} = A \sigma^{0.25} \mu_L^{0.25} \rho_L^{0.125} d_0^{0.5} \rho_A^{-0.25} \Delta P_L^{-0.375} \tag{2.100}$$

where A is an empirical constant and is generally set as 2.25.

Jones [43] took into account the influence of detailed parameters of centrifugal injectors:

$$\text{SMD} = 2.25\sigma^{0.25}\mu_L^{0.25}m_L^{0.25}\rho_A^{-0.25}\Delta P_L^{-0.5}\left(\frac{l_0}{d_0}\right)^{0.03}\left(\frac{L_s}{D_s}\right)^{0.07}\left(\frac{A_p}{D_sd_0}\right)^{-0.13}\left(\frac{D_s}{d_0}\right)^{0.21} \quad (2.101)$$

Juan Liu [45] *et al.* applied the Malvern laser-diffractometer to measure the commonly-used injectors in gas generators and obtained an empirical correlation for the mean spray diameter of the injector outlet angle, straight section length, and swirling diameter ratio:

$$\text{SMD} = B\sigma^{0.25}\mu_1^{0.25}\rho_1^{0.125}d_0^{0.5}\rho_g^{-0.25}\Delta P_1^{-0.375}f\left(\frac{D_s}{d_0},\frac{l_0}{d_0},\theta\right) \quad (2.102)$$

$$f = \left(\frac{D_s}{d_0}\right)^{0.33}\left(\frac{l_0}{d_0}\right)^{0.122}(1+\tan\theta)^{1.38} \quad (2.103)$$

where $B = 0.5536$ and θ is the angle of the injector divergence section.

2.4.3 Impinging-stream Injectors

For an impinging-stream injector, the axes of two or more straight-flow injector holes meet at one point. Impinging-stream injectors can be categorized into unlike doublet injectors and self-impinging injectors. For unlike doublet injectors, the fuel and oxidizer streams impinge upon each other. For self-impinging injectors, one fuel stream impinges on another fuel stream and one oxidizer stream impinges on another oxidizer stream.

Self-impinging injectors have two patterns: doublet and triplet. The axes of jets are all on one plane, which is called the jet plane. The advantages of self-impinging injectors are: (i) The momentum is utilized to break up the liquid column, which promotes atomization. (ii) The spray fan is formed after the jets impinge upon each other and the liquid is distributed in an extensive cross section. Therefore, the total mixing efficiency is enhanced because the mixing between neighboring spray fans is improved. (iii) Self-impinging injectors show good combustion stability. Moreover, a near-wall layer with a low residual oxygen coefficient can be conveniently formed, which can prevent ablation of the combustion chamber wall. (iv) The configuration is simple and easy to machine. Owing to the above advantages, self-impinging injectors are extensively applied. For a doublet self-impinging injector, the droplet mass median diameter indicates the atomization performance. The empirical formula proposed in Reference [46] is given below:

$$\bar{D}_{60°} = 4.5 \times 10^3 \frac{d_h^{0.57}}{v^{0.85}} \quad (2.104)$$

where $\bar{D}_{60°}$ stands for the droplet mass median diameter (μm), d_h for the injector hole diameter (mm), and v for the jet speed (m s^{-1}).

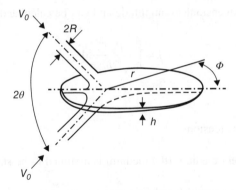

Figure 2.25 Impinging atomization.

When the angle between jets is θ, the mass median diameter is:

$$\bar{D}_\theta = (1.44 - 0.00734\theta)\bar{D}_{60^\circ}.$$ (2.105)

where θ stands for half of the angle between jets, and $\bar{D}_\theta$ for the droplet mass median diameter (μm) at an angle of θ.

Many researchers have investigated the atomization of two equal diameter jets between which the angle is 2θ (Figure 2.25) and obtained calculation formulas for liquid membrane thickness. Among the formulas, the Hasson [47] model is the most extensively applied.

The Hasson model assumes that the cross section of any jet or liquid membrane parallel to the symmetry plane of two jets is elliptic. The impinging point is one focus of the elliptic liquid membrane:

$$\frac{hr}{R^2} = \frac{\sin^3\theta}{(1 - \cos\phi\cos\theta)^2}$$ (2.106)

Shaoping Shi [48] argued that the shape of the liquid membrane should be cardioid, and the following correlation can be obtained from momentum and mass conservation:

$$\frac{hr}{R^2} = \frac{\beta e^{\beta(1-\phi/\pi)}}{e^\beta - 1}$$ (2.107)

where β can be solved from the following equation:

$$1 + (\pi/\beta)^2 - \left(1 + \frac{2}{e^\beta - 1}\right)/\cos\theta = 0$$ (2.108)

The diameter of ligaments formed from the liquid membrane can be calculated from the following equation:

$$d_1 = 2\left(\frac{4}{3f}\right)^{1/3}\left(\frac{K^2\sigma^2}{\rho_g\rho_1 u^2}\right)^{1/6}\left[1 + 2.6\mu^3\sqrt{\frac{K\rho_g^4 u^8}{6f\rho_1^2\sigma^5}}\right]^{1/5}$$ (2.109)

where f stands for the dimensionless amplitude and can be calculated from:

$$f = \int_0^t \frac{\rho_g u^2}{\sqrt{2h\rho_l\sigma}}dt \tag{2.110}$$

where:

σ is the the liquid surface tension,
μ is the liquid viscosity,
ρ is the density of ambient medium (the medium is assumed to be static),
ρ_l is the liquid density,
u is the speed at which the liquid membrane extends outside.

Assume that the liquid membrane thickness h and time t follow a hyperbolic function with the constant of Z:

$$ht = Z \tag{2.111}$$

Then the liquid membrane is assumed to be in inversely proportion to r, namely:

$$h = Z_1/r \tag{2.112}$$

The following correlation can then be derived:

$$Z = ht = Z_1 t/r = Z_1/u \tag{2.113}$$

From Equations 2.105 and 2.112, the following formulation can be obtained:

$$Z_1 = \frac{R^2 \sin^3\theta}{(1-\cos\phi\cos\theta)^2} \tag{2.114}$$

From Equations 2.106 and 2.112, the following formulation can be obtained:

$$Z_1 = \frac{\beta R^2 e^{\beta(1-\phi/\pi)}}{e^\beta - 1} \tag{2.115}$$

Equations 2.114 and 2.115 indicate that Z_1 is a function of the azimuth angle (ϕ) when R and θ are set to a certain value.

Equation 2.112 is first substituted into Equation 2.109, and f is set to be 12 according to Reference [16], and then u is approximated by substitution with v_0, resulting in the following formulation:

$$d_1 = 0.9614 \left[\frac{K_1^2 \sigma^2}{\rho_g \rho_1 v_0^4} \right]^{1/6} \left[1 + 2.6\mu^3 \sqrt{\frac{K_1 \rho_g^4 v_0^7}{72\rho_1^2 \sigma^5}} \right]^{1/5} \tag{2.116}$$

When the amplitude is assumed to be equal to the diameter of the ligament, the ligament of one wavelength produces a droplet. The diameter of the resulting droplet is:

$$d_d = \left(\frac{3\pi}{\sqrt{2}}\right)^{1/3} d_1 \left[1 + \frac{3\mu}{(\rho_1 \sigma d_1)^{1/2}}\right]^{1/6} \tag{2.117}$$

With the development of modern flow field display technologies, the atomization process of impinging streams can be further revealed through observation using optical methods. Yoon and Jeung [24] investigated the atomization for a self-impinging injection unit with a diameter of 0.7 mm. Flow is classified as laminar and turbulent according to the speed at which the liquid is injected. Figure 2.26 demonstrates liquid membrane atomization in laminar flow at different ambient pressures and injection speeds. When:

$$We_g = \frac{\rho_g u_j^2 d_0}{\sigma} < 1$$

the length of liquid membrane in laminar flow before breakup increases with the increasing mass flux while the aerodynamic force has no influence on liquid membrane breakup. When $We_g > 1$, the liquid membrane in laminar flow breaks up under the action of aerodynamic forces.

When the injected liquid flow is turbulent, a shadowgraph method can be used to capture the atomization process (Figure 2.27). The closed liquid membrane is smaller and the aerodynamic breakup happens earlier than the case of laminar flow at the same jet speed and ambient pressure. Depending on the We, the liquid membrane atomization is divided into three cases:

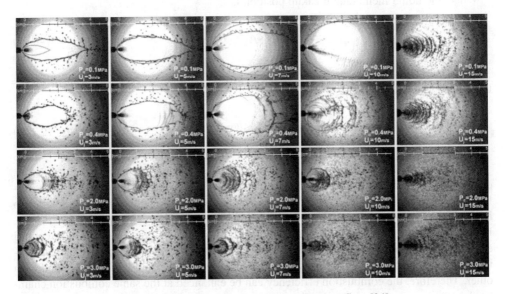

Figure 2.26 Impinging atomization in laminar flow [24].

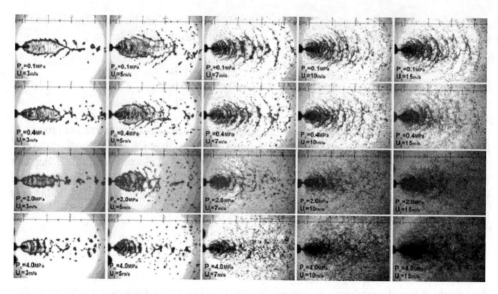

Figure 2.27   Impinging atomization in turbulent flow [24].

(i) When $\mathrm{We_g} < 2$, the expansion of the liquid membrane is determined by the liquid flux and the aerodynamic force is negligible. No periodic breakup appears. (ii) When $2 < \mathrm{We_g} < 50$, liquid membrane breakup is controlled by surface tension and periodic breakup appears. (iii) When $\mathrm{We_g} > 50$, the breakup wavelength is difficult to measure, and atomization happens immediately at the outlet of the injector.

Based on a large amount of experiment results, the empirical formulas of liquid membrane breakup position and droplet diameters are fitted as shown in Figure 2.28.

Where the liquid membrane breakup position is:

$$x_\mathrm{b}/d_\mathrm{o} = 9.4\rho^{-2/3}\mathrm{We}_j^{-1/3} \tag{2.118}$$

where $\mathrm{We}_j = e_\mathrm{L}u_j^2 d_\phi/\delta$
the droplet diameter is:

$$\frac{d_\mathrm{D}}{d_\mathrm{o}} = 1.64\rho^{-1/6}\mathrm{We}_j^{-1/3} \tag{2.119}$$

There are several patterns of unlike impingement injection units, generally doublet, triplet, quadruplet, and quintuplet. Doublet impinging injectors are extensively applied and the triplet and quadruplet impinging injectors are commonly used together with the doublet to improve the combustion efficiency, while quintuplet impinging injectors are rarely used. Impinging injection units are generally applied in apogee engines and attitude control engines. The combustion preparation process is shortened because the atomization and mixing processes happen simultaneously as the oxidant jet and the fuel jet impinge upon each other. Therefore, the combustion efficiency can be enhanced at the same combustion chamber length.

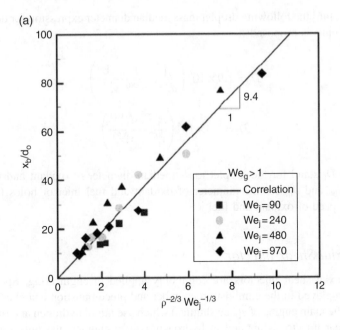

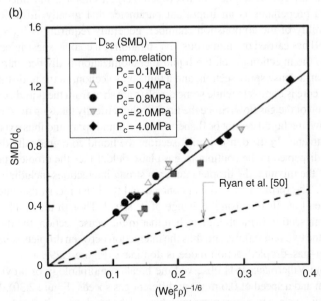

Figure 2.28 Atomization model of an impinging stream injector [24]: (a) position of liquid membrane breakup; (b) liquid droplet diameter.

In Reference [46] the following droplet mass median diameter expression for doublet impinging injection units is proposed:

$$\bar{D}_o = 1.9 \times 10^6 \left(\frac{1}{d_{h_o}^{0.78}} \cdot \frac{1}{v_o^{0.86}} \cdot \frac{1}{v_f^{1.19}} \right) \tag{2.120}$$

$$\bar{D}_f = 1.0 \times 10^4 \left(\frac{d_{hf}^{0.27}}{v_f^{0.74}} \cdot \frac{d_{h_o}^{0.023}}{v_o^{0.33}} \right) \tag{2.121}$$

where $\bar{D}_o$ and $\bar{D}_f$ stand for the droplet mass median diameter of oxidant and fuel injectors respectively, d_{h_o} and d_{hf} for the diameter of oxidant and fuel injector holes (mm), and v_o and v_f for the speed of oxidant and fuel jets.

2.4.4 Coaxial Shear Injector

Gas–liquid coaxial shear injectors are commonly applied in engineering. Specifically, the injectors are employed in the combustion chamber and precombustion chamber of cryogenic engines (e.g., the main engine of space shuttles) which use liquid hydrogen and oxygen. When the coaxial shear injector is applied in hydrogen/oxygen engines, the injection speed ratio between the two propellants is an important parameter that greatly influences the specific impulse and stability of the combustion chamber, generally requiring $v_f/v_o > 10$.

Glogowski [49] has carried out numerous investigations on coaxial shear injectors. According to the distance to the injection panel, the liquid fuel atomization is divided into an undisturbed liquid core section, a dense spray section, and a thin spray section. In the undisturbed liquid core section the spray core is complete while some droplets are shed from the liquid core surface under the stripping action of the gas flow. Since the liquid volumetrically takes up most of the mixture of the two phase flow, the liquid cannot be dispersed into a liquid spray and thus exists in the form of thin slices or ligaments. In the dense droplet section, the liquid core gradually disintegrates into droplets that are dispersed in the continuous gas phase field. Since the droplets take up a considerable volume of the mixture, the droplets undergo strong interactions including "collision" and "quasi-collision". The former means direct contact, and the latter means that one droplet passes through the wake of another, causing a change of its speed. Though the number density of the droplets in the thin section is much higher than that in the dense section, the droplets take up a small volume of the two phase flow, and thus the interaction between droplets is negligible. In this case, the "droplet–gas–droplet" action mode is dominant (Figure 2.29).

Marmottant and Villermaux [51] observed the breakup morphology of a cylinder jet with a diameter of 8 mm and a speed of 0.6 m s^{-1} at different gas speeds (Figure 2.30). With increasing gas speed, the liquid surface becomes more unstable and the breakup position is closer to the outlet.

2.4.5 Coaxial Centrifugal Injectors

Gas–liquid coaxial centrifugal injectors possess the advantages of centrifugal injectors and pneumatic injectors and are extensively applied in cryogenic propellant engines such as LOX/LH engines. In comparison with gas–liquid coaxial straight-flow injectors, the gas–liquid

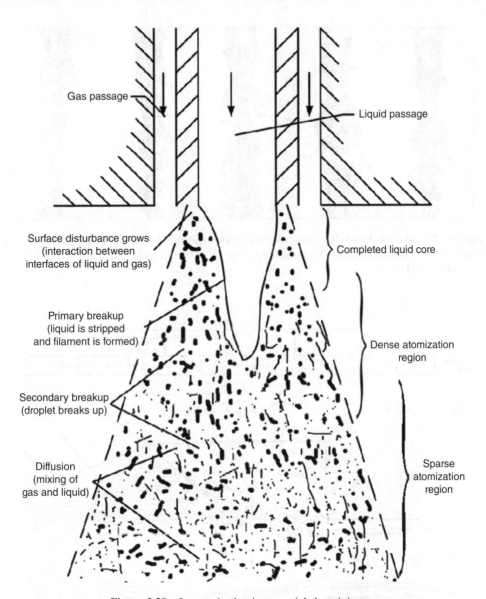

Figure 2.29   Jet atomization in a coaxial shear injector.

coaxial centrifugal injectors have a small flux coefficient, perform well in atomization when the liquid flux is high, and are insensitive to combustion instability.

Hulka [52] *et al.* have described the spray and combustion process of a gas–liquid coaxial centrifugal injector based on observations of the windowed combustion chamber in a single-nozzle rocket engine (Figure 2.31). They divided the spray combustion process into primary atomization, secondary atomization, evaporation, mixing, combustion, and expansion.

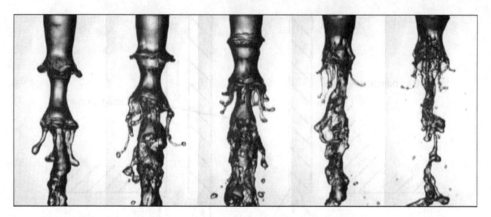

Figure 2.30 Breakup morphology of a cylinder jet with a diameter of 8 mm and a speed of 0.6 m s^{-1} at different gas speeds (from left to right the speed is 20 m s^{-1}, 30 m s^{-1}, 40 m s^{-1}, 50 m s^{-1}, and 60 m s^{-1}, respectively) [51].

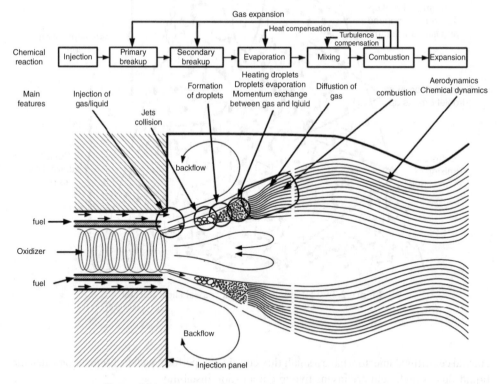

Figure 2.31 Process and characteristics of spray combustion.

In the 1990s, Howell [53] et al. (Pratt & Whitney Institute and Aerojet Institute) carried out numerous experiments on gas–liquid coaxial centrifugation injectors, which are extensively applied in LOX/LH rocket engines, and proposed several improvement methods. The results of Kalitan [54] from the University of Pennsylvania implied that J and We are two important parameters for a gas–liquid coaxial centrifugation injector:

$$J = \frac{\rho_g v_g^2}{\rho_l v_l^2}, \quad We = \frac{\rho_g (v_g - v_l)^2 d_g}{\sigma} \tag{2.122}$$

where ρ_g and v_g stand for the density and speed of gas, respectively, ρ_l and v_l for the density and speed of liquid, respectively, and d_g for the diameter of the circular seam.

Inamura [55] analyzed theoretically the flow inside and outside the injector and measured the liquid membrane thickness at the outlet of the injector with contacting probes. From the experimental results they obtained an empirical equation for the breakup length and liquid membrane cone angle including the influence of the mass flux ratio of the gas and liquid. The equation indicates that droplet diameter is smaller with a larger gas flux.

The formula for liquid membrane thickness is as follows:

$$L_{bu} = C \left\{ \frac{\rho_l \sigma \ln(\zeta/\zeta_0) \delta_0 \cos \alpha_R}{\rho_g^2 U_0^2} \right\}^{0.3} \tag{2.123}$$

where:

ζ stands for the amplitude at the position where the liquid membrane breaks up,
C is an empirical constant with a value of 0.2175,
δ_0 is the liquid membrane thickness at the outlet of swirl chamber,
U_1 is the liquid speed at the outlet,
α_R is the angle of the tapered liquid membrane at the outlet.

The empirical formula of the atomization taper angle is as follows:

$$2\alpha = 2\alpha_R \frac{a}{\sqrt{\dfrac{L}{\delta_0 \cos \alpha}}} \exp\left(-\frac{b}{Re} \right) \tag{2.124}$$

$$\alpha_R = \tan^{-1} \left\{ \frac{k}{\sqrt{1-k^2}} \right\} \tag{2.125}$$

where:

α is the modified atomization taper angle;
a and b are constants obtained by comparing the experiments and empirical equations and are given as 18.9 and 670, respectively, in Reference [55];
L is the distance from the inlet to the outlet of the injector;
K is the constant related to the injector configuration.

The empirical correlation for the droplet mean diameter is:

$$D_{32} = W \left(\frac{\sigma}{\rho_g U_r^2} \right)^{\frac{2}{3}} h^{\frac{1}{3}} \left[1 + \frac{0.01}{GLR^2} \right] \tag{2.126}$$

where W is the correction coefficient obtained by comparing the calculations and experiments and is given as 25 in Reference [25], GLR is the gas–liquid mass ratio.

Fengchen Zhuang [1] established the criterion for liquid membrane breakup after the liquid is injected out from the gas–liquid coaxial centrifugal injector. In his theory, the liquid starts to breakup when the liquid membrane thickness is less than the minimum liquid membrane thickness (t_{min}) which is a function of the disturbance wavelength, λ_{opt}:

$$\lambda_{opt} = \frac{k_1 \sigma v_r^{k_2}}{P_g u_t} \tag{2.127}$$

where k_1 and k_2 are both constants; in the theory $k_1 = 1$, $k_2 = 0.21$, $v_r = v_g / v_l$, $u_r = v_g - v_l$ is the gas–liquid relative speed, and σ is the surface tension coefficient.

The diameter of the droplet resulting from the liquid membrane breakup is:

$$d_l = k_3 \sqrt{t_{min} \lambda_{opt}} \tag{2.128}$$

where k_3 is a constant.

The gas–liquid relative speed is employed to judge whether the droplets resulting from the primary atomization undergo aerodynamic breakup. The formula for the minimum gas–liquid relative speed required for aerodynamic breakup is:

$$u_{min} = \left(\frac{256 C_1^2 \sigma^2}{81 G''(0) \rho_g \mu_g d_l} \right) \tag{2.129}$$

where C_1 is a constant that is approximately equal to 1, and $G''(0)$ is the value of the second derivative of the dimensionless boundary layer equation when $\eta = 0$.

Jinxiang Wu [56] considered the influence of the gas–liquid ratio, gas initial pressure, back pressure, outlet diameter of the injector, and the gas circular seam area on the droplet diameter for a gas–liquid coaxial centrifugal injector, resulting in the following fitted correlation:

$$\frac{SMD}{\delta} = C_1 \left(1 + \frac{1}{GLR} \right)^{C_2} \left(\frac{P_{g1} - P_0}{P_0} \right)^{C_3} \left(\frac{X}{d_0} \right)^{C_4} \tag{2.130}$$

where C_{1-4} are constants and d_0 is the diameter of the injector.

2.5   Numerical Simulation of Liquid Propellant Atomization

In liquid rocket engines, liquid fuel is injected into the combustion chamber through the injectors, and then undergoes an atomization process that consists of primary breakup and secondary breakup. The atomization process is considerably affected by the nozzle size, manufacture precision, liquid properties, and gas–liquid fuel, making it difficult to use a uniform formula to predict the droplet size distribution. Numerical simulation methods can be used to simulate the atomization process in order to better understand the liquid fuel atomization process and reveal the atomization mechanism.

2.5.1   Theoretical Models of Liquid Propellant Atomization

2.5.1.1   Models of Primary Breakup

In the primary breakup process, the continuous liquid column is disintegrated into ligaments and large droplets. The Kelvin–Helmholtz (K-H) model and Rayleigh–Taylor (R-T) model [57] are commonly used to describe the primary breakup.

K-H Model

K-H instability refers to the phenomenon whereby the liquid surface becomes unstable under the shear from the gas flow. When a cylindrical jet with a radius of a is injected into the ambient gas, the liquid jet is subject to first-order linear vibration.

The displacement of axisymmetric perturbations can be described in the following form:

$$\eta = \eta_0 e^{ikx + wt} \tag{2.131}$$

The pressure acting on the interface is:

$$P_g = -\rho_g \left(W - i\frac{\omega}{k} \right)^2 k\eta \frac{K_0(ka)}{K_1(ka)} \tag{2.132}$$

where:

ω is the growth rate of the surface wave,
k is the wave number of the surface wave,
η is the displacement of the disturbance wave,
η_0 is the initial amplitude of the surface wave,
W is the amplitude of the gas–liquid relative velocity,
K_0 and K_1 are, respectively, the zero order and first order modified Bessel functions of the
     second kind.

Since $\eta \ll a$, the kinetic energy, shear stress, and normal stress at the interface can be expressed as:

$$v_1 = \frac{\partial \eta}{\partial t}, \frac{\partial u_1}{\partial r} = -\frac{\partial v_1}{\partial x} \tag{2.133}$$

$$-p_1 + 2v_1\rho_1 \frac{\partial v_1}{\partial r} - \frac{\sigma}{a^2}\left(\eta + a^2\frac{\partial^2\eta}{\partial x^2}\right) + p_g = 0 \tag{2.134}$$

where r is the radial coordinate of the cylindrical jet, v_1 and u_1 are, respectively, the radial and axial velocity at the interface; the other parameters are physical properties of the liquid. The following dispersion equation is derived:

$$\omega^2 + 2v_1k^2\omega\left[\frac{I_1'(ka)}{I_0(ka)} - \frac{2k\varsigma}{k^2+\varsigma^2}\frac{I_1(ka)I_1'(\varsigma a)}{I_0(ka)I_1(\varsigma a)}\right]$$
$$= \frac{\sigma k}{\rho_1 a^2}\left(1-k^2a^2\right)\left(\frac{\varsigma^2-k^2}{\varsigma^2+k^2}\right)\frac{I_1(ka)}{I_0(ka)} + \frac{\rho_g}{\rho_1}(W-i\omega/k)^2k^2\left(\frac{\varsigma^2-k^2}{\varsigma^2+k^2}\right)\frac{I_1(ka)K_1(ka)}{I_0(ka)K_0(ka)} \tag{2.135}$$

where $\varsigma = \sqrt{k^2 + \omega/v_1}$, I_n and K_n are linear functions of the n_{th} order modified Bessel function of the first kind.

Reitz [10] used a numerical method to solve Equation 2.135. The results indicate the presence of a maximum growth rate (Ω) and the corresponding wavelength is Λ:

$$\frac{\Lambda}{a} = 9.02\frac{\left(1+0.45Z^{0.5}\right)\left(1+0.4T^{0.7}\right)}{\left(1+0.87We_g^{1.67}\right)^{0.6}} \tag{2.136}$$

$$\Omega\left[\frac{\rho_1 a^3}{\sigma}\right]^{0.5} = \frac{\left(0.34+0.38We_g^{1.5}\right)}{(1+Z)(1+1.4T^{0.6})} \tag{2.137}$$

where $Z = We_1^{0.5}/Re_1$, $Re_1 = Wa/v_1$, $We_1 = \rho_1W^2a/\sigma$, $We_g = \rho_gW^2a/\sigma$, T is a Taylor number, and $T = ZWe_g^{0.5}$.

Reitz also derived the single droplet breakup model. He assumes that spherical droplets the size of the nozzle diameter are injected into the ambient gas. Droplets break up due to the disturbance of unstable surface waves, and the development of the drop size is determined by the characteristic length scale and time scale of the surface waves. This model is called the Blob model. The governing equation is:

$$\frac{da}{dt} = -\frac{a-r_w}{\tau} \tag{2.138}$$

where r_w is the radius of the children droplets:

$$r_w = \begin{cases} B_0\Lambda & (B_0\Lambda \le a) \\ \min\left[(3\pi a^2W/2\Omega)^{0.33}, (3a^2\Lambda/4)^{0.33}\right] & (B_0\Lambda > a) \end{cases} \tag{2.139}$$

$$\tau = 3.726B_1a/\Lambda\Omega \tag{2.140}$$

Here, $B_0 = 0.61$, $B_1 = 10$

Equation 2.138 can be written as:

$$\frac{da}{dt} = -\left[\frac{a}{\tau} - C_a\frac{L_w}{\tau_w}\right], \quad \text{when} \quad B_0\Lambda \ll a \tag{2.141}$$

where $L_w = \Lambda$, $\tau_w = a/\Lambda\Omega$; L_w and τ_w are the characteristic length scale and time scale of the surface waves, respectively. The coefficient C_a can be computed by:

$$C_a = \frac{B_0}{3.726B_1}$$

In summary, the primary breakup is entirely controlled by the surface waves in the Blob model, with the viscosity, surface tension, and aerodynamic forces playing a major role.

Rayleigh–Taylor Model

In a Rayleigh–Taylor (R-T) model, the breakup time and breakup morphology are determined by the fastest growing instability wave. The wavelength of the fastest growing R-T wave (Λ_{RT}) is:

$$\Lambda_{RT} = 2\pi C_1 \sqrt{\frac{3\sigma}{a_p(\rho_l - \rho_g)}} \tag{2.142}$$

Here, a_p is the acceleration of the liquid film and C_1 is a parameter depending on the nozzle geometry. The corresponding breakup time can be calculated by the frequency of the fastest growing R-T wave:

$$\tau_{RT} = C_2 \sqrt{\frac{\sigma^{0.5}(\rho_l - \rho_g)}{2}\left[\frac{3}{a_p(\rho_l - \rho_g)}\right]^{1.5}} \tag{2.143}$$

The radius of the children droplets produced by R-T breakup is:

$$r = 0.5\Lambda_{RT} \tag{2.144}$$

In the calculation, the reduction of the parent droplet mass is tracked, and new child droplets are produced when the reduced mass reaches 3% of the initial droplet mass.

2.5.1.2 Secondary Atomization Models

In a spray, secondary atomization refers to the process whereby large droplets disintegrate into small droplets. This process determines the initial condition in the combustion. The secondary atomization is commonly described by the Taylor analogy breakup (TAB) [58] model, where the movement of droplets is compared to the spring oscillator movement.

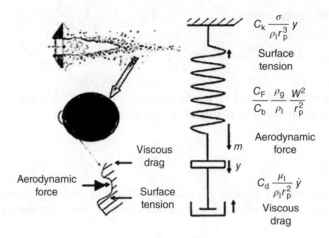

Figure 2.32 Harmonic vibration system.

For the Harmonic vibration system shown in Figure 2.32, the governing linear differential equation can be written as:

$$m\ddot{x} = F + F_t - kx - d.x \qquad (2.145)$$

where:

m is the mass of the droplet,
x is the deviation distance of the droplet central section from its equilibrium position, F is the interaction force between the gas and liquid,
kx is the surface tension force,
$d\dot{x}$ is the viscous drag.

The coefficients in Equation 2.145 can be calculated by:

$$\frac{F}{m} = C_F \frac{\rho_g W^2}{\rho_1 r_p}, \frac{k}{m} = C_k \frac{\sigma}{\rho_1 r_p^3}, \frac{d}{m} = C_d \frac{\mu_1}{\rho_1 r_p^2} \qquad (2.146)$$

Here, r_p is the droplet radius; $C_F = \frac{1}{3}$, $C_k = 8$, $C_d = 5$;. W is the amplitude of the gas–liquid relative velocity. After dimensionless displacement x is obtained, $y = x/C_b r_p$, where C_b is a constant. Equation 2.145 can be written in the following form:

$$\ddot{y} = \frac{C_F \rho_g}{C_b \rho_1} \frac{W^2}{r_p^2} - C_k \frac{\sigma}{\rho_1 r_p^3} y - C_d \frac{\mu_1}{\rho_1 r_p^2} \dot{y} \qquad (2.147)$$

The analytical solution of Equation 2.147 is:

$$y(t) = \frac{C_F}{C_k C_b} We + e^{-t/td} \left[\left(y_0 - \frac{C_F}{C_k C_b} We \right) \cos\omega t + \frac{1}{\omega} \left(\dot{y}_0 + \frac{y_0 - \frac{C_F}{C_k C_b} We}{t_d} \right) \sin\omega t \right] \quad (2.148)$$

where:

$$We = \rho_g W^2 r_p / \sigma,$$
$$\frac{1}{t_d} = \frac{C_d}{2} \frac{\mu_l}{\rho_l r_p^2},$$
$$\omega^2 = C_k \frac{\sigma}{\rho_l r_p^3} - \frac{1}{t_d^2},$$

$\dot{y}_0$ is the first derivative at time 0.

The droplet size can be calculated by the energy balance relation during the drop breakup. The energy before the droplet breakup includes the surface tension energy, oscillation energy, and deformation energy:

$$E_{surf} = 4\pi r_p^2 \sigma \quad (2.149)$$

$$E_{osc} = K \frac{4\pi}{5} \rho_l r_p^3 (\dot{x}^2 + \omega^2 x^2) = K \frac{\pi}{5} \rho_l r_p^5 (\dot{y}^2 + \omega^2 y^2) \quad (2.150)$$

Here K is a constant with a value of 10/3. The total energy before the droplet break is:

$$E_{par} = E_{surf} + E_{osc} = 4\pi r_p^2 \sigma + K \frac{\pi}{5} \rho_l r_p^5 (\dot{y}^2 + \omega^2 y^2) \quad (2.151)$$

Assuming that there is no distortion or deformation in the children droplets after the breakup, and thus only the minimum surface tension energy and kinetic energy (speed components of the children droplets that are perpendicular to the velocity direction of the parent droplet) exist. Therefore, the total energy of the droplet after droplet breakup is:

$$E_{pro} = 4\pi r_p^2 \sigma \frac{r_p}{r_{32}} + \frac{\pi}{6} r_p^5 \rho_l \dot{y}^2 \quad (2.152)$$

Since the total energy after droplet breakup is equal to that before the breakup, the following correlation can be derived from the above two equations:

$$r_{32} = \frac{4\pi r_p^3 \sigma}{E_{par} - \frac{\pi}{6} r_p^5 \rho_l \dot{y}^2} \quad (2.153)$$

2.5.2 Quasi-fluid Models

Quasi-fluid models include the single-fluid model (no-slip model), small-slip model, and two-fluid model. This method treats both vapor phase and liquid droplets as statistical continuous phases, and is an Euler–Euler method. The advantages of this method are: the vapor and liquid droplets can be handled by a uniform numerical method; turbulent transportation processes for the droplet phase can be fully taken into account; detailed information in the droplet phase can be obtained and thus compared with experiments. However, a lot of work is needed to deal with complex changes in droplet phase, such as droplet evaporation and combustion. When the number of droplet groups divided by droplet characteristics is large, the required memory storage in the calculation is extremely high, resulting in calculation difficulty. Moreover, pseudo-diffusion can be induced when an Euler method is used to treat the droplet phase, resulting in considerable numerical error.

2.5.2.1 Single-fluid Model

In the early 1970s, a single-fluid model (i.e. no-slip model) for gas–particle two-phase flow was developed on the basis of the single-phase turbulence model; its basic assumptions are:

1. The time-average velocity of each particle group characterized by a particle size is equal to the local time-averaged gas velocity (momentum balance, i.e., no-slip).
2. The particle temperature is set to be a constant (energy freeze) or to be equal to the local gas temperature (energy equilibrium).
3. The particle phase is treated as a gas-phase component, and diffuses at the same rate as other gaseous components (diffusion equilibrium).
4. The local size distribution: Since no-slip is assumed, the particle continuity equation and diffusion equation are required while the particle momentum and energy equations are deleted from particle phase equations.

When particles are grouped by the initial particle, we have the equation:

$$\frac{\partial \rho_p}{\partial t} + \frac{\partial}{\partial x_j}\left(\rho_p v_j\right) = \frac{\partial}{\partial x_j}\left(\frac{v_e}{\sigma_p} m_p \frac{\partial n_p}{\partial x_j}\right) + n_p \dot{m}_p \tag{2.154}$$

$$\frac{\partial n_p}{\partial t} + \frac{\partial}{\partial x_j}\left(n_p v_j\right) = \frac{\partial}{\partial x_j}\left(\frac{v_e}{\sigma_p} \times \frac{\partial n_p}{\partial x_j}\right) \tag{2.155}$$

Solution of the single-fluid model for gas–particle two-phase flow is almost the same as that for a single-phase fluid, except for the inclusion of several particle phase continuity equations similar to the gas phase component diffusion equation and the addition of the particles' source term in the gas phase equation. Therefore, the program employed for single-phase turbulent flow can be used by making some minor modifications.

2.5.2.2   Two-fluid Model

In the two-fluid model, the gas-phase is a continuous medium, and the atomized liquid phase is also treated as a continuous fluid. Assuming that there is a spatially continuous velocity and temperature distribution and equivalent transport properties in the two phases, similar two-phase governing equations can be obtained. Although the model can simulate well the turbulent diffusion of the spray, a very fine computational grid is required in order to reduce the numerical diffusion error, especially in the region near the spray holes where the grid scale is reduced to tens of micrometers. Furthermore, droplets of different sizes are treated as different phases, and thus the required memory and computational cost is huge in practical applications.

2.5.3   Particle Trajectory Models

Particle trajectory models include the deterministic trajectory model and the stochastic trajectory model. This method can give directly the physical properties of the droplets, obtain the development of droplet characteristics, and simultaneously compute the trajectories of droplets of different species with different diameters and temperature. Since the droplet phase is calculated by Lagrangian methods, the false diffusion can be removed and the collision and coalescence of droplets can be solved using the collision probability function. The governing equations for discrete droplets are usually ordinary differential equations, which is simpler scenario than for the governing equations in the quasi-fluid model. However, the disadvantage of the particle trajectory models is that the fluctuation of the droplet phase cannot be fully simulated.

2.5.3.1   Deterministic Trajectory Model

In the early development of the trajectory model, Crowe *et al.* proposed the deterministic trajectory model, which ignored the turbulent fluctuation of particles [59]. In the mid-1980s, Gosman [61] first proposed the stochastic trajectory model to account for the turbulence diffusion caused by the fluctuation of particles. In the deterministic trajectory model, the governing equations for the gas phase are the same as those in the two-fluid model, and the basic conservation equations for the particles are equivalent to the turbulent two-phase flow equations with the correlation term equal to zero, which is equal in form to the laminar flow equations and transient equations [1], and the continuity equation for the kth particle phase is:

$$\frac{\partial \rho_k}{\partial t} + \frac{\partial}{\partial x_j}(\rho_k v_{ki}) = S_k \qquad (2.156)$$

The momentum equation for the kth particle phase is:

$$\frac{\partial}{\partial t}(\rho_k v_{ki}) + \frac{\partial}{\partial x_j}(\rho_k v_{kj} v_{ki}) = \rho_k g_i + \frac{\rho_k}{\tau_{rk}}(v_i - v_{ki}) + v_i S_k \qquad (2.157)$$

The energy equation for the kth particle phase is:

$$\frac{\partial}{\partial t}(\rho_k c_k T_k) + \frac{\partial}{\partial x_j}(\rho_k v_{kj} c_k T_k) = n_k(Q_h - Q_k - Q_{rk}) + c_p T S_k \qquad (2.158)$$

where S_k is mass source term for the kth particle (droplet) group caused by evaporation, volatility, and reaction. The sum of the mass source terms for all particles should be equal to the source term S in the gas phase continuity equation:

$$S = -\sum_k S_k = -\sum_k n_k \dot{m}_k, \dot{m}_k = \frac{dm_k}{dt} \qquad (2.159)$$

Continuity, momentum, and energy equations for the particle phase can be written in Lagrangian coordinates:

$$\int_A n_k v_{kn} dA = N_k = \text{const} \qquad (2.160)$$

$$\frac{dv_{ki}}{dt_k} = \left(\frac{1}{\tau_{rk}} + \frac{\dot{m}_k}{m_k}\right)(v_i - v_{ki}) + g_i \qquad (2.161)$$

$$\frac{dT_k}{dt_k} = \frac{Q_h - Q_k - Q_{rk} + \dot{m}_k(c_p T - c_k T_k)}{m_k c_k} \qquad (2.162)$$

In the original deterministic trajectory model, it is assumed that the total number flux of particles along the trajectory remains the same and particle turbulent diffusion is ignored. However, experiments show that the particle turbulent diffusion is not negligible in many cases. To take into account its effect, Smoot et al. [60] introduced the "particle drift velocity", and assumed that the particle velocity consists of two parts:

$$v_{kj} = v_{kc,j} + v_{kd,j} \qquad (2.163)$$

where $v_{kc,j}$ is the particle convection velocity, which is determined by the particle momentum equation; $v_{kd,j}$ is the particle diffusion drift velocity, which is determined by the diffusion law in the form of Fick's law:

$$-\rho_k v_{kd,j} = -n_k m_k v_{kd,j} = D_k m_k \frac{\partial n_k}{\partial x_j} \qquad (2.164)$$

$$v_{kd,j} = -\frac{D_k}{n_k} \times \frac{\partial n_k}{\partial x_j} \qquad (2.165)$$

Obviously, this amendment introduces the concept of the two-fluid model. The time-average particle continuity equation is:

$$\frac{\partial n_k}{\partial t} + \frac{\partial}{\partial x_j}(n_k v_{kj}) = -\frac{\partial}{\partial x_j}\left(\overline{n'_k v'_{kj}}\right) \qquad (2.166)$$

The following correlation can be obtained by taking the gradient of diffusion mass flow:

$$-\overline{n'_k v'_{kj}} = -n_k v_{kd,j} = D_k \frac{\partial n_k}{\partial x_j} \tag{2.167}$$

Then the time-average particle continuity equation can be written in the following form:

$$\frac{\partial n_k}{\partial t} + \frac{\partial}{\partial x_j}(n_k v_{kj}) = \frac{\partial}{\partial x_j}\left(D_k \frac{\partial n_k}{\partial x_j}\right) \tag{2.168}$$

The total speed of the particles is assumed to be:

$$v_{kj,0} = v_{kj} + v_{kd,j} \tag{2.169}$$

The following equation can then be obtained:

$$\frac{\partial n_k}{\partial t} + \frac{\partial}{\partial x_j}(n_k v_{kj,0}) = 0 \tag{2.170}$$

This means that the concept of drift velocity is based on the diffusion term of the particle continuity equation in the two-fluid model in Eulerian coordinates, but this concept is not introduced for the particle momentum equation. The particle diffusion coefficient (D_k) and the gradient of particle number density ($\partial nk/\partial xj$) are needed to compute the particle diffusion drift velocity. These two variables cannot be given by the trajectory model itself, and thus we have to resort to the two-fluid model. The former is determined by the Hinze–Tchen formula:

$$\frac{v_p}{v_T} = \frac{D_p}{D_T} = \left(\frac{k_p}{k}\right)^2 = \left(1 + \frac{\tau_{r1}}{\tau_T}\right)^{-1} \tag{2.171}$$

For the latter, the following number density equation can be derived by assuming that the particle velocity is equal to the gas velocity from the single-fluid or no-slip concepts:

$$\frac{\partial n_k}{\partial t} + \frac{\partial}{\partial x_j}(n_k v_j) = \frac{\partial}{\partial x_j}\left(\frac{v_T}{\sigma_{kp}} \times \frac{\partial n_k}{\partial x_j}\right) \tag{2.172}$$

The gradient of the particle number density can thus be obtained; σ_{kp} is set as 0.35.

2.5.3.2 Stochastic Trajectory Model

The stochastic trajectory model was first developed in the mid-1980s by Gosman *et al.* [61], who took into account the turbulence diffusion caused by the fluctuation of particles. It is essentially a semi-direct simulation, which uses a direct numerical simulation method for particles and applies the general statistical turbulence model approach for the gas. The Stochastic trajectory model is based on the instantaneous particle momentum equation, including the action of the turbulent kinetic energy on the droplet. The trajectory of a droplet is calculated by:

$$\frac{du_p}{dt} = \frac{\bar{u} + u' - u_p}{\tau_r} \tag{2.173}$$

$$\frac{dv_p}{dt} = \frac{\bar{v} + v' - v_p}{\tau_r} + \frac{w_p^2}{r_p} + g \tag{2.174}$$

$$\frac{dw_p}{dt} = \frac{\bar{w} + w' - w_p}{\tau_r} + \frac{v_p w_p}{r_p} \tag{2.175}$$

Here, u_p, v_p, and w_p are components of the instantaneous particle velocity; $\bar{u}$, $\bar{v}$, and $\bar{w}$ represent the average velocity of the gas; and u', v', and w' represent the fluctuation velocity of the gas.

Assuming that the fluctuation velocity satisfies the local Gaussian distribution, the instantaneous fluctuation velocity u' is given by:

$$u_i' = \zeta \left(\overline{u_i'^2} \right)^{\frac{1}{2}} \tag{2.176}$$

where ζ is a random variable following the probability density function.

Then, Equation 2.176 is first substituted into Equations 2.173–2.175, and these equations are then integrated to obtain the particle trajectory equations. The integration interval is given by:

$$t_{int} = \min(t_e, t_r) \tag{2.177}$$

Here, t_r is the relaxation time for the particle, which is the ratio of the inertial force to the drag force:

$$t_r = d_p^2 \bar{\rho}_p / 18\mu \tag{2.178}$$

$$t_p = k/\varepsilon \tag{2.179}$$

The Monte-Carlo method is commonly used to solve the statistical trajectories. Since thousands or millions of trajectories need to be computed, the computational cost is huge.

In the particle trajectory models, droplets are assumed to be a sparse phase, ignoring the collisions and coalescence of droplets and the influence of the interaction force between neighboring droplets on droplet transportation. Droplets are assumed to be isolated in the calculation of drag and convection. For the atomization of liquid jets, the droplet size distribution and droplet trajectories calculated by the particle trajectory model are considerably different from experimental data. This is because the assumption of a sparse phase for the liquid is inappropriate in this case. To obtain more accurate simulation results, the interface tracking method is proposed and is detailed in the next section.

2.5.4   Simulation of Liquid Jet Atomization Using Interface Tracking Method

In this section, the two-phase LES formulation using the CLSVOF interface tracking method is described first, and associated numerical methods are then described. Finally, a detailed analysis of the simulation results and both a qualitative and quantitative comparison between numerical predictions and experiments will be given.

2.5.4.1   Governing Equations

In current studies of two-phase flow modeling, both liquid and gas are assumed to be incompressible and immiscible. Derivation of the two-phase flow LES formulation is described in detail by Xiao [62–65]. The thinking behind this formulation is as follows: the liquid/gas interface is resolved directly without modeling sub-grid scale (SGS) features; the usual spatially filtered LES formulation is employed in the single-phase flow regions; an appropriate treatment is adopted when discretizing the governing equations when interpolating cell face fluxes for cells that are intersected by the interface for both phases.

A coupled level set (LS) and VOF (volume of fluid) method (CLSVOF) is used here to capture and evolve the interface. The level set Φ is a signed distance function from the interface satisfying $\nabla \Phi = 0$. The interface is defined by $\phi = 0$, with $\Phi > 0$ representing liquid and $\Phi \leq 0$ representing air. The Φ is evolved by the simple advection equation using the resolved velocity field (note: in what follows a spatially filtered (resolved) quantity is indicated by an overbar) and ignoring the contribution of any sub-grid-scale (SGS) velocity effects:

$$\frac{\partial \bar{\phi}}{\partial t} + \bar{U}_i \frac{\partial \bar{\phi}}{\partial x_i} = 0$$

To maintain the signed-distance property, the reinitialization equation is also solved:

$$\frac{\partial \varphi}{\partial \tau} = S(\varphi_0)\left(1 - \sqrt{\frac{\partial \varphi}{\partial x_k}\frac{\partial \varphi}{\partial x_k}}\right) \quad S(\varphi_0) = \frac{\varphi_0}{\sqrt{\varphi_0^2 + d^2}}$$

where τ represents pseudo-time, $S(\varphi_0)$ is a modified sign function, $\varphi_0 = \varphi(x_i, \tau = 0) = \phi(x_i, t)$, and $d = max(\Delta x, \Delta y, \Delta z)$. After solving this equation to the steady state in the interface vicinity, $\bar{\phi}$ is replaced by φ.

The VOF function F is defined as the volume fraction occupied by the liquid in each computational cell. The resolved evolution of the VOF function is governed by:

$$\frac{\partial \bar{F}}{\partial t} + \bar{U}_i \frac{\partial \bar{F}}{\partial x_i} = 0$$

With residual or SGS stress tensor τ_{ij}^r modeled by a simple Smagorinsky eddy viscosity approach, the governing equations for the resolved velocity field are:

$$\frac{\partial \bar{U}_i}{\partial x_i} = 0$$

$$\frac{\partial(\bar{U}_i)}{\partial t} + \frac{\partial(\bar{U}_i\bar{U}_j)}{\partial x_j} = -\frac{1}{\rho}\frac{\partial\bar{P}}{\partial x_i} + \frac{1}{\rho}\frac{\partial(\bar{\tau}_{ij} + \tau_{ij}^{\mathrm{r}})}{\partial x_j} + g_i + \frac{1}{\rho}\overline{F_i^{\mathrm{ST}}}$$

$$\bar{\tau}_{ij} = 2\mu\overline{S}_{ij} \quad \tau_{ij}^{\mathrm{r}} = 2\mu_{\mathrm{r}}\overline{S}_{ij} \quad \overline{S}_{ij} = \frac{1}{2}\left(\frac{\partial\bar{U}_i}{\partial x_j} + \frac{\partial\bar{U}_j}{\partial x_i}\right)$$

$$\mu_{\mathrm{r}} = \rho(C_{\mathrm{S}}\Delta)^2\bar{S} \quad \bar{S} = \sqrt{2\overline{S}_{ij}\overline{S}_{ij}}$$

$$\rho = \rho_{\mathrm{G}} + (\rho_{\mathrm{L}} - \rho_{\mathrm{G}})H(\bar{\phi})$$
$$\mu = \mu_{\mathrm{G}} + (\mu_{\mathrm{L}} - \mu_{\mathrm{G}})H(\bar{\phi}) \qquad H(\bar{\phi}) = \begin{cases} 1 & \text{if } \bar{\phi} > 0 \\ 0 & \text{if } \bar{\phi} \le 0 \end{cases}$$

Here, g_i is gravitational acceleration, Δ represents the filter width, taken as the cube root of the local cell volume, and the value of the Smagorinsky constant (C_{S}) is set in all the calculations reported below to 0.1. The surface tension term F_i^{ST} is computed via:

$$\overline{F_i^{\mathrm{ST}}} = \sigma\bar{\kappa}\frac{\partial H}{\partial x_i} \quad \bar{\kappa} = \frac{\partial\bar{n}_i}{\partial x_i} \quad \bar{n}_i = -\frac{1}{\sqrt{\dfrac{\partial\bar{\phi}}{\partial x_k}\dfrac{\partial\bar{\phi}}{\partial x_k}}}\frac{\partial\bar{\phi}}{\partial x_i}$$

Here, σ is the surface tension coefficient, κ is the interface curvature, and n_i is interface normal vector pointing from the liquid phase to the gas phase. The SGS term arising from filtering the surface tension term is neglected. To maintain the implication of a sharp interface, fluid density and viscosity are in the present approach not considered as spatially filtered quantities, but are set to be the properties of liquid or gas depending on the local value of the resolved level set variable $\bar{\phi}$.

2.5.4.2 Numerical Methods

Coupled Level Set and VOF Method (CLSVOF)

A CLSVOF methodology for interface advection is adopted in the present approach since it offers an optimum combination of the good mass conservation property of the VOF approach and the convenient and accurate capability of LS for evaluation of interface geometrical properties. Coupling of LS and VOF variables is enforced in the interface reconstruction step where both LS and VOF information can be used to good effect:

- the interface normal vector is computed from LS (a smooth function) rather than VOF (a discontinuous function), since LS gives more accurate information on interface location and shape;
- the interface position in the cell is constrained by the VOF function, since this gives more accurate information on liquid volume conservation,

Based on such a reconstructed interface, the VOF function is evolved to the next time step, and the LS function is corrected. A flow chart of the CLSVOF method is shown in Figure 2.33.

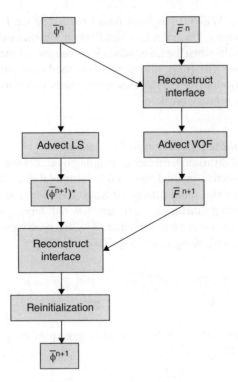

Figure 2.33   Flow chart of the CLSVOF method [62].

Coupling of the LS and VOF methods occurs during the interface reconstruction and LS-re-distance processes. The detailed algorithm of the present CLSVOF method is as follows:

- Initialize the LS and VOF functions at time step $n = 0$: $\overline{\phi}^n$ and $\overline{F}^n$.
- Reconstruct the interface in cells where $0 < \overline{F}^n < 1$. The interface normal vector n_i is calculated from the LS function, and the position of the interface within the cell is constrained by the VOF function.
- Advect the VOF function from $\overline{F}^n$ to $\overline{F}^{n+1}$ based on the reconstructed interface. Advect the LS function from $\overline{\phi}^n$ to $\overline{\phi}^{n+1,*}$. The operator split method is used to solve both equations (see Xiao [62] for details).
- Reconstruct and constrain the interface in cut cells using the new LS function $\overline{\phi}^{n+1,*}$ and the VOF function $\overline{F}^{n+1}$.
- Perform a reinitialization step on $\overline{\phi}^{n+1,*}$ to obtain the final level set function $\overline{\phi}^{n+1}$ with a recovered signed distance property.

In the current CLSVOF method, the normal vector is calculated directly by discretizing the LS gradient using a finite difference scheme. By appropriately choosing one of three finite difference schemes (central, forward, or backward differencing), it has been demonstrated that thin liquid ligaments can be well resolved (see Xiao [62]). Although a high order discretization scheme (e.g., fifth order WENO) has been found necessary for LS evolution in pure LS methods to reduce mass error, low order LS discretization schemes (second order is used here) can produce accurate results when the LS equation is solved and constrained as indicated above in a CLSVOF method (see Xiao [62]), since the VOF method maintains second order accuracy. This is a further reason to adopt the CLSVOF method, which has been used for all the following simulations of liquid jet primary breakup.

Discretization of Two-phase Flow Transport Equations

Since the convection and diffusion terms are discontinuous across the interface, a cautious first order forward-Euler projection method was used for temporal discretization of the two-phase flow governing equations (for more details see Xiao [62]). First, an intermediate velocity is computed from convection, diffusion, and gravitational terms (spatial discretization is described below). Note: the surface tension is treated via the pressure term using the ghost-fluid approach (Fedkiw *et al.* [66]; Kang *et al.* [67]; Liu *et al.* [68]):

$$\frac{\overline{U_i^*} - \overline{U_i^n}}{\delta t} = -\frac{\partial \left(\overline{U_i^n U_j^n} \right)}{\partial x_j} + \frac{1}{\rho^n} \frac{\partial \left(\overline{\tau_{ij}^n + \tau_{ij}^{r^n}} \right)}{\partial x_j} + g_i$$

Second, the intermediate velocity field is updated using a pressure gradient term to obtain the velocity at time step $n+1$:

$$\frac{\overline{U_i^{n+1}} - \overline{U_i^*}}{\delta t} = -\frac{1}{\rho^n} \frac{\partial \overline{P^{n+1}}}{\partial x_i}$$

Since the velocity field at time step $n+1$ must satisfy the continuity equation, a pressure Poisson equation may be derived by taking the divergence of the above equations to allow $\overline{P^{n+1}}$ to be calculated:

$$\frac{\partial}{\partial x_i} \left(\frac{1}{\rho^n} \frac{\partial \overline{P^{n+1}}}{\partial x_i} \right) = \frac{1}{\delta t} \frac{\partial \overline{U_i^*}}{\partial x_i}$$

The variables are arranged in a staggered manner in the current two-phase flow LES formulation: the pressure, LS, and VOF are located at the cell centre; velocity components are located at corresponding faces. In general, second order central methods are used to discretize spatial derivatives. Since the gas phase has a much smaller density and viscosity than the liquid phase, the velocity gradient in the gas phase is typically much larger than in the liquid phase. However, for discretization of the momentum equation for cells in the vicinity of the interface, it was found important to specify correctly which velocity should be used when calculating cell face fluxes associated with convection and diffusion terms. If the velocity in the adjoining gas phase cell was used to discretize the momentum equation for an interface cut liquid phase cell, large momentum errors were induced in the vicinity of the interface. To reduce this momentum error (arising from the density jump across the interface) an extrapolated liquid velocity field $\overline{U_i^L}$ was introduced for spatial discretization of convection and diffusion terms for interface cut cells.

A constant extrapolation technique (Fedkiw *et al.* [66]; Aslam [69]) was used here for liquid velocity extrapolation. For a detailed explanation of this technique, readers are referred to Xiao [62]. It was also demonstrated by Xiao [62] that: (i) a divergence free step for the extrapolated liquid velocity field was necessary to reduce the momentum error when the interface moved rapidly across a fixed grid; (ii) a more accurate capture of interface dynamics was obtained when the extrapolated liquid velocity field was also used in the VOF and LS transport equations.

Algorithm for Two-phase Flow LES

Figure 2.34 summarizes the final procedure of the developed two-phase flow LES. The detailed algorithm for the current two-phase flow LES is as follows:

- Based on the interface represented by the LS function $\overline{\phi}^n$, discretize the two-phase flow governing equations to solve for the velocity field at the next time step $\overline{U}_i^{n+1}$.
- Construct the extrapolated liquid velocity field $\overline{U}_i^{L,n+1}$ using the extrapolation technique described by Xiao [62].
- Ensure continuity for the extrapolated liquid velocity:

$$\left(\frac{\partial \overline{U}_i^{L,n+1}}{\partial x_i} = 0 \right)$$

in the gas phase by a divergence free step.

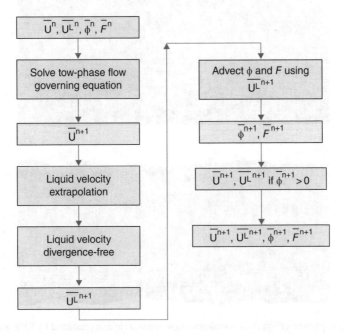

Figure 2.34 Procedure for current two-phase flow LES formulation [62].

- Based on $\overline{U_i^{L,n+1}}$ advect the LS and VOF functions to the next time step to obtain $\overline{\phi^{n+1}}$ and $\overline{F^{n+1}}$ using the CLSVOF algorithm shown in Figure 2.33.
- Set the velocity in CVs which change from gas to liquid (i.e., $\overline{\phi^n} \leq 0$ but $\overline{\phi^{n+1}} \geq 0$) to liquid velocity via $\overline{U_i^{n+1}} = \overline{U_i^{L,n+1}}$.
- Repeat for further time steps.

2.5.4.3   Results

Figure 2.35 compares the LES predicted interface topologies near the nozzle exit with the two different inflow treatments. In contrast to the predicted long smooth interface region observed when applying uniform and laminar inlet conditions, the interface is disturbed by turbulent eddies immediately after the jets exit the nozzles when using turbulent inflow conditions. Figure 2.36 shows the overall liquid jet structure predicted by LES when turbulent inflows are specified for both phases using R^2M. The growth of the surface instability in the liquid jet under aerodynamic forces is well reproduced by LES, and the breakup point of the liquid core now agrees well with the experiment. Ligaments and droplets are ejected from the resulting liquid clusters, which is consistent with shadowgraph observations. It is an interesting

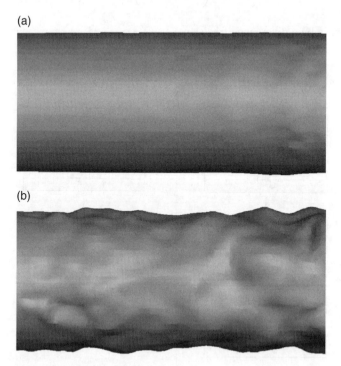

Figure 2.35   Comparison of interface topologies near the nozzle exit predicted by LES with: (a) uniform laminar inflow and (b) realistic turbulent inflow [62].

(a)

(b)

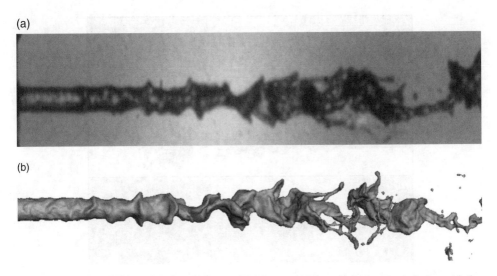

Figure 2.36 Comparison of LES predicted liquid jet structure and shadowgraph image: (a) from Charalampous *et al.* [70]; (b) LES with turbulent inflow by R^{2}M. (U_G = 47 m s^{-1}, U_L = 4 m s^{-1}) [62].

question as to whether it is liquid or gas flow turbulence (or both) that exerts the larger influence upon the interface; this will be investigated in the following section.

2.5.5   Liquid Jet Structure – Varying Flow Conditions

Figure 2.37 compares the liquid jet structures predicted by the current CLSVOF LES method with shadowgraphs taken by Charalampous *et al.* [70] for liquid jet primary breakup with different coaxial air velocities. The breakup morphologies of the continuous liquid jet are well reproduced by the current method for all conditions.

As the air velocity increases, the predicted location where drops and ligaments are first seen decreases, in good agreement with experimental observations. Owing to the increasing aerodynamic forces, the dimensions of predicted liquid ligaments and droplets resulting from the primary breakup become smaller. Owing to the relatively coarse mesh, the secondary breakup in the furthest downstream region is probably not well resolved for the high speed air velocity cases (U_G = 119 and 166 m s^{-1}). Figure 2.38 shows simulated primary breakup at three different liquid injection velocities with a fixed coaxial air flow of 70 m s^{-1}, together with shadowgraph images at the corresponding flow conditions. As the liquid velocity increases, the liquid jet interface before the breakup point becomes noticeably rougher due to the disturbing liquid eddies becoming more energetic as the Reynolds number of the liquid flow inside the nozzle increases from 5440 to 21770. This physical phenomenon is correctly captured by the current LES but obviously only when the R^{2}M approach is applied to generate the turbulent inflow boundary conditions. As the liquid injection velocity grows from 2 to 8 m s^{-1}, the liquid jet breakup position moves downstream, in good agreement with the experimental images.

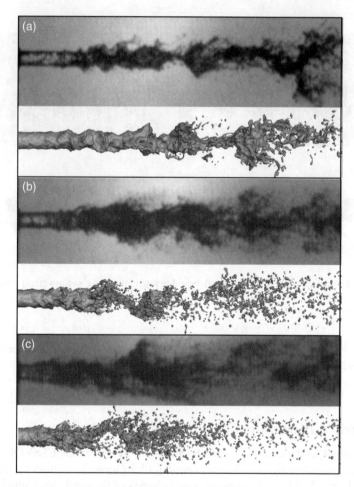

Figure 2.37   Liquid jet primary breakup, LES predictions and experiments; $U_L = 4$ m s^{-1}; air velocity $U_G =$ (a) 70 m s^{-1}, (b) 119 m s^{-1}, and (c) 166 m s^{-1} [65].

Figure 2.39 shows the liquid core length predicted by the current two-phase flow LES for flows in Figs. 2.36 and 2.37; U_L is 4 m s^{-1} in all cases, with the air velocity varied to produce different Weber numbers. When laminar inflows were used for both phases, the predicted liquid core length was (as expected from the discussion above) much larger than the experimental measurements of Charalampous *et al.* [71] for lower Weber numbers. When turbulent inflows were specified for both phases, the simulated core length agreed very well with the experimental value for all Weber numbers, confirming that the initial interface perturbations caused by liquid eddies play an important role in the resulting surface instability development and primary breakup process. For the cases with higher gaseous co-flow ($U_G = 119$ and 166 m s^{-1}), strong aerodynamic forces dominate the primary breakup process, and the liquid jet core is destroyed after a short distance and the difference of predicted core length between simulations with different inflow conditions is much smaller. Overall, liquid core lengths are well reproduced by

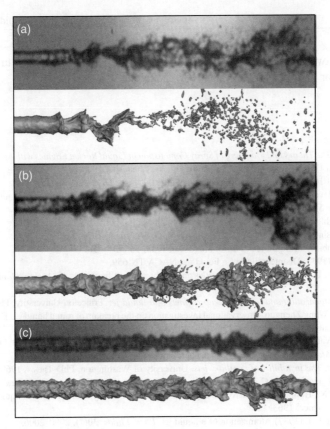

Figure 2.38 Liquid jet primary breakup, LES predictions and experiment; $U_G = 70$ m s^{-1}; water velocity $U_L =$ (a) 2 m s^{-1}, (b) 4 m s^{-1}, and (c) 8 m s^{-1} [65].

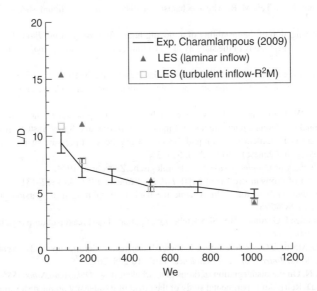

Figure 2.39 Comparison of the LES predicted liquid core length versus experimental data of Charalampous *et al.* [71] for different Weber numbers ($U_L = 4$ m s^{-1}) [65].

the developed two-phase flow LES method for all four flows when realistic turbulent inflows are provided for both phases using the R^2M technique.

References

[1] Zhuang F C. *The Theory and Modeling of Spray Combustion in Liquid Rocket Engines with Applications* (in Chinese), National University of Defense Technology Press, Changsha, 1995.

[2] Wikipedia (2016) RL10. http://en.wikipedia.org/wiki/RL-10-(rocket-engine).

[3] Du Y J. The development of liquid rocket engine, R10. *Journal of Rocket Propulsion (in Chinese)*, 2000, **2**: 49–58.

[4] Cao M J. *The Spray Theory* (in Chinese). Beijing: China Machine Press, 2005.

[5] Lord Rayleigh. On the instability of jets. Proc. London Math. Soc., 1878.

[6] Weber C. Disintegration of liquid jets. *Z. Angew. Math. Mechanik*, 1931, **11**(2):136–159.

[7] Tyler F. Instability of liquid jets. *Philos. Magazine*, 1933, **16**: 504–518.

[8] Haenlein A. (1932) Disintegration of a liquid jet. NACA TN 659.

[9] Ohnesorge W. Formation of drops by nozzles and the breakup of liquid jets. *Z. Angew. Math. Mechanik*, 1936, **16**: 355–358.

[10] Reitz R. D. *Atomization and other breakup regimes of a liquid jet*. Princeton Universtity, PhD Thesis, 1978.

[11] Liu J, Li Q L, Wang Zhenguo. Observational investigation on the breakup of round liquid jet. Presented at the 13th Annual International Conference on Liquid Atomization and Spray System—Asia, 2009.

[12] Taylor J. J. Water jet photography-techniques and methods. *Experimental Fluids*, 1983, **1**:113–120.

[13] Lefebvre H. *Atomization and Sprays*. Hemisphere Publishing Corporation, 1989.

[14] Sterling, A.M. *The instability of capillary jets*. University of Washington, PhD Thesis, 1969.

[15] Sterling, A.M. and Sleicher, C.A. The instability of capillary jets. *Journal of Fluid Mechanics*, 1975, **68**:477–495.

[16] Dombrowski N., Hopper P. The aerodynamics instability and disintegration of viscous liquid sheets. Chemical Engineering Science, 1963, 18.

[17] Lin S.P, Kang D.J. (1987) Atomization of a liquid jet. *Phys. Fluids*, **30**(7), 2000–2006.

[18] Bracco F.V., Reitz R.D. Mechanism of atomization of a liquid jet. *Physics Fluids (A)*, 1982, **25**(10), 1730–1742.

[19] Fraser R P. Liquid fuel atomization. *Sixth Symposium on Combustion*, Reinhold, New York, 1957, pp. 687–701.

[20] Eisenklam P. Recent research and development work on liquid atomization in Europe and the U.S.A, 5th Conference on Liquid Atomization, Tokyo, May, 1976.

[21] York J L, Stubbs H F, Tek M R. The mechanism of disintegration of liquid sheets. Trans. *ASME*, 1953, **75**:1279–1286.

[22] Squire H B. Investigation of the instability of a moving liquid film. *Br. J. Appl. Phys*, 1953, **4**:167–169.

[23] Hagerty W W, Shea J F. A study of the stability of plane fluid sheets. *Journal of Applied Mechanics*, 1955, **22**(4): 509–514.

[24] Youngbin Yoon, In-Seuk Jeung. Effects of Ambient Gas Pressure on the Breakup of Sprays in Like-Doublet and Swirl Coaxial Injectors. Presented at the International Symposium on Energy Conversion Fundamentals, Istanbul, 21–25 June 2004.

[25] Liu J, Li Q L, Liu W D. Influence of the swirl injector diameter on the process of conical sheet breakup. Presented at the 14th Annual Conference International on Liquid Atomization and Spray System -Asia, 2010.

[26] Liu J, Li Q L, Liu W D. Experiment on liquid sheet breakup process of pressure swirl injector. *Journal of Propulsion Technology* (in Chinese), 2011, **32**(4): 539–543.

[27] Lenard P. *Mechanism of Droplet Breakup*. Meteorologische Zeitschrift, 1904.

[28] Lane W R. Shatter of drops in streams of air[J]. *Ind. Eng. Chem.*, 1951, **43**:1312–1317

[29] Hinze J. O. Fundamentals of the hydrodynamic mechanism of splitting in despersion processes. *Journal of AICHE*, 1955, **1**(3):289–295.

[30] Hanson A R, Domich E G, Adams H S. Shock tube investigation of the breakup of drops by air blasts. *Phys Fluids*, 1963, **6**:1070–1080.

[31] Rumscheidt F D, Mason S G. Particle motion in sheared suspensions-deformation and burst of fluid drops in shear and hyperbolic flows. *Journal of Colloid Science*, 1967, **16**:238–261.

[32] Kolmogorov A N. On the disintegration of drops in a turbulent flow. *Dokl.Akad.Nauk SSSR*, 1949, **66**:825–828.

[33] Lee C.H., Rolf D. Reitz. An experimental study of the effect of gas density on the distortion and breakup mechanism of drops in high speed gas stream. *International Journal of Multiphase Flow*, 2000, **26**:229–244.

[34] Dinh T N, Li G J, Theofanous T G. An investigation of droplet breakup in a high Mach, low Weber number regime. Presented at the 41st Aerospace Sciences Meeting and Exhibit, Reno, Nevada, 6–9 January 2003.

[35] Fang T W. *The mechanism of liquid jet atomization and secondary breakup in supersonic flow (in Chinese)*. Changsha: National University of Defense Technology, 2010.

[36] Rizk N K, Lefebvre A H. Drop size distribution characteristics of spill-return atomizers. *Journal of Propulsion and Power*, 1985, **1**(3):16–22.

[37] Mugele R., Evans H. D. Droplet size distribution in sprays. *Journal of Industrial and Engineering Chemistry*, 1951, **43**(6):1317–1324.

[38] Van De Hulst. *Light scattering by small particles*. New York: Dover Publications, 1981.

[39] Bachalo W D. Method for measuring the size and velocity of spheres by dual-beam light-scatter interferometry. *Applied Optics*, 1980, **19**(3), 363–370.

[40] Bachalo W D., Houser M J. Phase/Doppler spray analyzer for simultaneous measurements of drop size and velocity distributions. *Optical Engineering* 1984, **23**(5):583–590.

[41] Dodge L G, Rhodes D J, Reitz R D. Drop-size measurement techniques for sprays:comparison of Malvern laser-diffraction and aerometrics phase/Doppler. *Applied Optics*, 1987, **11**(26):2144–2154.

[42] Rizk N.K., Lefebvre A.H. Internal flow characteristics of simplex swirl atomizers. *Journal of Propulsion and Power*, 1985, **1**(3):193–199.

[43] Jones A.R. Design optimization of a large pressure-jet atomizer for power plant. Proceedings of the 2nd International Conference on Liquid Atomization and Spray Systems: ICLASS-'82, Madison, Wisconson, 20–24 June 1982, pp. 181–185.

[44] Lefebvre A.H. *Gas Turbine Combustion*. Washington,D.C.: Hemisphere, 1983.

[45] Liu J, Zhang Xin-Qiao, Li Qing-Lian, and Wang, Z.G. Effect of geometrical parameters on spray cone angle in pressure swirl injector. *Proc. Institution Mech. Eng., Part G: J. Aerospace Eng.*, 2013, **227** (2), 342–353.

[46] Zhu N C at el. *The design of liquid rocket engines (part 1) (in Chinese)*. Beijing: China Astronautic Publishing House, 1994.

[47] Hasson D., Peck R.E. Thickness distribution in a sheet formed by impinging jets. Journal of AICHE, 1964:752–754.

[48] Shi S P, Zhuang F C. A mathematical model of impinging jets atomization at low Weber number. *Journal of Aerospace Power (in Chinese)*, 1994, **3**: 285–288.

[49] Glogowski M. and Micci M M. Shear coaxial injector spray characterization near the LOX post tip. Presented at The American Institute of Aeronautics and Astronautics Conference, July 1995.

[50] Ryan H M, Anderson W E, Pal S, et al. Atomization characteristics of impinging liquid jets[J]. Journal of Propulsion and Power, 1995, **11**(1): 135–145.

[51] Marmottant P., Villermaux E. On spray formation. *Journal of Fluid Mechanics*, 2004, **498**: 73–111.

[52] Hulka J., Makel D. Liquid oxygen hydrogen testing of a single swirl coaxial injector element in a windowed combustion chamber. Presented at the 29th Joint Propulsion Conference and Exhibit, Monterey, CA, 28–30 June 1993.

[53] D. Howell, E. Petersen, J. Clark. Performance characteristics of LOX-H2, tangential-entry, swirl-coaxial, rocket injectors. AIAA Journal, 1993, 93–0228.

[54] Kalitan D M, Salgues D, Mouis A G. Experimental liquid rocket swirl coaxial injector study using non-intrusive optical techniques. Presented at the 41st Joint Propulsion Conference & Exhibit, Tuscon, Arizona, 10–13 July 2005.

[55] T. Inamura, K. Miyata, H. Tamura. Spray characteristics of swirl coaxial injector and its modeling. AIAA Paper 2001–3570, 2001.

[56] Wu J X. *The theoretical and experimental study on atomization performances and two-phase flow of gas–liquid coaxial injectors with mass flow at different back pressures*. Changsha: National University of Defense Technology (in Chinese), 1993.

[57] Elgowainy A, Ashgriz N. The Rayleigh-Taylor instability of viscous fluid layers. *Physics of Fluids*, **1997**, 9, 1635.

[58] O'rourke P.J., Amsden A.A. The TAB Method for Numerical Calculation of Spray Droplet Breakup. SAE Technical Paper, 872089, 1987.

[59] Crowe C T, Sharma M P, Stock D E. The particle-source-in-cell method for gas-droplet flows. *J Fluid Eng*, 1997, **99** (2):325–332.

[60] Smoot L D, Smith P J. Coal Combustion and Gasification. Plenum Press, 1985.

[61] Gosman A D, Ioannides E. Aspects of computer simulation of liquid-fueled combustors, Vol **7**(6), 1983: 482–490 *Journal of Energy*.

[62] Xiao, F., *"Large Eddy Simulation of Liquid Jet Primary Breakup,"* PhD thesis, Loughborough University, (2012).

[63] Xiao, F., Dianat, M., McGuirk, J. J., "Large eddy simulation of single droplet and liquid jet primary breakup using a coupled Level Set/Volume of Fluid method", *Atomization and Sprays*, **24**:281–302 (2014).

[64] Xiao, F., Dianat, M., McGuirk, J. J., "Large Eddy Simulation of Liquid-Jet Primary Breakup in Air Crossflow", *AIAA Journal* **51**:2878–2893 (2013).

[65] Xiao, F., Dianat, M., McGuirk, J. J., "LES of turbulent liquid jet primary breakup in turbulent coaxial air flow", *International Journal of Multiphase Flow* **60**:103–118 (2014).

[66] Fedkiw, R., Aslam, T., Merriman, B., Osher, S., "A non-oscillatory Eulerian approach to interfaces in multimaterial flows (the ghost fluid method)", *J. Comp. Phys.* **152**:457–492 (1999).

[67] Kang, M., Fedkiw, R., Liu, X. D., "A boundary condition capturing method for multiphase incompressible flow", *J. Sci. Comput.* **15**:323–360 (2000).

[68] Liu, X. D., Fedkiw, R., Kang, M. A boundary condition capturing method for Poisson's equation on irregular domains. *J. Comp. Phys.*, 2000, **160**, 151–178.

[69] Aslam T. D., A partial differential equation approach to multidimensional extrapolation. *J. Comp. Phys.* 2003, **193**, 349–355.

[70] Charalampous, G., Hardalupas, Y., Taylor, A. M. K. P., Novel technique for measurements of continuous liquid jet core in an atomizer. *AIAA J.* 2009, **47**, 2605–2615.

[71] Charalampous, G., Hardalupas, Y., Taylor, A. M. K. P., Structure of the continuous liquid jet core during coaxial air-blast atomisation. *Int. J. Spray Combustion Dynamics.* 2009, **1**, 389–415.

3

Modeling of Droplet Evaporation and Combustion

As liquid propellants are atomized into droplets in a liquid rocket engine combustor, the evaporation and combustion of droplets plays an important role in the engine performance. In a practical rocket engine, the combustion of liquid propellants is carried out in the form of a spray or droplet cluster. However, it is very difficult to study the combustion process of the whole liquid spray theoretically as well as experimentally. Since the combustion of a single droplet is the fundamental of the spray combustion, some approximate estimates for the liquid spray combustion can be obtained from droplet combustion. The combustion process of a fuel droplet consists of several stages: fuel evaporation from the droplet surface, mixing with the surrounding medium, and reaction with the oxide gas in the medium.

This chapter elaborates on the evaporation of a single droplet under various conditions: in a static or convection environment at atmospheric pressure, at a high pressure, and in an oscillation environment. The evaporation of a single droplet of a multicomponent fuel will also be described.

3.1 Theory for Quasi-Steady Evaporation and Combustion of a Single Droplet at Atmospheric Pressure

As a classic case of droplet evaporation, the evaporation of an isolated spherical droplet in a static ambient at atmospheric pressure is widely studied. To derive an analytical solution in theoretical modeling, the droplet evaporation is commonly assumed to be a stationary process, though it is actually an unsteady process. It is also assumed that the droplets contain only one chemical component. Moreover, in most cases there is a velocity difference between the droplet

Internal Combustion Processes of Liquid Rocket Engines: Modeling and Numerical Simulations,
First Edition. Zhen-Guo Wang.
© 2016 National Defense Industry Press. Published 2016 by John Wiley & Sons Singapore Pte Ltd.

and the ambient gas during combustion, and thus the effect of convective heat and mass transfer on the droplet evaporation rate should be examined. This section first describes droplet evaporation and the combustion process in a static environment, and then analyzes droplet evaporation and combustion in the case of convection.

3.1.1 Quasi-Steady Evaporation Theory for Single Droplet in the Static Gas without Combustion

Consider the evaporation of an isolated droplet in a static gas without combustion (Figure 3.1). The initial droplet temperature is T_0 and the temperature of the ambient gas is T_∞.

To simplify the analysis, the following assumptions are introduced:

1. There is only spherically one-dimensional flow caused by the Stefan flow around the droplet, and the spherically symmetric center is the center of the spherical droplet.
2. The gas flow field is quasi-steady with a constant pressure.
3. The gas is immiscible with the liquid droplet.
4. Droplets contains only one component, the internal temperature is uniform and equal to the initial temperature, and the surface of the two phases is at equilibrium.
5. The gas phase Lewis number (Le) is 1.
6. The physical properties of the flow field (thermal conductivity coefficient (λ), specific heat capacity (c_p), and ρD where D is the gas mass diffusion coefficient) are assumed to be constant.

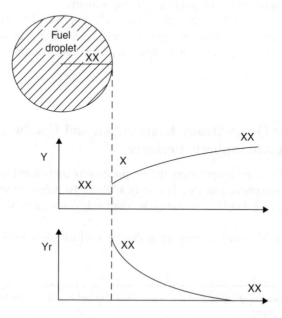

Figure 3.1   Quasi-steady evaporation model for a droplet in a static gas without combustion.

3.1.1.1 Gas Phase Conservation Equations

Based on the above assumptions, the mass, energy, and components conservation equations for droplet evaporation at spherical coordinates can be obtained [1]:

$$\dot{m} = 4\pi r^2 \rho v = 4\pi r^2 \rho_s v_s \tag{3.1}$$

$$r^2 \rho v c_p \frac{dT}{dr} = \frac{d}{dr}\left(r^2 \lambda \frac{dT}{dr}\right) \tag{3.2}$$

$$r^2 \rho v \frac{dY_F}{dr} = \frac{d}{dr}\left(r^2 \rho D \frac{dY_F}{dr}\right) \tag{3.3}$$

where:

$\dot{m}$ is the droplet evaporation rate;
r is the radius of spherical coordinates;
ρ and v are the density and velocity, respectively, of the mixture of ambient gas and droplet vapor at r;
T is the temperature of the gaseous mixture;
c_p is the specific heat capacity;
Y_F is the mass fraction of the droplet vapor in the mixture;
D is the diffusion coefficient;
the subscript s denotes the droplet surface.

3.1.1.2 Equation Solution

Distribution of Droplet Vapor Concentration
Assuming that:

$$b = \frac{Y_F}{Y_{Fs} - 1},$$

Equation 3.3 becomes:

$$r^2 \rho v \frac{db}{dr} = \frac{d}{dr}\left(r^2 \rho D \frac{db}{dr}\right) \tag{3.4}$$

The boundary conditions are as follows:
At $r = r_s$:

$$\left(\frac{db}{dr}\right)_s = \frac{v_s}{D_s} \tag{3.5}$$

At $r \to \infty$:

$$b = b_\infty = \frac{Y_{F\infty}}{Y_{Fs} - 1} \tag{3.6}$$

The following correlation can be derived by integrating Equation 3.4 over $[r_s, r]$:

$$r_s^2 \rho_s v_s (b - b_s) = r^2 \rho D \frac{db}{dr} - r_s^2 \rho_s D_s \left(\frac{db}{dr} \right)_s \tag{3.7}$$

Substituting Equation 3.5 into 3.7, the following equation can be obtained:

$$r_s^2 \rho_s v_s (b - b_s + 1) = r^2 \rho D \frac{db}{dr} \tag{3.8}$$

From assumption 6 above, Equation 3.8 can be written as:

$$-\frac{r_s^2 v_s}{D_s} d\left(\frac{1}{r} \right) = \frac{db}{(b - b_s + 1)} \tag{3.9}$$

Integration of Equation 3.9 over $[r_s, r]$ produces the following correlation:

$$\frac{r_s^2 v_s}{r D_s} = \ln \left(\frac{b_\infty - b_s + 1}{b - b_s + 1} \right) \tag{3.10}$$

Since the droplet surface parameters are known, the above equation describes the relation between the droplet vapor concentration and radius. The vapor concentration declines exponentially with the radius.

At $r = r_s$:

$$\frac{r_s v_s}{D_s} = \ln(b_\infty - b_s + 1) = \ln(1 + B_m) \tag{3.11}$$

where:

$$B_m = b_\infty - b_s = \frac{Y_{Fs} - Y_{F\infty}}{1 - Y_{Fs}}$$

which is called the Spalding transfer function.

The relation between the mixture velocity and vapor concentration on the droplet surface can be derived from Equation 3.11:

$$v_s = \frac{D_s \ln(1 + B_m)}{r_s} \tag{3.12}$$

The droplet mass consumption rate can be computed by:

$$\dot{m} = -4\pi r_s^2 \rho_1 \frac{dr_s}{dt} = -\frac{\pi}{4} d_s \rho_1 \frac{d(d_s^2)}{dt} \tag{3.13}$$

where ρ_1 is the density of the droplet. Since the droplet evaporation rate is equal to the consumption rate, the following correlation can be derived:

$$\frac{d\left(d_s^2\right)}{dt} = -\frac{8\rho_s D_s}{\rho_1}\ln(1+B_m) = -K \tag{3.14}$$

where K is the evaporation constant, which is proportional to the transport coefficient ρD. By integrating Equation 3.14, the following correlation can be obtained:

$$d_s^2 = d_0^2 - Kt \tag{3.15}$$

where d_0 is the initial droplet diameter.

Equation 3.15 shows that the square of the droplet diameter is a linear decreasing function of evaporation time, namely, the classic d^2 laws.

The time required for droplet evaporation can be computed by:

$$t_v = \frac{d_0^2}{K} \tag{3.16}$$

Therefore, a bigger droplet would take longer to finish the evaporation process.

Temperature Distribution of Droplet Vapor

For the energy conservation Equation 3.2 of the mixture around the droplet, the boundary conditions are:

At $r = r_s$, $T = T_s$:

$$\lambda\left(\frac{dT}{dr}\right)_s = \rho_s v_s [L_v + c_1(T_s - T_0)] \tag{3.17}$$

At $r \to \infty$, $T = T_\infty$.

Here:

λ is the heat conduction coefficient of the mixture;

L_v is the specific droplet vaporization heat (latent heat of evaporation) per unit mass at the temperature T_S;

c_1 is the specific heat;

T_0 is the initial droplet temperature;

T_∞ is the ambient temperature.

Integrating the energy Equation 3.2 over $[r_s, r]$, the following correlation can be obtained:

$$r^2 \rho v c_p (T - T_s) = r^2 \lambda \frac{dT}{dr} - r_s^2 \lambda \left(\frac{dT}{dr}\right)_s \tag{3.18}$$

Substituting Equations 3.17 and 3.1 into the above equation, the following correlation can be derived:

$$-\frac{r_s^2 \rho_s v_s c_p}{\lambda} d\left(\frac{1}{r}\right) = \frac{dT}{T - T_s + \frac{L_v}{c_p} + \frac{c_1}{c_p}(T_s - T_0)} \tag{3.19}$$

In turn, the following relation can be obtained by integrating the above equation over $[r, r_\infty]$:

$$\frac{r_s^2 v_s c_p}{r \alpha_s} = \ln \frac{c_p(T_\infty - T_s) + L_v + c_1(T_s - T_0)}{c_p(T - T_s) + L_v + c_1(T_s - T_0)} \tag{3.20}$$

where $\alpha_s = \lambda/\rho_s c_p$ is the thermal diffusion coefficient.

Equation 3.20 indicates that the mixture temperature declines exponentially with the radius. On the droplet surface, the velocity of the vapor is:

$$v_s = \alpha_s \frac{\ln(1 + B_T)}{r_s} \tag{3.21}$$

where:

$$B_T = \frac{c_p(T_\infty - T_s)}{L_v + c_1(T_s - T_0)} \tag{3.22}$$

From Equations 3.12 and 3.21, the following equation can be derived:

$$\text{Le} \ln(1 + B_m) = \ln(1 + B_T) \tag{3.23}$$

When Le=1, $B_m = B_T$, and thus the relation between concentration and temperature on the droplet surface can be obtained:

$$\frac{c_p(T_\infty - T_s)}{L_v + c_1(T_s - T_0)} = \frac{Y_{F\infty} - Y_{Fs}}{Y_{Fs} - 1} \tag{3.24}$$

Equilibrium between Droplets and Vapor

When two phases of the same material (such as the liquid and vapor on the droplet surface) are in the state of equilibrium at a constant temperature and pressure, the temperature, pressure, and Gibbs free energy of the two phases are equal. The Clausius–Clapeyron vapor pressure equation can be deduced from the phase equilibrium:

$$\frac{d \ln p_F}{dT} = \frac{\Delta H}{R_u T^2} \tag{3.25}$$

where p_F is the vapor partial pressure—when the evaporation is steady, p_F is the saturated vapor pressure at temperature T; ΔH is the heat absorbed by 1 mol material when it evaporates at constant temperature and pressure, namely, the molar vaporization heat; R_u is the universal gas constant.

Equation 3.25 describes the relation between the vapor partial pressure and the temperature. On integrating Equation 3.25, the following equation can be obtained:

$$\ln\frac{p_F}{p_{ref}} = \frac{\Delta H}{R_u}\left(\frac{1}{T_{ref}} - \frac{1}{T_s}\right) \tag{3.26}$$

where the subscript "ref" refers to a given reference value; p_{ref} is set to be atmospheric pressure and T_{ref} is the corresponding liquid boiling point.

Assuming that p is the pressure of the mixture around the droplet, the Dalton partial pressure law gives:

$$\frac{p_F}{p} = Y_{Fs}\frac{\bar{M}}{M_F} \tag{3.27}$$

where $\bar{M}$ is the molar mass. Based on the above equation and the relation between ΔH and L_v, the following relation can be obtained:

$$Y_{Fs} = \frac{M_F}{\bar{M}}\frac{p_{ref}}{p}\exp\left[\frac{L_v}{R}\left(\frac{1}{T_{ref}} - \frac{1}{T_s}\right)\right] \tag{3.28}$$

Y_{Fs} and T_s can be solved from Equations 3.24 and 3.28. Furthermore, the evaporation rate on the droplet surface can be calculated from Equation 3.1, the droplet diameter can be obtained as a function of time, and the distribution of the vapor concentration and temperature can be calculated from Equations 3.10 and 3.20, respectively.

3.1.2 Quasi-Steady Evaporation Theory for Droplet in a Static Gas with Combustion

Fuel droplet combustion in a static oxidant medium is analyzed in this subsection. The assumptions here are the same as those in Section 3.1.1. Droplet combustion is diffusion combustion, and the instantaneous reaction model is the simplest reaction model. In this model, the chemical reaction rate is regarded as infinite, and thus the reaction characteristic time is much shorter than the diffusion characteristic time, indicating that reaction finishes instantaneously at the flame discontinuity surface. Therefore, there will be a flame front dividing the flow field into two zones: zone A, which only contains fuel vapor and combustion products, and zone B, which only contains oxidants and combustion products. The quasi-steady evaporation model of a droplet combusting in a static medium is shown in Figure 3.2 [1].

3.1.2.1 Gas Phase Conservation Equations

According to the energy and component conservation in zones A and B and the relation between temperature and vapor concentration on the droplet surface, the gas phase governing equations for droplets combusting in a static medium can be derived as follows:

The component conservation equation in zone A is:

$$r^2\rho v\frac{dY_F}{dr} - \frac{d}{dr}\left(r^2\rho D\frac{dY_F}{dr}\right) = 0 \quad (r_s \leq r \leq r_f) \tag{3.29}$$

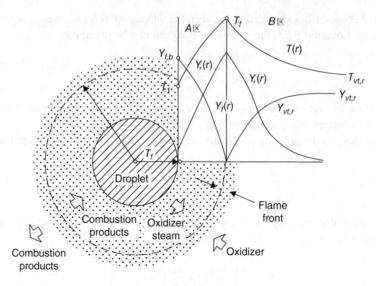

Figure 3.2   Quasi-steady evaporation model of a droplet combusting in a static medium.

The component conservation equation in zone B is:

$$r^2 \rho v \frac{dY_o}{dr} - \frac{d}{dr}\left(r^2 \rho D \frac{dY_o}{dr}\right) = 0 \quad (r_f \leq r \leq r_\infty) \tag{3.30}$$

The energy conservation equation is:

$$r^2 \rho v c_p \frac{dT}{dr} - \frac{d}{dr}\left(r^2 \lambda \frac{dT}{dr}\right) = 0 \tag{3.31}$$

The boundary conditions are as follows:
At $r = r_s$:

$$Y_F = Y_{Fs} \quad T = T_s$$

$$\rho_s v_s = \rho_{Fs} v_{Fs} = -\rho_s D_s \left(\frac{dY_F}{dr}\right)_s + \rho_s v_s Y_{Fs} \quad \text{(Stefan flow)} \tag{3.32}$$

$$\lambda \left(\frac{dT}{dr}\right)_s = \rho_s v_s [L_v + c_l(T_s - T_0)] \quad \text{(heat equilibrium)} \tag{3.33}$$

$$Y_{Fs} = \frac{M_F p_{ref}}{\bar{M} p} \exp\left[\frac{L_v}{R}\left(\frac{1}{T_{ref}} - \frac{1}{T_s}\right)\right] \quad \text{(phase equilibrium)} \tag{3.28}$$

At $r = r_\infty$:

$$Y_o = Y_{o\infty} \quad T = T_\infty$$

At $r = r_f$:

$$(Y_F)_f = (Y_o)_f = 0$$

$$\left(\rho D \frac{dY_F}{dr}\right)_f = -\beta\left(\rho D \frac{dY_o}{dr}\right)_f = -\frac{\dot{m}}{4\pi r_f^2} \tag{3.34}$$

$$\left(\lambda \frac{dT}{dr}\right)_{f,A} - \left(\lambda \frac{dT}{dr}\right)_{f,B} = Q_F\left(\rho D \frac{dY_F}{dr}\right)_f \tag{3.35}$$

where Q_F is the heat released by droplet combustion; subscripts "o" and "F", respectively, represent oxidant and fuel; f represents the flame front; β is the mass of fuel needed for complete combustion of 1 kg of oxidant.

The conditions at the flame front indicate that there is no fuel vapor or oxidant on the flame front; on both sides of the flame front the fuel to oxidant ratio is stoichiometric; the reaction heat of fuel is equal to the heat transferred from the flame front to zones A and B.

3.1.2.2 Equation Solution

Distribution of the Vapor Concentration and Temperature in Zone A

Integration of the fuel vapor component and energy equations in zone A over $[r_s, r]$ leads to:

$$r^2 \rho v Y_F - r^2 \rho D \frac{dY_F}{dr} = r_s^2 \rho_s v_s Y_{Fs} - r_s^2 \rho_s D_s \left(\frac{dY_F}{dr}\right)_s \tag{3.36}$$

$$r^2 \rho v T - r^2 \lambda \frac{dT}{dr} = r_s^2 \rho_s v_s T_s - r_s^2 \lambda \left(\frac{dT}{dr}\right)_s \tag{3.37}$$

Substitution of Equations 3.32 and 3.33 into the above two equations leads to:

$$\dot{m} = \dot{m} Y_F - 4\pi r^2 \rho D \frac{dY_F}{dr} \tag{3.38}$$

$$4\pi r^2 \lambda \frac{dT}{dr} = \dot{m}[L + c_1(T_s - T_o)] + \dot{m} c_p(T - T_s) \tag{3.39}$$

where the evaporation rate is $\dot{m} = 4\pi \rho_s v_s r_s^2 = \text{const}$

Integration of Equations 3.38 and 3.39 over $[r, r_f]$ leads to:

$$\dot{m} = \frac{4\pi \rho D}{\dfrac{1}{r} - \dfrac{1}{r_f}} \ln\left(\frac{1}{1 - Y_F}\right) \quad r_s \le r \le r_f \tag{3.40}$$

$$-\frac{1}{r} = \frac{4\pi \rho D}{\dot{m}} \ln\left[1 + \frac{c_p(T_f - T_s)}{L_v + c_1(T_s - T_o)}\right] - \frac{1}{r_f} \quad r_s \le r \le r_f \tag{3.41}$$

Equations 3.40 and 3.41 described the distribution of vapor concentration and temperature. At $r = r_s$, the relation between flame front temperature, vapor concentration, and temperature on the droplet surface can be obtained:

$$Y_{Fs} = \frac{c_p(T_f - T_s)}{c_p(T_f - T_s) + L_v + c_1(T_s - T_0)} \tag{3.42}$$

Distribution of the Oxidant Concentration and Temperature in Zone B

In zone B, integration of Equations 3.30 and 3.31 over $[r_f, r]$ leads to:

$$4\pi r^2 \rho D \frac{dY_o}{dr} - \dot{m} Y_o = \frac{\dot{m}}{\beta} \tag{3.43}$$

$$\dot{m} c_p(T - T_f) - 4\pi r^2 \lambda \frac{dT}{dr} = -4\pi r_f^2 \lambda \left(\frac{dT}{dr}\right)_{f,B} \tag{3.44}$$

By applying Equation 3.39 to the flame front $r = r_f$ with the boundary condition (Equation 3.35), the following equations can be obtained:

$$\left(\lambda \frac{dT}{dr}\right)_{f,B} = -Q_F \left(\rho D \frac{dY_F}{dr}\right)_f + \left(\lambda \frac{dT}{dr}\right)_{f,A}$$

$$= \frac{\dot{m}}{4\pi r_f^2} \left[Q_F + c_p(T_F - T_s) + L_v + c_1(T_s - T_0)\right] \tag{3.45}$$

Substitution of Equation 3.45 into 3.44 leads to:

$$\dot{m} c_p(T - T_f) - 4\pi r^2 \lambda \frac{dT}{dr} = \dot{m} \left[-c_p(T_f - T_s) - L_v - c_1(T_s - T_0) - Q_F\right] \tag{3.46}$$

Assuming that Le = 1 in zone B, integration of Equations 3.43 and 3.46 over $[r, r_\infty]$ leads to the relation of oxidant concentration and temperature with radius in zone B:

$$\frac{1}{r} = \frac{4\pi \rho D}{\dot{m}} \ln \frac{1 + \beta Y_{o\infty}}{1 + \beta Y_o} \tag{3.47}$$

$$-\frac{1}{r} = \frac{4\pi \rho D}{\dot{m}} \ln \frac{c_p(T - T_s) + L_v + c_1(T_s - T_0) + Q_F}{c_p(T_\infty - T_s) + L_v + c_1(T_s - T_0) + Q_F} \qquad r_f \le r \le r_\infty \tag{3.48}$$

Flame Front Temperature and Radius

At $r = r_f$ Equations 3.47 and 3.48 can be written as:

$$\frac{1}{r_f} = \frac{4\pi \rho D}{\dot{m}} \ln(1 + \beta Y_{o\infty}) \tag{3.49}$$

$$-\frac{1}{r_f} = \frac{4\pi\rho D}{\dot{m}}\ln\frac{c_p(T_f-T_s)+L_v+c_1(T_s-T_0)+Q_F}{c_p(T_\infty-T_s)+L_v+c_1(T_s-T_0)+Q_F} \tag{3.50}$$

Combination of the above two equations results in the following relation:

$$-Q_F = \frac{c_p(T_f-T_\infty)}{\beta Y_{o\infty}}+c_p(T_f-T_s)+L_v+c_1(T_s-T_0) \tag{3.51}$$

The equation indicates that the heat released from the reaction is used to increase the oxidant temperature from T_∞ to T_f, evaporate the droplet, increase the droplet temperature from T to T_s, and increase the temperature of fuel vapor from T_s to T_f.

According to Equation 3.51 the flame temperature is:

$$T_f = \frac{\beta Y_{o\infty}\left[-Q_F+c_pT_s-L_v-c_1(T_s-T_0)\right]+c_pT_\infty}{c_p(1+\beta Y_{o\infty})} \tag{3.52}$$

Combination of Equations 3.40 and 3.49 leads to:

$$\dot{m} = 4\pi\rho Dr_s \ln(1+B) \tag{3.53}$$

where:

$$B = \frac{1+\beta Y_{o\infty}}{1-Y_{Fs}}$$

From Equations 3.53 and 3.49, the flame front radius is given by:

$$\bar{r}_f = \frac{r_f}{r_s} = \frac{\ln(1+B)}{\ln(1+BY_{o\infty})} \tag{3.54}$$

Combination of Equations 3.47, 3.42, and 3.52 can give the vapor concentration and temperature on the droplet surface, and the flame temperature. Substitution of the calculation results in Equations 3.40 and 3.41 produces the distribution of the vapor concentration and temperature in zone A. In addition, substitution of the calculation results in Equations 3.47 and 3.48 results in the distribution of the vapor concentration and temperature in zone B.

3.1.3 Non-Combustion Evaporation Theory for a Droplet in a Convective Flow

Since droplet evaporation in a convective medium should be at least a 2D axisymmetric process with a finite boundary layer thickness, it is more complicated to study droplet evaporation in a convective flow than in a static environment. The converting film method is applied in engineering for an analytic solution. The idea is to convert the real 2D axisymmetric flow into an equivalent problem of molecular heat conduction and diffusion in a spherical membrane.

In addition, the relation between the converted film radius (r_∞) and the intensity of convective heat and mass transfer can be obtained.

According to the principle that the convective heat transfer of a solid ball is equivalent to the heat conduction of a hypothetical molecular, we have:

$$4\pi r_s^2 h^\circ (T_\infty - T_s) = \frac{4\pi \lambda_g (T_\infty - T_s)}{\dfrac{1}{r_s} - \dfrac{1}{r_\infty}} \tag{3.55}$$

and the converted film radius is:

$$r_\infty = r_s N_{uT}^\circ / \left(N_{uT}^\circ - 2 \right) \tag{3.56}$$

where h° is the non-evaporation convective heat transfer coefficient, λ_g is the gas molecular heat conduction coefficient, and N_{uT}° is the Nusselt number of non-evaporation heat convection.

$$N_{uT}^\circ = 2 + 0.6 \mathrm{Re}^{0.5} \mathrm{Pr}^{0.33} \tag{3.57}$$

where R_e is the Reynolds number and Pr is the Prandtl number.

Changing the integration boundary conditions of droplet evaporation in a static environment into:

$$r = r_s, \ b = b_s; \ r = r_\infty, \ b = b_\infty$$

leads to the droplet evaporation rate in a convective environment:

$$\dot{m}_s = \pi d_s \rho_s D_s N_{uT}^\circ \ln(1 + B_m) \tag{3.58}$$

The difference between Equation 3.58 and the droplet evaporation rate in a static environment is $N_{uT}^\circ/2$. The above equation indicates that the evaporation rate grows with increasing convection intensity in the non-combustion evaporation.

3.1.4 Evaporation Theory for a Droplet in a Convective Medium with Combustion

The converting film concept is applied because of the influence of convection. Since the flame front is very close to the free flow border, the approximation is:

$$
\begin{aligned}
r_f &= r_\infty \\
T_s &= T_b
\end{aligned}
\tag{3.59}
$$

where T_b is the boiling temperature. Consideration of the above factors leads to:

$$\dot{m} = 4\pi r_s \frac{\lambda}{c_p} \frac{N_{uT}^0}{2} \ln \left[1 + \frac{c_p (T_f - T_b)}{L_v + c_p (T_b - T_0)} \right] \tag{3.60}$$

$$d_o^2 - d^2 = K_c t \tag{3.61}$$

$$K_c = \frac{4\pi N_{uT}^o}{\rho_c} \frac{\lambda}{c_p} \ln\left[1 + \frac{c_p(T_f - T_b)}{L_v + c_p(T_b - T_o)}\right] = \frac{1}{2} N_{uT}^o K_c^o \tag{3.62}$$

where K_c is the evaporation constant of a combusting droplet in the forced convection.

According to Equation 3.62, K_c increases with the growth of the convection (N_u^o) intensity. When the convection velocity is high enough, the enveloping flame in front of the droplet can be blown out and the wake combustion appears. As the convection velocity grows further beyond a critical value, the wake combustion can also be blown out and pure evaporation appears.

3.2   Evaporation Model for a Single Droplet under High Pressure

The pressure in a modern liquid rocket engine combustor is extremely high. For example, the combustor pressure of RD-120 engines developed by the former Soviet Union is 16.28 MPa. The combustor pressure of the SSME is 22.05 MPa, and the pressure of the MMSE pre-combustion chamber can reach 35.67 MPa. Under these conditions the application of droplet evaporation theory deduced under atmospheric pressure leads to considerable errors. This is mainly due to the following aspects:

1. Evaporation theory under atmospheric pressure is based on the quasi-steady theory, which makes use of the fact that the liquid density is much larger than the gas density, making the liquid phase process slower than the gas phase process. However, at high pressures, the liquid and gas densities tend to be of the same order of magnitude. In addition, the transportations of the two phases are similar. Therefore, quasi-steady theory is no longer appropriate.
2. The critical pressure and temperature of the commonly-used propellants are not very high. For example, liquid hydrogen: $P_{crit} = 12.8$ atm, $T_c = 33.3$ K, liquid oxygen: $P_{crit} = 49.75$ atm, $T_c = 154.58$ K, and kerosene: RP-1 $P_{crit} = 23.73$ atm, $T_c = 679.0$ K. When the temperature of the droplet is approaching T_c, the evaporation rate can be calculated by following equations:

$$\dot{m} = 4\pi r_s \frac{N_u}{2} \frac{\lambda}{C_p} \ln(1 + B)$$

$$B = \frac{C_{p \cdot g}(T_\infty - T_o)}{L}$$

As $L \to 0$, $B \to 0$, which is clearly inappropriate.
3. Under a high pressure, gas can be partly dissolved in droplets, which is difficult to model. In addition, use of the state equation under atmospheric pressure induces considerable error because of the intensive effects of non-ideal gases.

For the above three reasons, there are significant differences between evaporation under high pressure and that under ordinary pressure. Many theoretical and experimental studies have been carried out, e.g., by T. Kadota [2], R.D. Sutton [3], R.L. Matlosz [4], K.C. Hsieh [5], and Yang [6].

In China, Professor Zhuang Fengchen [1, 7] obtained important theoretical and experimental results on high pressure evaporation. This section mainly describes the achievements of Professor Zhuang Fengchen. There are some similarities in the work of Matlosz, K.C. Hsieh and Yang. The relevant literature is recommended to interested readers.

3.2.1 ZKS Droplet High Pressure Evaporation Theory

The theory proposed by Zhuang Fengchen [7], T. Kadota [2], and R.D. Sutton [3] is based on intuitive physical concepts. An explicit formula for droplet evaporation rate at high pressure can then be obtained by applying converting film theory. Practical application of the theory is extensive and it is known as ZKS theory.

1. Modeling assumptions and considerations
    1) the droplet evaporates in the forced convection environment, the ambient gas is inert, and the converting film theory is applied to account for the convective heat transfer effect;
    2) the evaporation process is spherically symmetric;
    3) consideration that the droplet temperature changes over time; the temperature distribution inside the droplet is assumed to be uniform;
    4) consideration of the effect of droplet interface shrinkage;
    5) consideration of the effect of the non-ideal gas.
2. Gas phase conservation equations

Component Conservation Equation of Droplet Vapor
According to the modeling assumptions, the vapor produced from evaporation is partly taken away by Stefan flow, partly fills the space generated by droplet interface movement, and partly transfers through diffusion:

$$\dot{m}_v = Y_v \dot{m}_{\text{Stefan}} - 4\pi r_s^2 \rho_{v,s} \frac{dr_s}{dt} - 4\pi r^2 \rho D \frac{dY_v}{dr}$$

The Stefan flow is equal to the macro-material flow in the normal direction of the phase interface. In addition, the Stefan flow includes macro vapor flow and ambient flow filling the space generated by movement of the phase interface. Furthermore, the macro vapor flow is equal to the vapor flow evaporated from the droplet minus the amount filling the interface movement. Taking the outward normal direction of the phase interface as the positive:

$$\dot{m}_v = Y_v \dot{m}_{\text{Stefan}} - 4\pi r_s^2 \rho_{v,s} \frac{dr_s}{dt} - 4\pi r^2 \rho D \frac{dY_v}{dr}$$

Combination of the above two equations leads to:

$$\dot{m} + 4\pi r_s^2 \rho_{v \cdot s} \frac{dr_s}{dt} = \left(\dot{m} + 4\pi r_s^2 \rho_{v \cdot s} \frac{dr_s}{dt} + 4\pi r_s^2 \rho_{e \cdot s} \frac{dr_s}{dt} \right) Y_v - 4\pi r^2 \rho D \frac{dY_v}{dr} \tag{3.63}$$

where $\rho_{v \cdot s}$ and $\rho_{e \cdot s}$, respectively, represent the vapor density on liquid surface and ambient gas density; r_s is the droplet radius at time t.

Note:

$$A = 1 + \frac{4\pi r_s^2 \rho_{v \cdot s}}{\dot{m}} \frac{dr_s}{dt}, \quad B = 1 + \frac{\rho_{e \cdot s}}{\rho_{v \cdot s}} \left(1 - \frac{1}{A}\right) \tag{3.64}$$

and the boundary conditions are:

$$\begin{cases} r = r_s, & Y_v = Y_{v \cdot s} \\ r = r_\infty, & Y_v = Y_{v \cdot \infty} \end{cases} \tag{3.65}$$

where:

$$r_\infty = \frac{r_s \mathrm{Nu}_m^0}{\mathrm{Nu}_m^0 - 2}$$

is the converting film radius.

Then the droplet evaporation rate can be computed by:

$$\dot{m} = \frac{2\pi \rho D r_s \mathrm{Nu}_m^0}{AB} \ln \left[\frac{1 - BY_{v \cdot \infty}}{1 - BY_{v \cdot s}}\right] \tag{3.66}$$

When ignoring interface movement ($A = B = 1$), Equation 3.66 is the expression of the droplet evaporation rate at the atmospheric pressure.

Combining:

$$\dot{m} = -d\left(\frac{4}{3}\pi \rho_l r_s^3\right)/dt$$

and:

$$\frac{d\rho_l}{dt} = \frac{d\rho_l}{dT_l} \frac{dT_l}{dt}$$

the droplet radius variation over time can be obtained by:

$$\frac{dr_s}{dt} = -\left[\frac{\dot{m}}{4\pi r_s^2 \rho_l} + \frac{r_s}{3\rho_l} \frac{d\rho_l}{dT_l} \frac{dT_l}{dt}\right] \tag{3.67}$$

Energy Conservation Equation

$$q = 4\pi r^2 \lambda \frac{dT}{dr} - \dot{m}(T - T_s)\left[AC_{p \cdot v} + (A - 1)\frac{\rho_{e \cdot s}}{\rho_{v \cdot s}} C_{p \cdot e}\right] \tag{3.68}$$

where q is the heat transferred into the droplet.

The boundary conditions are:

$$\begin{cases} r = r_s, & T = T_s \\ r = r_\infty, & T_\infty = T_\infty \end{cases}$$
(3.69)

Integration of the above equations leads to:

$$q = 2\pi\lambda r_s N_{uT}^0 K \frac{T_\infty - T_s}{e^K - 1}$$
(3.70)

where:

$$K = \frac{\dot{m}\left[A c_{p,v} + (A-1)c_{p,e}\dfrac{\rho_{e \cdot s}}{\rho_{v \cdot s}}\right]}{2\pi\lambda r_s N_T^0}$$
(3.71)

The droplet heat equilibrium equation leads to:

$$\frac{dT_s}{dt} = \frac{1.5\lambda N u_T^0 Z}{\rho_1 C_p \cdot \lambda r_s^2}\left[\frac{T_\infty - T_s}{e^Z - 1} - \frac{\Delta H_v}{A c_{p \cdot v} + C_{p \cdot e}(A-1)\dfrac{\rho_{e \cdot s}}{\rho_{v \cdot s}}}\right]$$
(3.72)

where ΔH_v is the droplet latent heat of phase change.

Real Gas State Equation

Under high pressure, a gas differs from the ideal gas. Therefore, the ideal gas state equation is no longer appropriate. For nonpolar gases such as hydrocarbons, nitrogen, and oxygen, the R-K double parameter equation is applied to describe the gas in engineering [8]:

$$\left[P + \frac{a}{T^{0.5}V(V+a)}\right](V-b) = RT$$
(3.73)

where a and b are the physical parameters and can be obtained by making the first and second derivative at the critical point of constant temperature line equal to zero:

$$a = 0.427R^2 T_c^{2.5}/p_c, \quad b = 0.0867RT_c/P_c$$
(3.74)

where the units of, R, P, T_c, V are respectively $cm^3 \, atm \, gmol^{-1} \, K^{-1}$, atm, K and $cm^3 \, gmol^{-1}$. Equation 3.73 is difficult to solve, and thus the compression factor Z is usually applied to indicate the degree to which the real gas deviates from an ideal gas:

$$pV = ZRT$$
(3.75)

The R-K equation can be converted into the following form with the compression factor Z:

$$Z = \frac{1}{1-h} - \frac{A^2}{B}\frac{h}{(1-h)} \tag{3.76}$$

where $A^2 = a/(R^2 T^{2.5})$, $B = b/(RT)$, $h = Bp/Z$. When the above equation is applied to a mixture of gases:

$$A = \sum y_i A_i, \quad B = \sum Y_i B_i \tag{3.77}$$

Z can be calculated by iteratively solving Equations 3.76 and 3.77.

Vapor Relative Mass Concentration on the Droplet Surface

In the case of atmospheric pressure, the vapor relative mass concentration on the droplet surface can be calculated from the droplet surface equilibrium conditions and vapor pressure equation. However, the thermodynamic equilibrium conditions on the droplet surface at a high pressure are different from those at atmospheric pressure. Not only are the pressure and temperature of two phases required to be equivalent but also the fugacity (chemical potential) of the same component in two phases is supposed to be equivalent:

$$p_c = p_v, T_1 = T_v, f_i^l = f_i^v \tag{3.78}$$

$$f_i^l = f(T, p, x_i = 1) = p_i(T)\phi_i^l(T)\exp\left[V_i^l(p - p_i)/RT\right] \tag{3.79}$$

where P_i is the saturation vapor pressure of liquid component i; φ_i^l is the fugacity coefficient of the saturated vapor under the system temperature; V_i^l is specific heat capacity of the liquid component i under the system temperature; x_i is the relative molar concentration of the liquid component i.

In addition, the following empirical formula is proposed by Zhuang Fengchen [1]:

$$\rho_1 = \rho_c \left[1 + \sum_{i=1}^{4} K_j(1 - T_r)^{j/3}\right]\left[1 + \frac{9z_c N(p - p_i)}{P_c}\right]^{1/9} \tag{3.80}$$

where:

$$K_1 = 17.4425 - 214.578Z_c + 989.625Z_c^2 - 1522.06Z_c^3;$$

$$K_2 = -3.28257 + 13.6377Z_c + 107.4844Z_c^2 - 384.211Z_c^3 \quad (Z_c \leq 0.26);$$

$$K_2 = 60.2091 - 402.063Z_c + 501.0Z_c^2 + 641.0Z_c^3 \quad (Z_c > 0.26);$$

$$K_3 = 0;$$

$$K_4 = 0.93 - K_2;$$

$$N = (1.0 - 0.89\omega)\left[\exp\left(6.9547 - 76.2853T_r + 191.306T_r^2 \right.\right.$$
$$\left.\left. - 203.5472T_r^3 + 82.7631T_r^4\right)\right]$$

where ω is the eccentric factor, Z_c is the critical compression factor, and T_r is the non-dimensionalized temperature by the critical temperature.

The gas phase fugacity of the component I is:

$$f_i^v = py_i\varphi_i^v \tag{3.81}$$

where p is the system pressure, y_i is the relative molar concentration of component i in the gas phase, and φ_i^v is the fugacity coefficient of component i in a gas mixture.

The R-K equation can be written in the following form:

$$\ln\varphi_i = \ln\frac{v}{v-b} + \frac{b_i}{v-b} - \frac{2\sum_j y_j a_{ij}}{bRT^{1.5}}\ln\frac{v+b}{v} - \ln Z + \frac{ab_i}{b^2RT^{1.5}}\left(\ln\frac{v+b}{v} - \frac{b}{v+B}\right) \tag{3.82}$$

where y_j is the relative molar concentration of component j in the gas mixture.

Because the chemical potentials of component i in the gas and liquid phase are equivalent, the relative molar concentration of component i in the gas mixture on the droplet surface is:

$$y_i|_s = \frac{p_i(T)\varphi_i^l(T)}{p\varphi_i^v}\exp\left[V_i^l\cdot(p-p_i)/RT\right] \tag{3.83}$$

According to the concentration conversion, the relative mass concentration of vapor on the droplet surface is:

$$Y_{v\cdot s} = \frac{y_{i\cdot s}\cdot M_v}{y_{i\cdot s}\cdot M_v T(1-y_{i\cdot s})M_e} \tag{3.84}$$

where M_v and M_e are, respectively, the molar mass of droplet vapor and ambient inert medium.

Partial Molar Phase Change Heat (ΔH_v)

The partial molar phase change heat in liquid evaporation refers to the heat needed for 1 mole of component i in the solution to be transferred into 1 mole of vapor component i in the gas mixture. In evaporation theory under the condition of atmospheric pressure, ΔH_v is replaced by the latent heat of evaporation. The latent heat of evaporation refers to the heat absorbed by the pure substance when it is converted from liquid into gas under standard conditions. Assuming that the droplet is pure with a single component by ignoring the dissolution effect, the following equations are derived:

$$\Delta H_v = L + \Delta\bar{H}_i$$

$$\frac{\Delta\bar{H}_i}{RT^2} = -\frac{B_i}{BT}\left(I + \frac{1.5K}{j+1}\right) + \frac{1}{T(Z-BP)}\left(I + \frac{1.5K}{j+1} + BP\right) - \frac{1.5A^2}{BT}\left(\frac{2A_i}{A} - \frac{B_i}{B}\right)In\left(1 + \frac{BP}{Z}\right)$$

$$-\frac{A^2P}{ZT}\left(\frac{2A_i}{A} - \frac{B_i}{B}\right)\frac{\left(I + Z + \dfrac{1.5K}{j+1}\right)}{Z+BP}$$

where:

$$j = Z/BP, \quad K = A^2/B$$

$$l = \frac{j\left[j(j+1)^3 + 0.5K(j-1)(j+1)^2 - Kj(j+1)(j-1)^2 - 0.5K^2(j-1)^2\right]}{(j^2-1)\left[K(j-1)^2(j+1) + Kj(j-1)^2 - j^2(j+1)^2\right]}$$

Under the system temperature, the liquid latent heat can be calculated by the Watson formula:

$$L_2 = L_1 \left(\frac{1 - T_{r,2}}{1 - T_{r,1}}\right)^{0.38}$$

The above equations together with Equations 3.66 and 3.72 can give the development of the droplet diameter and temperature over time. The results of one study [7] on dodecane droplets of different diameters at different ambient temperatures, pressure, and Nu^0 show that the theory agrees very well with experimental findings.

3.2.2   Application of the Liquid Activity Coefficient to Calculate the Gas–Liquid Equilibrium at a High Pressure

It was found experimentally that the dissolution should be considered in droplet evaporation at a high pressure. The dissolution of gas will obviously influence the thermal properties and thus the evaporation. However, the ZKS theory did not account for this factor. Formulas derived in this section for the droplet evaporation rate and the variation of droplet temperature and radius are the same as those in ZKS theory. The feature of the model in this section is the application of an improved R-K state equation, including consideration of the gas dissolution and adoption of the corresponding thermal phase equilibrium relation.

Application of the conventional R-K equation at high pressures can induce considerable errors. The following modification is proposed in Reference [9]:

$$P = \frac{RT}{V-b} - \frac{a}{T^{0.5}V(V+b)} \tag{3.85}$$

To apply the above equation to mixtures, the following mixing rules are used:

$$b = \sum_1^n y_i b_i; \quad b_i = \frac{\Omega_{bi} R T_{ci}}{P_{ci}}; \quad a = \sum_1^n \sum_1^n y_i y_j a_{ij}$$

$$a_{i,j} = \frac{(\Omega_{ai} + \Omega_{aj}) R^2 T_{cij}^{2.5}}{2 P_{cij}}; \quad P_{cij} = \frac{Z_{cij} R T_{cij}}{V_{cij}}$$

$$V_{cij}^{1/3} = \frac{1}{2}\left(C_{ci}^{1/3} + V_{ci}^{1/3}\right); \quad T_{cij} = \sqrt{T_{ci} T_{cj}}\left(1 - K_{ij}\right)$$

$$Z_{cij} = 0.291 - 0.08 \left(\frac{\omega_i + \omega_j}{2} \right)$$

The Ω_{ai} of commonly-used materials and the interaction constant K_{ij} of two components are given in Reference [9].

According to thermodynamic phase equilibrium theory, the chemical potential of each component in the vapor is equal to that in the liquid under the condition of thermodynamic equilibrium:

$$f_i^v = f_i^l$$

The gas-phase fugacity is a function of pressure and vapor concentration:

$$f_i^v = \phi_i p y_i \tag{3.86}$$

where ϕ_i is the fugacity coefficient, which is related to parameters of the state equation by the thermal relation, and can be calculated from Equation 3.82. The definition of liquid fugacity (f_i^l) for subcritical condensable components is different from that for supercritical noncondensable components as it relates to the choice of standard state and normalized conditions.

For the subcritical condensable components:

$$f_1^l = r_1^{(p^r)} f_1^{v(p^r)} x_1 \exp \left(\int_{p^r}^{p} \frac{\bar{V}_1^l}{R_u T} dp \right) \tag{3.87}$$

where $r_1^{(p^r)}$ is the activity coefficient at the reference pressure p^r, x_1 is the mole fraction of component 1, and $f_1^{v(p^r)}$ is the standard-state fugacity of pure liquid 1 at p^r

For supercritical noncondensable components:

$$f_2^l = \gamma_2^{*(p^r)} H_{2(1)}^{(p^r)} x_2 \exp \left(\int_{p^r}^{p} \frac{\bar{V}_2^l}{RT} dP \right) \tag{3.88}$$

where $H_{2(1)}^{(p^r)}$ is the Henry's constant of component 2 in solution 1 at the reference pressure (p^r) and temperature (T); $\gamma_1^{(p^r)}$ and $\gamma_2^{*(p^r)}$ are activity coefficients that can be calculated from the excess Gibbs energy [9]:

$$\ln \left(\gamma_1^{(p^r)} \right) = \left[\frac{\partial \left(n g^{E^*} RT \right)}{\partial n_1} \right]_{T,P,n_2} \tag{3.89}$$

$$\ln \left(\gamma_2^{*(p^r)} \right) = \left[\frac{\partial \left(n g^{E^*} RT \right)}{\partial n_2} \right]_{T,P,n_1} \tag{3.90}$$

$$g^{E^*} / RT = (x_1 q_1 + x_2 q_2) a_{22(1)} \phi_2^2 \tag{3.91}$$

where:

n is the total mole number and equals $n_1 + n_2$;
g^E is the excess Gibbs energy;
q_i is the effective size of molecule i;
$a_{22(1)}$ is the self-interaction constant of molecule 2 in the ambient of molecule 1;
ϕ is the effective volume fraction.

The values of q_1, q_2, ϕ_2, and $a_{22(1)}$ can be obtained by empirical formulae and from published tables [10].

The standard state fugacity $f_1^{v(p^r)}$ and Henry's constant $H_{2(1)}^{p^r}$ can be calculated by following equations:

$$f_1^{v(p^r)} = \phi_1^s p_1^s \exp\left(\int_{p_1^s}^{p^r} \frac{\bar{V}_1^l}{RT} dP \right) \tag{3.92}$$

$$H_{2(1)}^{p^r} = H_{2(1)}^{p_1^s} \exp\left(\int_{p_1^s}^{p^r} \frac{\bar{V}_2^\infty}{RT} dP \right) \tag{3.93}$$

where p_1^s is the saturated vapor pressure of the solvent (condensable component); $\bar{V}_2^\infty$ is the partial molar volume of the solute (noncondensable component) after being infinitely dissolved in solvent; $H_{2(1)}^{(p_1^s)}$ is the value of $\ln (f_2/x_2)$ extrapolated to $x_2 = 0$ on the curve varying over x_2.

In the droplet and gas medium binary system, the concentrations of different components at the given pressure and temperature can be computed by the above equations. After obtaining the solubility of the gas medium in the droplet, the evaporation process of the droplet can be calculated by substituting it into other corresponding equations.

3.3 Subcritical Evaporation Response Characteristics of Propellant Droplet in Oscillatory Environments

Liquid rocket engines are prone to unstable combustion characterized by pressure oscillation, which can result in serious damage to the engines. Generally, unstable combustion is caused by coupling between the spray combustion and acoustic oscillation in a combustor. However, the coupling mechanism has not been established.

In an oscillatory environment, the droplet evaporation is no longer a stationary process. Oscillating liquid parameters can induce heat release fluctuations and then the generated energy is transmitted to the disturbance waves, forming a feedback loop. A pressure wave with high amplitude can cause acoustic instability. Acoustic instability studies focus on: (i) revealing the detailed response mechanism; (ii) finding the transfer function of the wave-energy feedback process. For linear instability, the analytical solution is an explicit combustion response function of state and flow variables [11–15]. In early qualitative stability theories, the open loop response function was applied to describe the droplet combustion response. After the simplified spherically symmetric droplet combustion model [16], the following models have been proposed: unsteady droplet evaporation models [17, 18], an oscillation evaporation model in convection flow [19], and an evaporation-controlling combustion model with thermal wave [20]. Owing to the significant improvement of computing speed, droplet evaporation in oscillation can be reproduced and analyzed using numerical methods. One study [21] numerically analyzed droplet evaporation in the subcritical or supercritical state and the pressure-coupled evaporation response, and gave the response curves under different conditions. The droplet evaporation model and gas model have been solved using the numerical method [22] and the response characteristics of droplet evaporation to ambient pressure variation obtained. In addition, the open-loop response of evaporating droplets to high frequency pressure waves has been studied [23]. It was found that the evaporation response strongly depends on the environment pressure. This section describes the unsteady evaporation model of a propellant droplet in an inert gas. The droplet evaporation response to the ambient pressure is obtained by numerically solving droplet model equations and gas phase model equations.

3.3.1   Physical Model

When a droplet is suddenly set in a static inert gas, there are pressure fluctuations in the ambient gas (Figure 3.3). In the static inert gas, the velocity of Stefan flow caused by evaporation is small. Furthermore, a pressure disturbance propagates at the acoustic speed, and the relaxation time for the flow field around the droplet is much smaller than the pressure oscillation period. Therefore, the concerned domain can be considered as a constant pressure field. The temperature distribution inside the droplet is calculated by heat conduction equations without considering the effect of internal circulation on droplet temperature distribution.

To solve the mass, momentum, energy, and component conservation equations in the gas phase, the following assumptions are made:

1. mass transfer and thermal diffusion are spherically symmetric;
2. there is no forced convection, volume force, viscous dissipation, or thermal radiation;

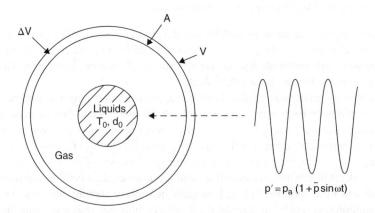

Figure 3.3   Model of an evaporating droplet disturbed by pressure in a static medium.

3. the chemical reaction can be ignored, and the thermal equilibrium is satisfied at the interface of the gas and liquid phases;
4. environmental stress is constant, ignoring the Soret effect and the Dufour effect.

Owing to the decrease of droplet diameter in the droplet evaporation process, coordinate transformation for the one-dimensional unsteady equations ($\eta = r/r_s$) is carried out to account for the interface movement. At the surface of a droplet, the dimensionless diameter is always $\eta = 1$. The equations after transformation are:

$$\rho_l c_{p,l} \frac{\partial T_l}{\partial t} = \frac{1}{r_s^2 \eta^2} \frac{\partial}{\partial \eta} \left(\eta^2 \lambda_l \frac{\partial T_l}{\partial \eta} \right) \tag{3.94}$$

$$\frac{\partial \rho_g}{\partial t} + \frac{1}{r_s \eta^2} \frac{\partial}{\partial \eta} \left(\eta^2 \rho_g u_g \right) = 0 \tag{3.95}$$

$$\rho_g c_{p,g} \frac{\partial T}{\partial t} + \rho_g u c_{p,g} \frac{1}{r_s} \frac{\partial T}{\partial \eta} = \frac{1}{r_s^2 \eta^2} \frac{\partial}{\partial \eta} \left(\eta^2 \lambda_g \frac{\partial T}{\partial \eta} \right) \tag{3.96}$$

$$\rho_g \frac{\partial Y_f}{\partial t} + \rho_g u \frac{1}{r_s} \frac{\partial Y_f}{\partial \eta} = \frac{1}{r_s^2 \eta^2} \frac{\partial}{\partial \eta} \left(\eta^2 \rho_g D_g \frac{\partial Y_f}{\partial \eta} \right) \tag{3.97}$$

$$p = \bar{p}(1 + \hat{p} \sin \omega t) \tag{3.98}$$

where:

subscripts "g", "l", and "s", respectively, represent the gas phase, liquid phase, and droplet surface;
R_0 is the universal gas constant;
$\bar{p}$ is the time-averaged pressure of the environment;
$\hat{p}$ is the amplitude of the pressure disturbance;
ω is the acoustic frequency.

The boundary conditions for the above equations are:

$$\eta = 0, \ \frac{\partial T_l}{\partial \eta} = 0 \tag{3.99}$$

$$\eta = 1, \ T_l = T_g = T_s, \ Y_f = Y_{f,s} \tag{3.100}$$

$$\eta = r_\infty / r_s, \ Y_f = 0.0 \tag{3.101}$$

The modeling equations are discretized with an implicit central difference scheme that has the spatial accuracy of secondary order. The key to solving the equations is to determine the droplet surface temperature. Given the initial conditions, the gas equations and the droplet heat conduction equation are sequentially solved to obtain the temperature, component concentration distribution, and droplet evaporation rate. The droplet surface temperature is calculated from the liquid–gas interface equilibrium:

$$4\pi r_s^2 \lambda_g \frac{\partial T_g}{\partial r}\bigg|_s = 4\pi r_s^2 \lambda_1 \frac{\partial T_1}{\partial r}\bigg|_s + \dot{m}\Delta H_v \tag{3.102}$$

$$\dot{m} = 4\pi r_s^2 \rho_{g,s} D_g \frac{\partial Y_f}{\partial r}\bigg|_s + \dot{m}Y_{f,s} \tag{3.103}$$

$$\dot{m} = 4\pi r^2 \rho_g u_g \tag{3.104}$$

where ΔH_v is the latent heat of evaporation; the relation between $Y_{f,s}$ and T_s is obtained from the surface vapor pressure equation [24].

On the basis of the flow field at the previous time step, model equations with the current outside boundary conditions and droplet surface temperature are solved to obtain the new temperature and concentration fields and then the new droplet surface temperature. The above process is repeated until the relative error of surface temperature satisfies the requirement, and then the flow field at the next time step is calculated.

3.3.2 Examples and the Analysis of Results

n-Heptane droplet evaporation has been simulated with the above droplet evaporation models [22]. In the calculation, the initial droplet diameter is $d_0 = 80$, 120, and 160 μm, respectively. The initial droplet temperature is $T_0 = 300$ K. The ambient temperature is $T_\infty = 500$ K. The pressure oscillation pattern is $P_\infty(t) = P_0(1 + \alpha \sin \omega t)$, where α and ω are, respectively, the pressure amplitude coefficient and angular frequency.

Figure 3.4 demonstrates the radial distribution of the droplet internal temperature and ambient gas temperature without pressure oscillation. The distribution of the droplet internal temperature has a considerable gradient. This implies that the heat transferred by the gas phase is mainly used to heat the droplet. When it reaches equilibrium, droplet temperature continues to rise with a uniform temperature distribution. The gas temperature gradient is high at the droplet surface, indicating that the gas heat conduction is intense.

Figure 3.5 shows the development of the droplet diameter and surface temperature in the droplet evaporation without pressure oscillation. The figure shows that the droplet surface temperature rapidly rises at the beginning of the evaporation and then slowly approaches the equilibrium temperature. However, the evaporation process finishes before reaching the critical temperature. The diameter decreases slowly in the early stages of heating and then more rapidly with a nearly constant speed.

Figure 3.6 illustrates the evaporation rates of droplet with the same diameter and pressure oscillation amplitude but different pressure oscillation frequencies. With increasing pressure oscillation frequency, the oscillation amplitude of the droplet evaporation rate rises gradually. As the pressure oscillation frequency increases further, the oscillation amplitude of the evaporation rate no longer increases. In conclusion, there is a wide frequency range rather than a single natural frequency making the droplet evaporation response strongly. Moreover, at the

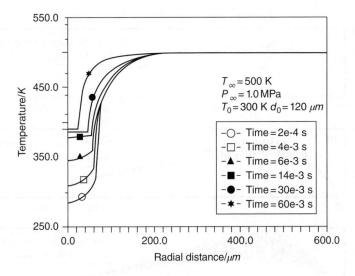

Figure 3.4 Radial distribution of droplet internal temperature and ambient gas temperature without pressure oscillation.

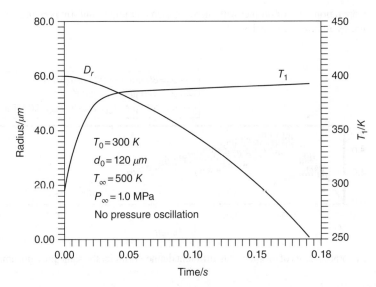

Figure 3.5 Droplet diameter and surface temperature variation curve over time without pressure oscillation.

same pressure oscillation frequency, the evaporation responds most intensely when the droplet temperature just reaches the equilibrium temperature.

Figure 3.7 shows the evaporation rates of droplets with different diameters under the same pressure amplitude and frequency. It is shown that the absolute evaporation rate is faster and the

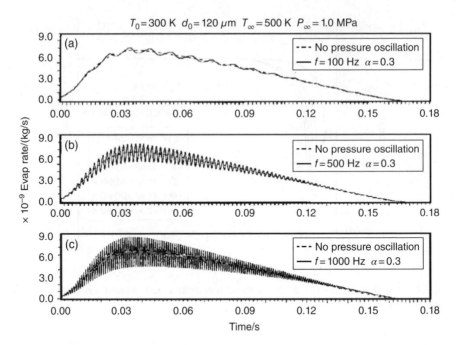

Figure 3.6 Evaporation rates of droplet with the same diameter under uniform pressure amplitude and different pressure oscillation frequencies.

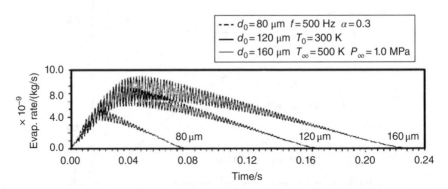

Figure 3.7 Evaporation rates of droplets with different diameters under the same pressure amplitude and frequency.

response to high-frequency pressure oscillation is more intense for a droplet with a larger diameter. It takes longer for the large droplet to reach the strongest response.

Figure 3.7 shows that evaporation varies synchronously with the pressure and that both have the same frequency. This is mainly because equilibrium is assumed at the liquid–gas interface and the ambient pressure has a direct influence on the heat transfer and diffusion. Therefore, the evaporation rate response frequency is the same as the pressure oscillation frequency.

3.4   Multicomponent Fuel Droplet Evaporation Model

The commonly-used fuels in aerospace engineering and industry such as kerosene, diesel oil, and gasoline are actually a multicomponent mixture. The evaporation of a multicomponent droplet is a complex process involving the transient heat transfer on the droplet surface, multicomponent phase transition, and component diffusion in the droplet. The volatility difference of different components leads to variation of droplet components, making it difficult to study the evaporation and heat transfer of a multicomponent droplet. The complexity of the multicomponent droplet evaporation and the lack of experiments on the multicomponent droplet evaporation rate makes research more difficult.

Figure 3.8 is a diagram of the component and temperature distributions in the evaporation of multicomponent droplets. In the droplet evaporation there are different evaporation characteristics for different components. Therefore, the average droplet concentration is higher than the volatile components and lower than the nonvolatile components.

Most multicomponent droplet evaporation models are developed from the single-component droplet evaporation model by including multicomponent effects. Droplet evaporation models can be categorized into discrete and continuous models.

When discretizing a multicomponent droplet evaporation model, the distribution of each component in a droplet is calculated separately. This kind of model is to be applied to droplets with a few components. The commonly-used multicomponent fuels such as kerosene, diesel oil, and gasoline consist of hundreds of different hydrocarbons. Furthermore, the products from different manufactures or batches are also different. Even ignoring the differences between products, the calculation requirement is still beyond the reach of available computers.

To simplify the analysis of multicomponent fuel evaporation, a continuous multicomponent droplet evaporation model has been developed. The components of a droplet can be defined by a continuous distribution function of physical properties such as molar mass. The function of droplet components gradually varies in the process of evaporation. Therefore, the computation cost can be reduced while considering the multicomponent characteristics in the droplet evaporation.

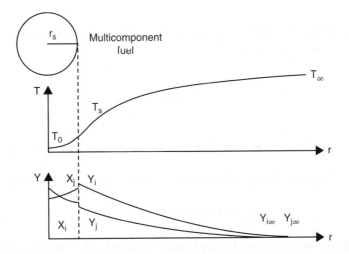

Figure 3.8   Component and temperature distributions in the evaporation of multicomponent droplets.

3.4.1   Simple Multicomponent Droplet Evaporation Model

3.4.1.1   Basic Assumptions [25, 26]

1. The droplet is spherical, and the evaporation is a spherically symmetric process.
2. The droplet is stationary in the ambient gas, and there is no other form of flow except the Stefan flow caused by evaporation.
3. The quasi-steady assumption is applied in the droplet and gas phase around the droplet, and the droplet surface radial movement caused by evaporation is ignored.
4. The component distribution in the droplet is uniform at every moment.
5. The pressure is constant in the flow field.
6. The gas phase Le is assumed to be 1.
7. The gas phase is regarded as an ideal gas.
8. In the gas phase flow field a physical property is constant and is equal to a particular average value calculated at a reference temperature and concentration with the Sparrow 1/3 rule.
9. Gas dissolution in the droplet is ignored.
10. The influence of surface tension on the phase equilibrium is ignored.
11. The radiation heat transfer between droplets and the environment is ignored.

3.4.1.2   Liquid–Gas Equilibrium of Multicomponent Gas

At the gas–liquid equilibrium on the solution surface, the ratio of the concentration x_i of component i in the liquid to the concentration y_i of component i in the gas is denoted as the equilibrium constant under the temperature and pressure [27]:

$$K_i = \frac{y_i}{x_i} = \frac{\gamma_i P_i^\circ}{P} \tag{3.105}$$

where γ_i is the activity coefficient of component I, and when solution is ideal, $\gamma_i = 1$; P is the total pressure at gas–liquid equilibrium:

$$P = \sum_{i=1}^{n} \gamma_i P_i^\circ x_i + P_O \tag{3.106}$$

where P_O is the partial pressure of environment gas and P_i° is the saturated vapor pressure of component i.

When the solution is a binary solution, the activity coefficient [28] of component i is:

$$\ln \gamma_i = A_{1,2}/\left(1 + A_{1,2} X_1/A_{2,1} X_2\right)^2 \tag{3.107}$$

where $A_{1,2}$ and $A_{2,1}$ are constants; X_1 and X_2 are, respectively, the molarity of the volatile and nonvolatile components in solution.

The saturated vapor pressure of component i can be calculated with the Antoine saturated vapor pressure equation [29]:

$$\ln P_i^{\circ} = A - \frac{B}{T_s + C} \tag{3.108}$$

where A, B, and C are constants and T_s is the liquid surface temperature (K).

The concentration $Y_{f,s}$ of the gas component resulting from evaporation on the droplet surface can be calculated from Equations 3.106–3.108:

$$Y_{f,s} = \frac{\sum_{i=1}^{n} \gamma_i P_i^{\circ} x_i M_i}{\sum_{i=1}^{n} \gamma_i P_i^{\circ} x_i M_i + \left(P - \sum_{i=1}^{n} \gamma_i P_i^{\circ} x_i \right) M_o} \tag{3.109}$$

where M_i is the molar mass of component i; P is the ambient pressure.

The physical property of a gas mixture is given by:

$$X = \sum_{i=1}^{n} y_i X_i \tag{3.110}$$

where X represents density (ρ), molar mass (M), specific heat capacity (C_g), diffusion coefficient (D), and heat transfer coefficient (α_g).

3.4.1.3 Conservation of Components on Liquid Surface

The component ratio is assumed to be constant in the process in which the fuel gas generated from the multicomponent droplet evaporation is diffused to the flame front. The fuel gas mixture can be taken as an equivalent fuel. When considering just the Stefan flow, the following equation is satisfied:

$$\dot{m}_s'' Y_{f,R} = \dot{m}_s'' Y_{f,s} - \rho_g D_g \frac{dY_f}{dr}\bigg|_{r_s} \tag{3.111}$$

where Y is the mass percent concentration; subscript "f" refers to the liquid; "s" refers to the liquid surface; "g" refers to the gas; "i" refers to the i-th component; "o" refers to the ambient gas and "R" refers to the inside of the droplet; and:

$$Y_f + Y_o = 1, \quad Y_f = \sum_{i=1}^{n} Y_i, \quad Y_{f,R} = 1$$

Rearrangement of Equation 3.111 leads to:

$$\dot{m}_s'' = \rho_g D_g \frac{d}{dr} \left(\frac{Y_f}{Y_{f,s} - Y_{f,R}} \right) \Big|_{r_s} \tag{3.112}$$

A dimensionless concentration is introduced:

$$b_D = \frac{Y_f - Y_{f,\infty}}{Y_{f,s} - Y_{f,R}} \tag{3.113}$$

where the subscript "∞" refers to the far field.

Combination of Equations 3.112 and 3.113 leads to component boundary conditions on the liquid surface:

$$\dot{m}_s'' = \rho_g D_g \frac{db_D}{dr} \Big|_s \tag{3.114}$$

3.4.1.4 Energy Conservation on Liquid Surface

The energy equilibrium condition on the droplet surface is that the energy needed for evaporation is equal to the transferred heat:

$$\dot{m}_s''[L + C_l(T_s - T_0)] = \lambda_g \frac{dT}{dr} \Big|_s, \quad (\lambda_g = \rho_g C_g \alpha_g) \tag{3.115}$$

where:

$$L = \sum \frac{Y_i L_i}{Y_f}, \quad C_l = \sum \frac{Y_i C_{l,i}}{Y_f}, \quad C_g = \sum \frac{Y_i C_{g,i}}{Y_f}$$

A dimensionless temperature is introduced:

$$b_T = \frac{C_g(T - T_\infty)}{L + C_l(T_s - T_0)} \tag{3.116}$$

The energy boundary condition on the liquid surface can be obtained from Equation 3.115:

$$\dot{m}_s'' = \rho_g \alpha_g \frac{db_T}{dr} \Big|_s \tag{3.117}$$

3.4.1.5 Multicomponent Droplet Evaporation Model

The mass conservation equation is as follows:

$$\dot{m} = 4\pi r_s^2 \dot{m}_s'' = 4\pi r^2 \dot{m}'' \tag{3.118}$$

where $\dot{m}$ is the mass flow on the droplet surface (kg s^{-1}), r is the radius with the origin at the droplet center, and $\dot{m}''$ is the mass flow per unit area at radius r.

The energy conservation equation is:

$$\frac{d}{dr}\left(K_g 4\pi r^2 \frac{dT}{dr}\right) - \frac{d}{dr}\left(\left[4\pi r^2 \dot{m}''\right]C_{p,g}T\right) = 0 \tag{3.119}$$

The component conservation equation is as follows:

$$\frac{d}{dr}\left(\rho_g D_g 4\pi r^2 \frac{dY_{f,i}}{dr}\right) - \frac{d}{dr}\left(\left[4\pi r^2 \dot{m}''\right]Y_{f,i}\right) = 0 \tag{3.120}$$

From equation (3.120), the following equation can be derived:

$$\frac{d}{dr}\left(\rho_g D_g 4\pi r^2 \frac{dY_f}{dr}\right) - \frac{d}{dr}\left(\left[4\pi r^2 \dot{m}''\right]Y_f\right) = 0 \tag{3.121}$$

Substitution of Equations 3.113 and 3.116 into 3.119 and 3.121 leads to:

$$\rho_g a_g \frac{d}{dr}\left(r^2 \frac{db_T}{dr}\right) - \left[r_s^2 \dot{m}_s''\right]\frac{db_T}{dr} = 0 \tag{3.122}$$

$$\rho_g D_g \frac{d}{dr}\left(r^2 \frac{db_D}{dr}\right) - \left[r_s^2 \dot{m}_s''\right]\frac{db_D}{dr} = 0 \tag{3.123}$$

The boundary conditions on the droplet surface are:

$$r = r_s, \quad \begin{cases} b_T = b_{T,s} = \dfrac{C_g(T_s - T_\infty)}{L + C_l(T_s - T_0)} \\[3mm] b_D = b_{D,s} = \dfrac{Y_{f,s} - Y_{f,\infty}}{Y_{f,R} - Y_{f,s}} \end{cases} \tag{3.124}$$

$$r = r_\infty, \quad \begin{cases} b_T = b_{T,\infty} = 0 \\[2mm] b_D = b_{D,\infty} = 0 \end{cases}$$

When Le = 1, Equation 3.122 is the same as Equation 3.123. Omission of the subscripts "T" and "D" leads to a uniform equation:

$$\rho_g D_g \frac{d}{dr}\left(r^2 \frac{db}{dr}\right) - \left[r_s^2 \dot{m}_s''\right]\frac{db}{dr} = 0 \tag{3.125}$$

Integration of Equation 3.125 with boundary conditions given in Equation 3.124 leads to:

$$\ln\left(\frac{b_\infty - b_s + 1}{b - b_s + 1}\right) = \frac{\left[\dot{m}_s'' R^2\right]}{\rho_g a_g} \cdot \frac{1}{r} \tag{3.126}$$

Substitution of boundary conditions $r = R$, $b = b_s$ into Equation 3.126 leads to:

$$\dot{m}_s'' = \frac{\rho_g a_g}{R}\ln(b_\infty - b_s + 1) \tag{3.127}$$

where R is the droplet radius (m).

The term B is used to denote the difference of the b value between the free flow and the droplet surface:

$$B_T = \frac{C_g(T_\infty - T_s)}{L + C_1(T_s - T_0)}$$ (3.128)

where subscript "0" refers to the initial value:

$$B_D = \frac{Y_{f,s} - Y_{f,\infty}}{Y_{f,w} - Y_{f,s}}$$ (3.129)

Rearrangement of Equation 3.127 leads to:

$$\dot{m}_s'' = \frac{\rho_g a_g}{R} \ln(1 + B)$$ (3.130)

B is necessary for the solution of Equation 3.130, and when $B_T = B_D$:

$$\frac{Y_{f,\infty} - Y_{f,s}}{Y_{f,s} - Y_{f,R}} = \frac{C_g(T_\infty - T_s)}{L + C_1(T_s - T_0)}$$ (3.131)

Combination of Equations 3.109, 3.130, and 3.131 leads to $P_{f,s}$, $Y_{f,s}$, B, and the droplet evaporation rate. Though the droplet concentration varies over time in the solution, the component concentration on the droplet surface is assumed to be constant within each time step.

3.4.1.6 Influence of Convection on Evaporation Model

When a droplet is placed in a convective medium, the heat and mass transfer between the droplet and the environment is enhanced. Reference [1] gives a modified form of Equation 3.130:

$$\dot{m}_s'' = \frac{\rho_g a_g}{R} \frac{Nu_T^0}{2} \ln(1 + B)$$ (3.132)

where Nu_T^0 is defined as:

$$Nu_T^0 = 2 + 0.6 Re^{0.5} Pr^{0.33}$$ (3.133)

where Re is the Reynolds number and Pr is the Prandtl number:

$$Pr = \frac{c_{pg}\mu_g}{k_g}$$ (3.134)

$$Re = \frac{\rho_g u_{l,g}(t)d}{\mu_g}, \quad u_{l,g}(t) = |u_l - u_g|$$ (3.135)

3.4.1.7   Influence of Component Diffusion with a Finite Rate on the Evaporation Model

Since component and energy diffusion with a finite rate is not considered in the above derivation, the component and temperature in a droplet is assumed to be uniform within each time step, and the developed model is referred to as the equilibrium evaporation model (also called the infinite evaporation model or batch distillation evaporation model). This model is applicable to droplet evaporation with a low evaporation rate such as is the case in the low temperature area near the jet panel in a combustor. Another limit model is the freezing evaporation model where the component diffusion rate is small enough to be ignored. Ignoring the component diffusion inside the droplet, the component ratio of the fuel vapor is equal to that of the liquid fuel in the droplet evaporation process. This kind of evaporation model can be applied in evaporation with a high evaporation rate such as the case of a droplet in a combustion or high temperature region.

The practical droplet evaporation is between the two limit evaporation models. The evaporation rate of volatile components is higher than other components, and thus the concentration of the volatile components on the droplet surface is lower than that at the droplet core. In addition, this leads to diffusion of the volatile components from the droplet core to the surface at a finite rate.

When considering the finite diffusion rate in the droplet, the component distribution inside the droplet is not uniform. Therefore, the component conservation equation on the droplet surface needs to be modified. The component concentration on the droplet surface (x_i') is introduced to substitute the average droplet concentration in Equation 3.106:

$$x_i' = \xi_i x_i \qquad (3.136)$$

The concentration of volatile components on the droplet surface is lower than the average concentration inside the droplet. The difference between them is determined by the droplet evaporation rate and the component diffusion rate inside the droplet. When the environment temperature is high, the evaporation rate is high, and the component diffusion inside the droplet has less influence on the component concentration on the droplet surface, resulting in a large difference between the component concentration on the droplet surface and the average concentration. This evaporation process is close to the freezing evaporation model. When the environment temperature is low, the evaporation rate is low, and the component diffusion inside the droplet has more influence on the component concentration on the droplet surface, resulting in a small difference between the component concentration on the droplet surface and the average concentration. This evaporation process is close to the infinite diffusion rate evaporation model. Therefore, a bias factor (ξ_i) can be defined as:

$$\xi_i = \xi_i(D_i, (y_i - x_i), B) \qquad (3.137)$$

3.4.1.8   Calculation Results and Analysis [26]

Influence of Environmental Temperature on the Multicomponent Droplet Evaporation
Figures 3.9–3.12 show the simulation results obtained when the influence of environment temperature (T) on the droplet average concentration is taken into account. The environment temperature is, respectively, 400, 1200, and 2000 K. The pressure of the static oxygen

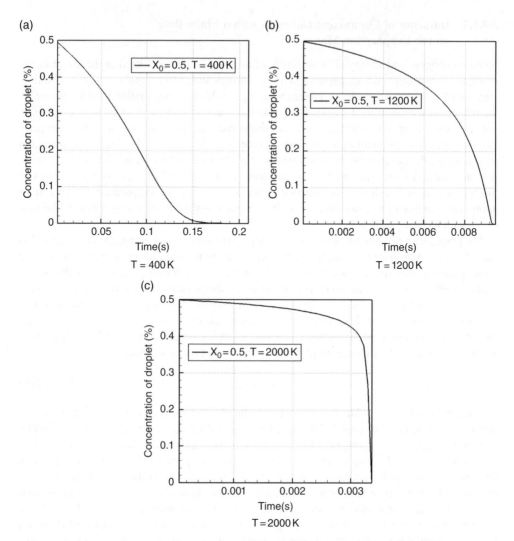

Figure 3.9 Droplet average concentration variation over time at different environment temperatures.

environment is 0.1 MPa. A solution with 50% alcohol concentration was used in the simulation. The initial droplet diameter was 60 μm; x_0 is the initial droplet concentration.

Figure 3.9 indicates that droplet average concentration declines faster in the early stages of the evaporation process with a lower environment temperature and vice versa, which is in agreement with practical evaporation. When the environment temperature is high, the component diffusion inside droplet has less influence on the component concentration on the droplet surface due to the high evaporation rate. Therefore, the difference between droplet surface concentration and average concentration is large. This evaporation process is close to the freezing evaporation model. When the environment temperature is low, the component diffusion inside droplet has more influence on the component concentration on the droplet surface due to the low evaporation rate. Therefore, the difference between droplet surface concentration and the average concentration is small. This evaporation process is close to the equilibrium evaporation model.

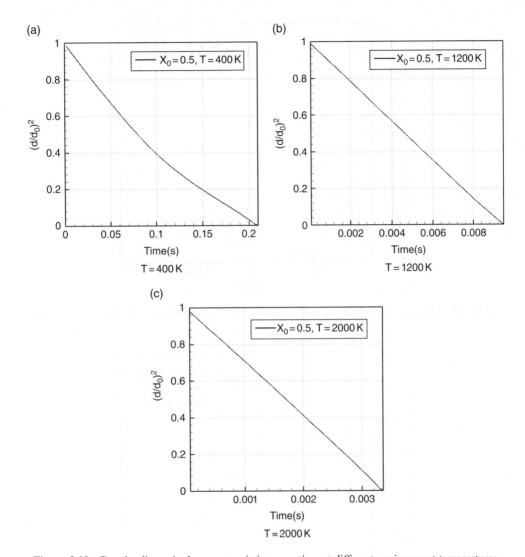

Figure 3.10 Droplet dimensionless area variation over time at different environment temperatures.

Figure 3.10 shows that the dimensionless area $(d/d_0)^2$ of the droplet is no longer linearly correlated with time in an environment with a temperature of 400 K. The decline rate of the dimensionless droplet area is high in the early stages and low in the later stages. This is because the evaporation rate reduces as the volatile component concentration inside the droplet gradually decreases in the evaporation process. Since the volatile component concentration shows less variation in the evaporation process in an environment with a higher temperature, the relation between the dimensionless area and the time approaches a linear function.

Figure 3.11 indicates that droplet surface temperature increases gradually during the evaporation. In addition, the droplet temperature is higher with a higher environment temperature. There is a plateau in the development of the surface temperature before the end of the evaporation for the case with an environment temperature of 400 K.

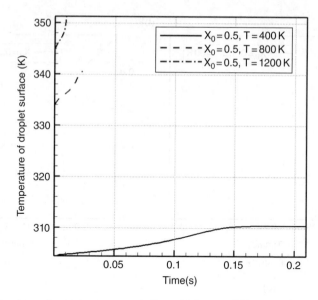

Figure 3.11   Droplet surface temperature variation over time at different environment temperatures.

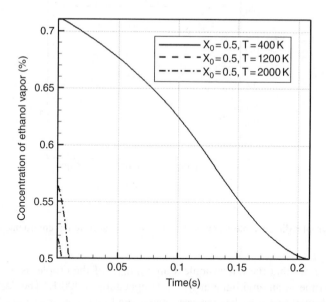

Figure 3.12   Variation of alcohol concentration on the droplet surface over time at different environment temperatures.

Figure 3.12 shows that the initial alcohol vapor concentration is smaller for the case with a higher environment temperature. At the end of the evaporation, the alcohol vapor concentration drops to the initial value.

Figures 3.9–3.12 indicate that the evaporation of a droplet with 50% alcohol is closely related with the environment temperature. In an environment with a lower temperature, the droplet evaporation rate is low, and the influence of component diffusion inside the droplet becomes stronger. The evaporation is closer to the equilibrium evaporation. The droplet evaporation rate is higher with a higher ambient temperature, and the influence of component diffusion inside the droplet is weaker. The evaporation is closer to the freezing evaporation.

Influence of Environmental Pressure on Multicomponent Droplet Evaporation

The pressure in a practical combustion chamber is usually a few MPa or higher, but a droplet evaporation experiment is commonly carried out under atmospheric conditions. Therefore, it is necessary to study the influence of the pressure on droplet evaporation. Equation 3.132 suggests that the pressure affects the evaporation rate mostly via the gas density. The pressure also influences properties related to heat transfer such as the thermal conductivity and specific heat capacity of the gas, and thus affects the evaporation rate.

Figures 3.13–3.16 illustrate the evaporation of a droplet (60 μm diameter) consisting of 30% alcohol in an environment with a temperature of 400 K but different pressures. Figure 3.13 shows that a lower environment pressure leads to a lower initial surface temperature and thus a faster increase of surface temperature in the evaporation process. Figure 3.14 shows that the total evaporation time is shorter with a lower environment pressure. Figures 3.15 and 3.16 indicate that a lower environment pressure leads to a higher initial alcohol vapor concentration and thus to a slower decrease of the alcohol vapor concentration and droplet concentration in the evaporation.

According to the above analysis, the vapor concentration and droplet concentration vary in the multicomponent droplet evaporation process. This result cannot be obtained by assuming that the multicomponent droplet is composed of one virtual component. For the combustion of a solution with low alcohol concentration, control of the atomization and evaporation performance can make the alcohol vapor concentration in the ignition area or combustion area higher

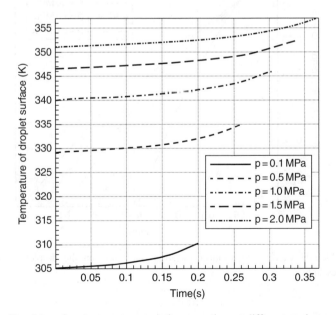

Figure 3.13   Droplet surface temperature variation over time at different environment pressures.

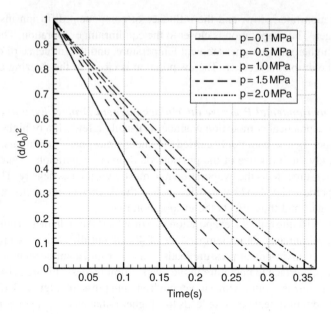

Figure 3.14 Droplet dimensionless area variation over time at different environment pressures.

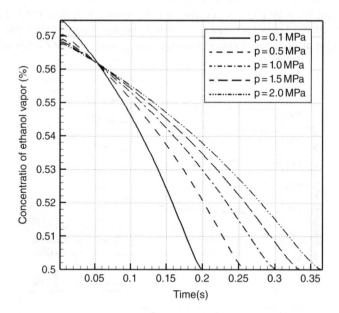

Figure 3.15 Alcohol vapor concentration variation over time on the droplet surface at different environment pressures.

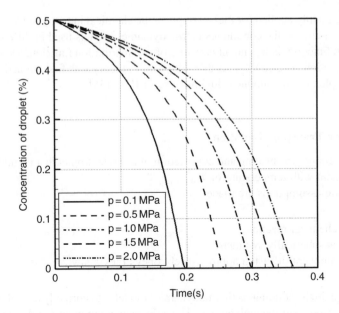

Figure 3.16   Droplet concentration variation over time at different environment pressures.

than the average alcohol vapor concentration, which can contribute to the ignition or flame stability. Moreover, control of the atomization and evaporation performance can make the alcohol vapor concentration in the wall region close to the average alcohol vapor concentration, thereby avoiding wall surface damage caused by a high temperature.

3.4.2   Continuous Thermodynamics Model of Complex Multicomponent Mixture Droplet Evaporation

A real fuel may be composed of hundreds or thousands of components. Since properties such molecular weight and volatility of components are significantly different, the evaporation characteristics cannot be accurately described by one or several substitute components. Moreover, the calculation cost increases as the number of components used in the discrete component evaporation model grows. To reduce the calculation cost of the complex multicomponent mixture evaporation, a continuous thermodynamics model of the multicomponent mixture droplet evaporation was developed. In this model, the multicomponent mixture is regarded as a continuous probability distribution function of its typical thermodynamic properties. Therefore, the main difference between the discrete multicomponent model and the continuous thermodynamic model is that a statistical distribution function of the thermal properties (such as molecular mass, boiling point, and number of carbon atoms, etc.) is applied in the continuous thermodynamic model to describe the mixture while the mole or mass fraction of the components is used in the discrete multicomponent model.

The idea of describing a complex mixture by a continuous function instead of discrete components first appeared about 60 years ago [30, 31]. As the development of a continuous thermodynamics technique was slow, for a long time it was rarely applied to a multicomponent mixture. It was mainly applied for fuel mixtures in the chemical industry in the 1980s [32–37]. The differences between the continuous thermodynamics models used in different studies are the distribution function, independent variables of the distribution function, and the liquid–gas equilibrium equation. The continuous thermodynamics theory was first applied to the evaporation of a complex multicomponent fuel mixture in 1995 [38].

3.4.2.1   Vapor Transport Equation

This section examines the heating and evaporation of a single droplet that is suddenly exposed to an environment with temperature of T_∞.

The following assumptions are made:

1. there is no chemical reaction;
2. the droplet is spherically symmetric;
3. the multicomponent diffusion approximately obeys Fick's law.

The main task in the continuous thermodynamics model is to correctly describe the diffusion of the mixture vapor into the ambient gas. For a discrete component i in the mixture, the diffusion is described by the following equation:

$$\frac{\partial}{\partial t}(cy_i) + \nabla \cdot (cv^* y_i) = \nabla \cdot (cD_{im} \nabla y_i) \tag{3.138}$$

where:

c is the molar density;
y is the molar fraction;
v^* is the molar average velocity;
D_{im} is the effective diffusivity of component i.

In the continuous thermodynamics theory, the molar concentration of components is described by the probability density function $f(I)$. The molar fraction of component i is:

$$y_i = f(I)_i \Delta I_i \tag{3.139}$$

where ΔI_i is the domain interval width of distribution variable I. The distribution variable I can represent a certain property such as compound molar mass, boiling point, or number of carbon atoms; here I is the molar mass.

The most commonly-used probability density distribution function is the Γ distribution function for the complex multicomponent mixture:

$$f(I) = \frac{(I-\gamma)^{\alpha-1}}{\beta^\alpha \Gamma(\alpha)} \exp\left[-\frac{I-\gamma}{\beta}\right] \tag{3.140}$$

where γ is the initial value; α and β are the shape control parameters; $\Gamma(\alpha)$ is the function of Γ, and its mean and variance are respectively:

$$\theta = \alpha\beta + \gamma; \quad \sigma^2 = \alpha\beta^2 \tag{3.141}$$

The property of the probability density distribution function is:

$$\int_0^\infty f(I)\,dI = 1 \tag{3.142}$$

For the evaporation of a single droplet, the molar fraction of the fuel vapor is y_F, and the remaining material is the ambient gas A:

$$y_A = 1 - y_F \tag{3.143}$$

The molar fraction of component i in vapor phase is defined as:

$$y_i = y_F f(I)_i \Delta I_i \tag{3.144}$$

Substituting the above equation into Equation 3.138 and assuming that ΔI_i approaches an infinitely small value, the transport equation of fuel vapor can derived by integrating the equation in $[0, \infty)$:

$$\frac{\partial}{\partial t}(c y_F) + \nabla \cdot (c v^* y_F) = \nabla \cdot \left\{ c\bar{D}\nabla y_F + y_F \nabla (c\bar{D}) \right.$$
$$\left. - y_F \int_0^\infty f(I)\nabla(c D_m(I))\,dI \right\} \tag{3.145}$$

where the average diffusivity $(\bar{D})$ is defined as:

$$\bar{D} = \int_0^\infty D_m(I)f(I)\,dI \tag{3.146}$$

Here, $D_m(I)$ is the component diffusivity, which is a function of variable I.

Equation 3.143 is averaged by the weight of I and integrated in $[0, \infty)$, leading to the transport equation of mean θ:

$$\frac{\partial}{\partial t}(c y_F \theta) + \nabla \cdot (c v^* y_F \theta) = \nabla \cdot \left\{ c\tilde{D}\nabla (y_F \theta) + y_F \theta \nabla (c\tilde{D}) \right.$$
$$\left. - y_F \int_0^\infty f(I)I\nabla(c D_m(I))\,dI \right\} \tag{3.147}$$

where the second order of diffusivity is defined as:

$$\tilde{D}\theta = \int_0^\infty D_m(I)f(I)I\,dI \tag{3.148}$$

Finally, Equation 3.143 is averaged by a weight of I^2 and integrated in $[0, \infty)$, resulting in:

$$\frac{\partial}{\partial t}(cy_F\Psi) + \nabla \cdot (cv^*y_F\Psi) = \nabla \cdot \left\{ c\hat{D}\nabla(y_F\Psi) + y_F\Psi\nabla(c\hat{D}) \right.$$

$$\left. -y_F\int_0^\infty f(I)I^2\nabla(cD_m(I))\,dI \right\} \tag{3.149}$$

Another definition of the mean diffusivity is:

$$\hat{D}\Psi = \int_0^\infty D_m(I)f(I)I^2\,dI \tag{3.150}$$

where the variable Ψ is the second order moment of the distribution function and is defined as:

$$\Psi = \int_0^\infty f(I)I^2\,dI = \theta^2 + \sigma^2 \tag{3.151}$$

where σ^2 is the distribution variance.

More torque equations can be obtained in the same way. However, for a two-parameter distribution the above equations are sufficient.

In Equations 3.145, 3.147, and 3.149, parameters such as fuel molar fraction (y_F), mean θ and second order moment Ψ (or variance σ^2) allow transport property to vary over time and space. Therefore, the above equation can reflect the temporal and spatial transport properties of vapor phase components. The boundary conditions of the droplet evaporation are:

$$r = R : y_F = y_{FR}; \quad \theta = \theta_R; \quad \sigma^2 = \sigma_R^2$$

$$r \rightarrow \infty : y_F = (y_F\theta) = (y_F\Psi) = 0 \tag{3.152}$$

where the subscript "R" represents the droplet surface.

In the transport equations, density (c) and diffusion coefficient (D) are linked together, and thus the influence of the property variation is weakened. The reason for this is that the cD of an ideal gas is not related to the pressure and is less sensitive to the temperature than D. Since the last two terms in Equations 3.145, 3.147, and 3.149 have a similar expression and contribute little to the equations, they are abandoned to simplify the transport equations:

$$\frac{\partial}{\partial t}(cy_F) + \nabla \cdot (cv^*y_F) = \nabla \cdot (c\bar{D}\nabla y_F) \tag{3.153}$$

$$\frac{\partial}{\partial t}(cy_F\theta) + \nabla \cdot (cv^*y_F\theta) = \nabla \cdot \left(c\tilde{D}\nabla(y_F\theta)\right) \tag{3.154}$$

$$\frac{\partial}{\partial t}(cy_F\Psi) + \nabla \cdot (cv^*y_F\Psi) = \nabla \cdot \left(c\hat{D}\nabla(y_F\Psi)\right) \tag{3.155}$$

If the diffusivity is constant throughout the space, $\nabla(c\bar{D}) \cdot \nabla y_F$ can be eliminated and then the diffusion item on the right-hand side of equation can be further simplified into $c\bar{D}\nabla \cdot \nabla y_F$.

Calculations indicate that $\nabla(c\bar{D}) \cdot \nabla y_F$ is 20% of $c\bar{D}\nabla \cdot \nabla y_F$ on the droplet surface. $\nabla(c\bar{D}) \cdot \nabla y_F$ has little influence on liquid properties but has a certain effect on the y_F and θ distribution of the vapor phase. Since the calculation cost is not reduced significantly by ignoring $\nabla(c\bar{D}) \cdot \nabla y_F$ it is retained in Equations 3.153–3.155 to account for the spatial variation of the diffusivity.

To calculate the droplet heat transfer, the energy equation needs to be converted into the continuous thermodynamics form. The energy equation of a mixture with discrete components is:

$$\bar{C}_p \frac{\partial}{\partial t}(cT) + \bar{C}_p \nabla \cdot (cv^*T) = \nabla \cdot \lambda \nabla T - \sum_{i=1}^{n} J_i^* \cdot \nabla \bar{h}_i \tag{3.156}$$

where T is the temperature, λ is the thermal conductivity, and $\bar{C}_p$ is the specific heat capacity of the mixture and is defined as:

$$\bar{C}_p = y_F \int_0^{\infty} C_p(I)f(I)\,dI + (1 - y_F)C_{pA} \tag{3.157}$$

The last term of Equation 3.156 represents the energy transport caused by the component cross diffusion. Though this term is often omitted, it can have a large influence when the specific heat capacity of the vapor is significantly different from that of the ambient gas. For the ideal gas, the gradient of enthalpy is:

$$\nabla \bar{h}_i = C_{pi} \nabla T \tag{3.158}$$

The flux of the component diffusion is:

$$J_i^* = -cD_{im} \nabla y_i \tag{3.159}$$

Introducing the distribution function, integration of it results in the cross diffusion term:

$$-\sum_{i=1}^{n} J_i^* \cdot \nabla \bar{h}_i = \left\{ \int_0^{\infty} C_p(I)cD_m(I)\nabla[y_Ff(I)]\,dI - J_A^* C_{pA} \right\} \cdot \nabla T \tag{3.160}$$

The continuity condition demands that:

$$\sum_{i=1}^{n} J_i^* = 0 \tag{3.161}$$

The diffusion flux of the ambient gas is:

$$J_A^* = \int_0^{\infty} cD_m(I)\nabla(y_Ff(I))\,dI \tag{3.162}$$

Therefore, the cross diffusion term can be simplified to:

$$-\sum_{i=1}^{n} J_i^* \cdot \nabla \bar{h}_i = \left\{ \int_0^{\infty} [C_p(I) - C_{pA}]cD_m(I)\nabla[y_Ff(I)]\,dI \right\} \cdot \nabla T \tag{3.163}$$

where $C_p(I)$ can be calculated by the Chou and Prausnitz linear empirical formula [28]:

$$C_p = a_C + b_C I \tag{3.164}$$

where a_C and b_C reflect the influence of temperature on the specific heat capacity. Integration of the right-hand side of Equation 3.164 leads to

$$\begin{aligned}&\{\left[\left(a_C - C_{pA}\right)c\bar{D} + b_C\theta c\tilde{D}\right]\nabla y_F + y_F\left[\left(a_C - C_{pA}\right)\nabla(c\bar{D}) + b_C\nabla\left(c\tilde{D}\theta\right)\right]\\&- y_F\int\left[a_C + b_C I - C_{pA}\right]f(I)\nabla[cD(I)]dI\}\cdot\nabla T\end{aligned} \tag{3.165}$$

Similar to the diffusion equations, finite difference approximations for the last two items of Equation 3.165 are the same and thus are deleted. Therefore, the final form of the energy equation is:

$$\bar{C}_p\frac{\partial}{\partial t}(cT) + \bar{C}_p\nabla\cdot(cv^*T) = \nabla\cdot\lambda\nabla T + \left[\left(a_C - C_{pA}\right)c\bar{D} + b_C\theta c\tilde{D}\right]\nabla y_F\cdot\nabla T \tag{3.166}$$

3.4.2.2 Liquid Phase Equilibrium Equation

The liquid phase is assumed to be completely mixed, and the droplet temperature distribution is approximately uniform. Under these assumptions, the molar flux on the droplet surface is:

$$N = -\frac{1}{A}\frac{d}{dt}(c_L V) \tag{3.167}$$

where c_L is the droplet molar density; A and V are, respectively, the droplet surface area and volume.

The variation rate of the droplet radius is:

$$\frac{dR}{dt} = -\frac{1}{c_L}\left(N + \frac{R\,dc_L}{3\,dt}\right) \tag{3.168}$$

The molar flux (N) is presumed to be matched with the component flux. For a discrete component i in the liquid phase, its evaporation rate is:

$$N_i = -\frac{1}{A}\frac{d}{dt}(x_i c_L V) = x_i N - c_1\frac{R\,dx_i}{3\,dt} \tag{3.169}$$

where x is the molar fraction of the liquid phase. In the vapor phase the molar flux of the component is:

$$N_i = N y_{iR} - cD_{im}\frac{\partial y_i}{\partial r}\bigg|_R \tag{3.170}$$

Combination of Equations 3.169 and 3.170 leads to:

$$N(x_i - y_{iR}) = -cD_{im}\frac{\partial y_i}{\partial r}\bigg|_R + \frac{c_L R\,dx_i}{3\,dt} \tag{3.171}$$

Introducing the distribution function in the liquid and gas phases, integration of the above equation in $[0, \infty)$ leads to the expression of the total molar flux N:

$$N(1-y_{FR}) = \left[-c\bar{D}\frac{\partial y_F}{\partial r} - y_F\frac{\partial}{\partial r}(c\bar{D}) + y_F\int_0^\infty f(I)\frac{\partial}{\partial r}(cD(I))\,dI \right]_R \tag{3.172}$$

Equation 3.171 is, respectively, averaged by a weight of I and I^2 and then is integrated in $[0, \infty)$, resulting in the variation rate of the liquid phase distribution parameters:

$$\frac{d\theta_L}{dt} = \frac{3}{c_L R}\left[N(\theta_L - y_F\theta) + c\tilde{D}\frac{\partial}{\partial r}(y_F\theta) + y_F\theta\frac{\partial}{\partial r}(c\tilde{D}) \right.$$
$$\left. - y_F\int_0^\infty f(I)I\frac{\partial}{\partial r}(cD(I))\,dI \right]_R \tag{3.173}$$

$$\frac{d\Psi_L}{dt} = \frac{3}{c_L R}\left[N(\Psi_L - y_F\Psi) + c\hat{D}\frac{\partial}{\partial r}(y_F\Psi) + y_F\psi\frac{\partial}{\partial r}(c\hat{D}) \right.$$
$$\left. - y_F\int_0^\infty f(I)I^2\frac{\partial}{\partial r}(cD(I))\,dI \right]_R \tag{3.174}$$

The equations can be simplified by deleting the last two terms of the equations:

$$N(1-y_{FR}) = -c\bar{D}\frac{\partial y_F}{\partial r}\bigg|_R \tag{3.175}$$

$$\frac{d\theta_L}{dt} = \frac{3}{c_L R}\left[N(\theta_L - y_F\theta) + c\tilde{D}\frac{\partial}{\partial r}(y_F\theta) \right]_R \tag{3.176}$$

$$\frac{d\Psi_L}{dt} = \frac{3}{c_L R}\left[N(\Psi_L - y_F\Psi) + c\hat{D}\frac{\partial}{\partial r}(y_F\Psi) \right]_R \tag{3.177}$$

where the liquid molar density is:

$$c_L = \rho_L/\theta_L \tag{3.178}$$

where ρ_L is the liquid density and is assumed to be a constant.

The droplet heat transfer equation is:

$$\frac{dT_L}{dt} = \frac{3}{C_{pL}c_L R}(q - Nh_{fg}) \tag{3.179}$$

where q is the heat flow rate including both the heat conduction and thermal radiation, C_{pL} is the liquid specific heat capacity, and h_{fg} is the evaporation enthalpy.

3.4.2.3 Phases Equilibrium Equation

In continuous thermodynamics, the phase equilibrium is commonly described by Raoult's law and the Clausius–Clapeyron equation. For a mixture with discrete components, Raoult's law is:

$$y_i = x_i(P_{vi}/P) \tag{3.180}$$

where P_{vi} is the component vapor pressure and P is the total pressure.

Normalization of the liquid continuous distribution function leads to the vapor molar fraction:

$$y_F = \int_0^\infty f_L(I)(P_v(I)/P)\,dI \tag{3.181}$$

The above equation is averaged, respectively, by a weight of I and $(I-\theta)^2$, and is integrated in $[0,\infty)$, leading to the vapor mean and variation:

$$y_F\theta = \int_0^\infty f_L(I)(P_v(I)/P)I\,dI \tag{3.182}$$

$$y_F\sigma^2 = \int_0^\infty f_L(I)(P_v(I)/P)(I-\theta)^2\,dI \tag{3.183}$$

The component vapor pressure is calculated by the Clausius–Clapeyron equation:

$$P_v(I) = P_{ATM}\exp\left[\left(s_{fg}/\bar{R}\right)\left(1 - T_B(I)/T\right)\right] \tag{3.184}$$

where s_{fg} is the evaporation entropy; $\bar{R}$ is the universal gas constant; T_B is the boiling point which can be approximately calculated by a linear function of the molar mass:

$$T_B(I) = a_B + b_B I \tag{3.185}$$

Trouton's law is applied to calculate s_{fg}, leading to the following expression of the vapor pressure:

$$P_v(I) = P_{ATM}\exp[A(1-BI)] \tag{3.186}$$

where:

$$\begin{aligned} A &= \left(s_{fg}/\bar{R}\right)(1 - a_B/T) \\ B &= b_B/(T - a_B) \end{aligned} \tag{3.187}$$

Substitution of Equations 3.186 and 3.142 into Equations 3.181–3.183 leads to the relation between the liquid and vapor distribution parameters:

$$\theta - \gamma = \frac{\theta_L - \gamma}{1 + AB\sigma_L^2/(\theta_L - \gamma)} \tag{3.188}$$

$$\sigma^2 = \sigma_L^2 [(\theta - \gamma)/(\theta_L - \gamma)]^2 \qquad (3.189)$$

The initial value of γ for the vapor and liquid is the same. Therefore, if one phase follows a Γ distribution, so does the other.

3.4.2.4 Results and Analysis of Calculation

The droplet evaporation of a kerosene and gasoline mixture is investigated here [38]. The distribution parameters of the initial mixture components are set by:

$$\text{Kerosene}, \alpha_L = 18.5, \ \beta_L = 10, \ \gamma = 0, \ \theta_L = 185, \ \sigma^2 = 1850$$

$$\text{Gasoline}, \alpha_L = 5.7, \ \beta_L = 15, \ \gamma = 0, \ \theta_L = 85.5, \ \sigma^2 = 1282.5$$

Figure 3.17 shows the distribution functions. It is assumed that a droplet with a diameter of 100 µm evaporates in an environment with $T_\infty = 700$. Figures 3.18 and 3.19 illustrate the development of the fuel droplet liquid-phase properties and evaporation characteristics. It is demonstrated that the kerosene first experiences a transient heat transfer. When the temperature is high enough, a significant amount of vapor can be produced. Since the gasoline is volatile and its initial surface vapor pressure is slightly higher than the quasi-steady vapor pressure, the gasoline vapor concentration drops in the evaporation process. As both fuel components reach the quasi-steady evaporation state, the liquid temperature and average molar mass grow almost

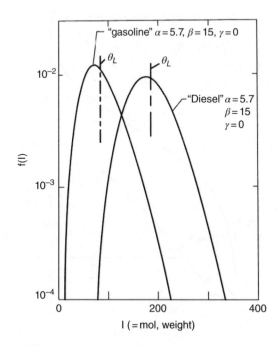

Figure 3.17 Distribution used to simulate kerosene and gasoline.

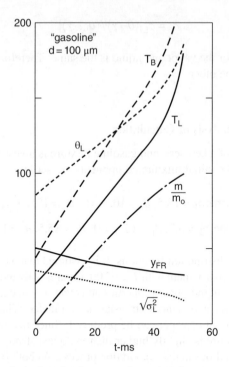

Figure 3.18   Variation of gasoline droplet properties and surface vapor mass fraction over time.

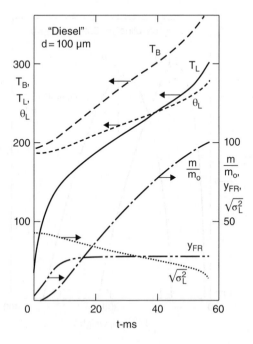

Figure 3.19   Variation of kerosene droplet properties and surface vapor mass fraction over time.

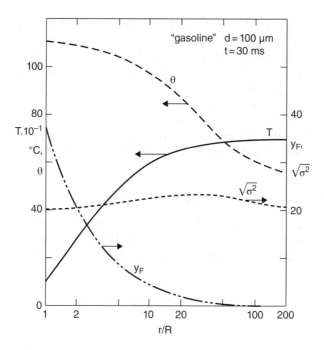

Figure 3.20   Development of gasoline vapor properties over r/R at $t = 30$ ms.

linearly. Components with small molar mass are evaporated and thus the vapor concentration approximately maintains a constant value. Since the liquid temperature is much lower than the boiling temperature, the vapor molar concentration on the droplet surface is very low. The reason for this is that the continuous transit heat transfer of the mixture droplet absorbs a part of heat for evaporation. In the later stages of the droplet life, only components with a large molar mass remain in the liquid phase and thus the liquid average molar mass and droplet temperature increase rapidly. During the whole evaporation process, the variation of the distribution function continuously decreases with the distillation of the light components.

Figure 3.20 shows the development of gasoline vapor properties in the radial direction at $t = 30$ ms. It is shown that the vapor average molar mass decreases with increasing radius. This reflects that θ_L increases over time. There is a small hump on the line of the vapor variation, meaning that the outward diffusion of the vapor on the droplet surface causes the increase of distribution deviation. This verifies the characteristics of a non-uniform distribution of vapor components in space.

3.5   Droplet Group Evaporation

The evaporation and combustion of a single droplet is calculated and analyzed above. However, there are a huge number of droplets in the spray and combustion process in a liquid rocket engine, and thus it is impractical to calculate the evaporation of every single droplet. Since a large number of droplets form a droplet group, it is necessary to study the evaporation and combustion of the droplet group.

In the existing spray and combustion models for a liquid rocket engine, the spray evaporation and combustion rate is assumed to be the sum of evaporation and combustion rate of every single droplet. Therefore, the study of overall spray combustion characteristics can be simplified into the study of single droplet evaporation and combustion. However, this assumption is only valid in a sparse spray. In the spray field of a liquid rocket engine, the spray is very dense so that droplets may be aggregated or broken due to collision. The evaporation and combustion of droplets cause the rapid change of gas parameters, which in turn influences the evaporation and combustion of the droplets. The interaction between the droplets and the gas makes the droplet group combustion in the spray significantly different from the combustion of an isolated droplet.

3.5.1 Definition of Group Combustion Number

Chiu and his colleagues first proposed the concept of group combustion [39] and introduced a group combustion number (G) to measure the gas–liquid interaction degree in the droplet group combustion. Its definition is the ratio of the gas–liquid two phase heat transport to the gas heat transport [1]:

$$G = \frac{n\left(1 + 0.276 \mathrm{Re}^{1/2} \mathrm{Sc}^{1/3}\right)}{\rho D} \frac{4\pi K r_1 R_b^2}{C_p} \tag{3.190}$$

or:

$$G = 3\left(1 + 0.276 \mathrm{Re}^{1/2} \mathrm{Sc}^{1/3}\right) \mathrm{Le} N^{2/3} \left(\frac{r_1}{d}\right) \tag{3.191}$$

where:

R_b is the group radius;
ρ is the gas density;
D is the diffusion coefficient;
K is the thermal conductivity;
C_p is the specific heat capacity at a constant pressure;
N is the number of droplets in the droplet group;
n is the droplet number density;
r_1 is the droplet radius;
d is the droplet spacing;
Re, Sc, Le are, respectively, the Reynolds number, Schmidt number, and Lewis number.

Equations 3.190 and 3.191 indicate that, as the droplet group grows dense or the droplet size increases, the group combustion number G increases, which means more intense interaction between the gas and liquid.

3.5.2 Droplet Group Combustion Model

Chiu and his colleagues pointed out that there were several modes of droplet group combustion depending on the group combustion number. These modes can coexist in practical spray

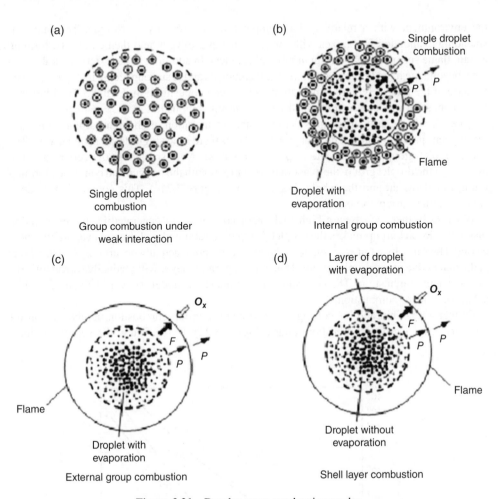

Figure 3.21   Droplet group combustion modes.

combustion and mutual conversion happens under certain conditions. Figure 3.21 illustrates several group combustion modes at different group combustion numbers.

For a thin liquid spray, the corresponding group combustion number is much smaller than one, i.e., $G \ll 1$. Under this condition droplet spacing is much larger than droplet size. The evaporation and combustion of a droplet has little influence on the gas parameters. Every single droplet is surrounded by separate flame (Figure 3.21a). This is called group combustion under weak interaction.

When the group combustion number (G) increases gradually, the interaction between the gas and the liquid becomes stronger. In addition, the evaporation and combustion of the droplet group causes significant changes in the gas component and temperature inside the droplet group, especially at the droplet group center. Analysis by Chiu *et al.* indicated that the entire droplet group is divided by the diffusion flame into a central pre-heating zone and an external combustion zone at $10^{-2} < G < 1$. In the preheating zone, droplets undergo pure evaporation in the environment with a relatively low temperature and scarcity of oxygen. The outward-spreading evaporated fuel mixes with inward-spreading oxygen and then combusts, forming a main flame surrounding a certain number of droplets. In addition, in the external combustion zone beyond the main flame, each droplet still combusts and is surrounded by its separate flame because of the relatively high temperature and oxygen concentration. Figure 3.21b illustrates this combustion mode, which is called internal group combustion.

When $G > 1$, the droplet group is denser. Rapid evaporation of abundant droplets prevents the oxygen spreading into the droplet group, and thus there is no single droplet combustion. The entire droplet group is in the state of pure evaporation. After spreading to a certain distance away from the droplet group, the evaporated fuel mixes with the oxygen and combusts, forming a uniform flame around the whole droplet group (Figure 3.21c). This combustion mode is called external group combustion.

When G further increases, a high fuel vapor concentration generated by the evaporation inside the droplet group makes the droplet group core saturated and droplet evaporation there ceases. The study by Chiu *et al.* suggests that the evaporation actually occurs only in a thin layer at the outer edge of the droplet group. This thin evaporation layer will gradually spread towards the center in combustion. The corresponding combustion model, shown in Figure 3.21d, is called shell layer combustion.

Furthermore, Chiu and his colleagues presented the group combustion number region for different droplet group combustion modes (Figure 3.22). In the figure, the vertical ordinate

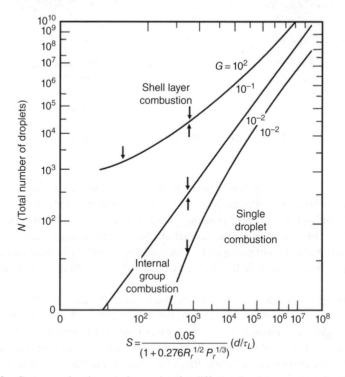

Figure 3.22   Group combustion number region for different droplet group combustion models.

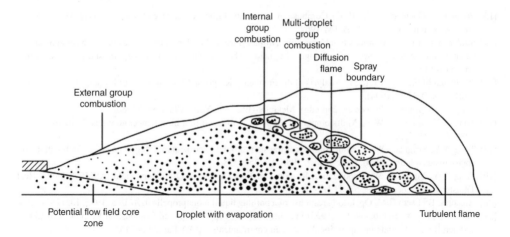

Figure 3.23 Spray combustion field.

N is the total droplet number and the horizontal ordinate S is the dimensionless droplet spacing. In addition, they also divided the real spray combustion field into several different zones (Figure 3.23). In the center of the spray, near the exit of injector, there is a potential flow field core zone. In addition, outside is the evaporation of droplets and the flame of the outer group combustion. The multi-droplet internal group combustion and the gas turbulent flame zone are downstream of the spray.

Readers interested in the droplet group evaporation and combustion model are referred to the literature [1].

References

[1] Zhuang F C. *The Theory and Modeling of Spray Combustion in Liquid Rocket Engines with Applications* (in Chinese), National University of Defense Technology Press, Changsha, 1995.

[2] Kadota T, Hiroyasa H. Evaporation of Single Droplet at Elevated pressures and Temperatures. *Bull. JSME*, 1976, **138**.

[3] Sutton R D. (1974) Operating Manual for Coaxial Injection Combustion Model. NASA CR-129031.

[4] Matlosz R L, Leipziger S. Investigation of liquid drop evaporation in a high temperature and high pressure environment. *Int. J. Heat Mass Transfer*, 1972(**15**):831–852.

[5] Hsieh K C, Shuen J S. Droplet vaporization in high pressure environments. *Combustion Science and Technology*, 1991, **76**, 111–132.

[6] Yang Y S. (1993) Evaporation of LOX under Super Critical and Subcritical Conditions. AIAA Paper 93-2188.

[7] Zhuang F C, Chen X H. Theoretical investigation of fuel droplet vaporization at high pressures. *J. Eng. Thermophysics (in Chinese)*, 1982, **3**(3).

[8] Reid R C. *The Properties of Gases and Liquid*. McGraw-Hill Book Company, 1977.

[9] Prausnits J M, Chuech P L. Calculation of high-pressure vapour-liquid equilibrium. *Ind. Eng. Chem*, 1968, **60**(3), 34–52.

[10] Prausnits J M, Chuech P L. *Computer Calculations for High-Pressure Vapour-Liquid Equilibria*. Prentice-Hall, Inc., 1968.

[11] Lee K W, Chae J W, Lee J Y. Analysis of high-pressure drop vaporization with flash vapor-liquid equilibrium calculation. *Int. Commun. Heat Mass Transfer*, 2003, **30**(5), 633–641.

[12] Yang V, Hsiao C C. (1991) Pressure-Coupled Vaporization and Combustion Responses of Liquid Fuel Droplets in High Pressure Environments. AIAA Paper 91-2310.

[13] Priem R J, Guentert D C. (1962) Combustion Instability Limits Determined by a Nonlinear Theory and a One-dimensional Model. NASA TM 1962.

[14] Strahle W C, Eriodic Solutions to a Convective Droplet Burning Problem. Proceedings of Tenth Symposium on Combustion, August 1964, Cambridge, England. The Combustion Institute, Pittsburgh. PA, 1965, pp. 1315–1325.

[15] Heidmann M F, Wieber P R. An Analysis of the Frequency Response Characteristics of Propellant Vaporization. NASA TN D-3749 (1994).

[16] King C J. *Separation Processes* (2nd edn). McGraw-Hill, New York, 1980, pp. 64–80.

[17] Tong A Y, Sirignano W A. Multicomponent droplet vaporization in a high temperature gas. Combust. *Flame* 1986, **66**, 221–235.

[18] Tong A Y, Sirignano W. A. Multicomponent transient droplet vaporization: integral equation formulation and approximate solution. Num. *Heat Transf.* 1986, **10**, 253–278.

[19] Tong A Y, Sirignano W. Vaporization Response of Fuel Droplet in Oscillatory Field. ASME National Heat Transfer Conference. Paper No.87-HT-58, 1987.

[20] Allison C B , Faeth G M. Open-loop response of a burning liquid monopropellant. *AIAA J.* 1975, **13**, 1287–1294.

[21] Lafon P, Yang V, Habiballah M. (1995) Pressure-coupled Vaporization and Combustion Responses of Liquid Oxygen (LOX) droplets in supercritical hydrogen environments. AIAA Paper 95–2432.

[22] Liu W D, Zhou J, Wang Z G. Response of propellant droplet subcritical evaporation to environmental pressure oscillation. *Journal of Aerospace Power (in Chinese)*, 2001, **16**(1), 52–54.

[23] Lee G Y, Kim S Y, Yoon W S. Oscillatory vaporization and acoustic response of droplet at high pressure. Int. *Commun. Heat Mass Transfer*, 2008, **35**, 1302–1306.

[24] Ehlers, J.G., Gordon, S., Heimel, S., and McBride, B.J. (1963) Thermodynamic Properties to 6000K for 210 Substances Involving the First 18 Elements. NASA-SP-3001.

[25] Tian Z F, Tao Y J, Su L Y at el. Experimental study on the evaporation of the multi-component ethanol droplet. Journal of Propulsion Technology (in Chinese), 2006.

[26] Tian Z F. Study on spray combustion in hydrogen peroxide gas generator with low concentration ethanol. Changsha: National University of Defense Technology (in Chinese), 2007.

[27] Jia S Y, Chai C J. *Chemical mass transfer and separation process*. Beijing: Chemical Industry Press (in Chinese), 2001.

[28] Qiao P at el. *Chemical calculation manual*. Beijing: Chemical Industry Press (in Chinese), 1988.

[29] Tong J S. *Thermal physical properties of the fluid*. Beijing: China Petrochemical Press, 1996 (in Chinese).

[30] Katz D L, Brown G G. (1933) Vapor pressure and vaporization of petroleum fractions. *Ind. Eng. Chem.* **25**, 1373–1384.

[31] Bowman J R. Distillation of an indefinite number of components. *Ind. Eng. Chem.* 1949, **41**, 2004–2007.

[32] Hoffman E J. Flash Calculations for petroleum fractions. *Chem. Eng. Sci.* 1968, **23**: 957–964.

[33] Whitson C H. Characterizing Hydrocarbon Plus Fractions. *Soc. Petrol. Eng. J.* 1983, **23**: 683–694.

[34] Chou G F, Prausnitz J M. Adiabatic flash calculations for continuous or semicontinuous mixtures using an equation of state. *Fluid Phase Equilibria* ,1986, **30**, 75–82.

[35] Peng D Y, Wu R S, Batycky J P. Application of continuous thermodynamics to oil reservoir fluid systems using an equation of State. *AOSTRA J.* 1987 (**3**), 113–122.

[36] Willman B, Teja A S. Prediction of dewpoints of semicontinuous natural gas and petroleum mixtures 1: Characterization by use of an effective carbon number and ideal solution prediction. *Ind. Eng. Chem.* 1987,**26**, 948–952.

[37] Cotterman R L, Prausnitz J M. (1990) Application of continuous thermodynamics to natural-gas mixtures. *Rev. Inst. Francais Petroleum* **45**, 633–643.

[38] Tamin J, Hallett W L H. A continuous thermodynamics model for multicomponent droplet vaporization. *Chem. Eng. Sci.*, 1995, **50**(18), 2933–2942.

[39] Chiu H H, Liu T M. Group combustion of liquid droplets. *Combustion Flame*, 1977, **17**, 127–142.

4

Modeling of Turbulence

The accurate prediction of turbulence is of great importance in the design and optimization of rocket combustors and nozzles, while turbulence modeling is actually one key factor in the uncertainty of the whole numerical simulation (N-S) process. Turbulent flow is a complex multi-scale system. However, the Kolmogorov scale of turbulence is much larger than the molecule's free path, with the continuity hypothesis still working, and the N-S equations are effective in describing the behavior of turbulence. In recent decades, numerous turbulence modeling methods have been proposed for theoretical studies and industrial applications.

Currently, the numerical simulation approaches on turbulence mainly include direct numerical simulation (DNS), large-eddy simulation (LES), Reynolds average method (RANS), and some other hybrid approaches. DNS based on the original N-S equation is expected to capture all scales contained in the turbulence field, which is capable of presenting all the flow information. However, the massive computational cost makes it only available for some theoretical studies; The RANS simulation approach starts from the Reynolds averaged N-S equation, with all kinds of turbulence models developed to enclose the high-order statistic terms in the governing equations. It is presently the most economical and popular approach for industrial applications [1]. The LES approach lies between the DNS and RANS simulations, with the large scale eddies resolved and small scale eddies modeled [2]. However, in the problems involving no-slip walls, the large computational cost is still a major challenge for industrial applications of LES; consequently, hybrid RANS/LES approaches have been developed. Aiming at industrial applications, we will mainly focus on the RANS and LES simulations in this chapter.

Internal Combustion Processes of Liquid Rocket Engines: Modeling and Numerical Simulations,
First Edition. Zhen-Guo Wang.
© 2016 National Defense Industry Press. Published 2016 by John Wiley & Sons Singapore Pte Ltd.

4.1 Turbulence Modeling in RANS

The Reynolds average could help to avoid resolving all the details of turbulence fluctuations and the high computational cost. However, it has proposed some additional high-order statistics of turbulent fluctuations. In the past 100 years, many researchers have been dedicated to the enclosure of these statistical terms in the frame of the Reynolds average. The idea is to establish some empirical or semi-empirical relationship between the high-order and low-order statistics or averaged flow parameters based on the theoretical analyses, experimental studies, and more detailed DNS simulations. The most popular turbulence models are the first-order (viscous eddy model) and second-order (Reynolds stress model) models. The second and higher order models need to solve more equations, which make them time consuming and numerically sensitive. The first order model is economical and robust, and is popular among engineers.

The density weighted Reynolds averaged N-S equation for compressible flow is mainly considered in this chapter:

Continuity equation:

$$\frac{\partial}{\partial t}\bar{\rho} + \frac{\partial}{\partial x_j}\left(\bar{\rho}\widetilde{u}_j\right) = 0 \tag{4.1}$$

Momentum equation:

$$\text{Boussinesq:} \frac{\partial}{\partial t}\left(\bar{\rho}\widetilde{u}_i\right) + \frac{\partial}{\partial x_j}\left(\bar{\rho}\widetilde{u}_i\widetilde{u}_j\right) = -\frac{\partial\bar{p}}{\partial x_i} + \frac{\partial}{\partial x_j}\left(\bar{t}_{ji} + \tau_{ji}\right) \tag{4.2}$$

Energy equation. Boussinesq:

$$\frac{\partial}{\partial t}\left(\bar{\rho}E\right) + \frac{\partial}{\partial x_j}\left(\bar{\rho}\widetilde{u}_j H\right) = \frac{\partial}{\partial x_j}\left(-q_{Lj} - q_{Tj}\right) + \frac{\partial}{\partial x_j}\left[\widetilde{u}_i\left(\bar{t}_{ij} + \tau_{ij}\right)\right] \tag{4.3}$$

where the total energy E and enthalpy H have the expression:

$$E = \widetilde{e} + \frac{1}{2}\widetilde{u}_i\widetilde{u}_i, \quad H = \widetilde{h} + \frac{1}{2}\widetilde{u}_i\widetilde{u}_i$$

The average molecule viscosity is given by:

$$\bar{t}_{ij} = 2\mu S_{ij} - \frac{2}{3}\mu\frac{\partial\widetilde{u}_k}{\partial x_k}\delta_{ij}$$

The state equation for the compressible flow is:

$$p = \bar{\rho}R\widetilde{T}$$

The heater flux terms are:

$$q_{Lj} = -\frac{\mu}{\mathrm{Pr}_L}\frac{\partial\widetilde{h}}{\partial x_j}, \quad q_{Tj} = -\frac{\mu_T}{\mathrm{Pr}_T}\frac{\partial\widetilde{h}}{\partial x_j},$$

where the molecular Prandtl number:

$$\mathrm{Pr_L} = \frac{C_p \mu}{k}$$

is 0.72 for air, and the turbulent Prandtl number ($\mathrm{Pr_T}$) can be fixed as 0.89–0.9 in the near wall region and 0.5 in the shear flow.

In the 1970s, the Boussinesq conception of eddy viscosity was proposed to analogize the viscous term in the laminar flow; consequently, a linear relationship between the shear stress and shear strain can be obtained:

$$\tau_{ij} = -\rho \overline{u'_i u'_j} = 2\mu_t \left(S_{ij} - S_{kk} \delta_{ij}/3 \right) - 2\rho k \delta_{ij}/3 \tag{4.4}$$

The averaged shear strain can be expressed as:

$$S_{ij} = \frac{1}{2} \left(\frac{\partial U_i}{\partial x_j} + \frac{\partial U_j}{\partial x_i} \right) \tag{4.5}$$

Here μ_t is the turbulent eddy viscosity coefficient, which is obtained from turbulence modeling. Based on this, all kinds of eddy viscosity models have been proposed, including algebraic models and one-equation and two-equation models.

4.1.1 Algebraic Model

Algebraic models are obtained directly, from the simple algebraic relationship between the Reynolds stress and averaged velocity properties, which do not contain any differential equations. The early semi-empirical turbulence models, such as Prandtl's blend-length scale theory, Taylor's vorticity transport theory, and von Karman's similarity theory, are all algebraic models. Popular algebraic models include the CS (Smith–Cebeci) [3] model, BL (Baldwin–Lomax) [4] model, half equation JK model (Johnson–King) [5], and so on [6]. The BL model [4] was a popular turbulence model in the 1980s; it will be described here. Despite its poor prediction accuracy in the shock/boundary layer interaction and boundary layer/wake mixed flow problems, its simplicity and economy makes it attractive in applications involving the attached flow. The BL model is developed from the CS model by taking into account the fact that the turbulence fluctuation is suppressed in the near wall region, which will induce a decrease in the mixing length scale while there is intermittent increase in the outer region of the boundary layer, with the mass, momentum, and energy transportation ability decreasing obviously. Thus, different mixing length scale assumptions are used in the inner and outer layers of the boundary layer. The viscous coefficient has a form

$$v_t = \begin{cases} (v_t)_{\mathrm{in}} & (y \le y_c) \\ (v_t)_{\mathrm{out}} & (y > y_c) \end{cases} \tag{4.6}$$

1. In the inner layer $(v_t)_{in} = l^2\Omega$:

The vorticity $\Omega = \left|\varepsilon_{ijk}U_{k,j}\right|$, and the length scale $l = \kappa y D$, with the Van Driest damp function:

$$D = [1 - \exp(-y^+/A^+)],$$

and the Karman constant $(\kappa) = 0.41$. The turbulence model constant A^+ can be set to 26. The non-dimensional wall–normal distance $y^+ = u_\tau y/v_w$, where the wall friction velocity $u_\tau = (\tau_w/\rho)^{1/2}$ (τ_w is the shear stress on the wall), y is the wall-normal distance, and v_w the viscous coefficient on the wall.

2. In the outer layer of the boundary layer:

$$(v_t)_{out} = F_{wake}F_{kelb}(y) \tag{4.7}$$

$$F_{wake} = \min\left(y_{max}F_{max}, C_{wk}y_{max}U_{dif}^2/F_{max}\right) \tag{4.8}$$

where F_{max} and y_{max} correspond to the maximum value and its location in the function:

$$F(y) = y\Omega[1 - \exp(-y^+/A^+)] \tag{4.9}$$

while U_{dif} denotes the gap between the maximum and minimum value of the average velocity. The Klebanoff intermittent function has a form:

$$F_{kleb}(y) = \left|1 + 5.5\left(\frac{C_{kelb} \cdot y}{y_{max}}\right)^6\right|^{-1} \tag{4.10}$$

The model constant $C_{kelb} = 0.3$ and $C_{wk} = 1.0$.

4.1.2 One-Equation Model

To improve the stability and computational efficiency, researchers developed the turbulent Reynolds number transport and eddy viscosity transport model, including the BB model (Baldwin & Barth) [7] and SA model (Spalart & Allmaras) [8]. The BB model starts from the $k - \varepsilon$ model, with the turbulence intensity (k) and dissipation (ε) combined into a turbulence Reynolds number, and the transportation function for the turbulent Reynolds number is established. In contrast, the SA model is developed based on empirical and dimension analyses, which avoid the empirical algebraic relationship. The SA model is widely used in industrial applications, especially in turbines. Compared to the two-equation models, the SA model is more economical and stable. The required grid resolution near the wall is comparable to that for the algebraic model.

4.1.2.1 BB Model

The BB (Baldwin–Barth) model [7] needs to resolve the turbulent Reynolds number transportation equation. A non-dimensional form is given below:

$$
\frac{\partial R}{\partial t} + u_j \frac{\partial R}{\partial x_j} = (C_{\varepsilon 2} f_2 - C_{\varepsilon 1}) \sqrt{RP}
$$

$$
+ \frac{Ma_\infty}{Re} \left(\nu + \frac{\nu_t}{\sigma_\varepsilon} \right) \frac{\partial^2 R}{\partial x_j^2} - \frac{Ma_\infty}{Re} \frac{1}{\sigma_\varepsilon} \frac{\partial}{\partial x_j} \left(\nu_t \frac{\partial R}{\partial x_j} \right)
$$

(4.11)

where:

$$
\frac{1}{\sigma_\varepsilon} = (C_{\varepsilon 2} - C_{\varepsilon 1}) \frac{\sqrt{C_\mu}}{\kappa^2}
$$

(4.12)

$$
\nu_t = C_\mu R D_1 D_2
$$

(4.13)

Here:

$$
D_1 = 1 - \exp\left(-\frac{y^+}{A^+} \right), \quad D_2 = 1 - \exp\left(-\frac{y^+}{A_2^+} \right)
$$

(4.14)

$$
f_2 = \frac{C_{\varepsilon 1}}{C_{\varepsilon 2}} + \left(1 - \frac{C_{\varepsilon 1}}{C_{\varepsilon 2}} \right) \left(\frac{1}{\kappa y^+} + D_1 D_2 \right) \{ \sqrt{D_1 D_2} +
$$

$$
\frac{y^+}{\sqrt{D_1 D_2}} \left[\frac{D_2}{A^+} \exp\left(-\frac{y^+}{A^+} \right) + \frac{D_1}{A_2^+} \exp\left(-\frac{y^+}{A_2^+} \right) \right]
$$

(4.15)

The relative model coefficient is set as:

$$
\left. \begin{array}{l} C_{\varepsilon 1} = 1.2, \quad C_{\varepsilon 2} - 2.0, \quad C_\mu - 0.09 \\ k = 0.41, \quad A^+ = 26, \quad A_2^+ = 10 \end{array} \right\}
$$

(4.16)

with a product term:

$$
P = 2\nu_t S_{ij} S_{ij} - \frac{2}{3} \nu_t \left(\frac{\partial u_k}{\partial x_k} \right)^2
$$

(4.17)

which has an approximate value of:

$$
P \cong \nu_t \Omega^2
$$

(4.18)

4.1.2.2   Spalart–Allmaras Model

The control parameter $\tilde{\nu}$ in the SA (Spalart–Allmaras) model [8] is used to denote the turbulent viscosity in the near wall region (except for the molecule viscosity dominated region):

$$\frac{\partial \tilde{\nu}}{\partial t} + u_j \frac{\partial \tilde{\nu}}{\partial x_j} = C_{b1} \tilde{S} \tilde{\nu} + \frac{1}{\sigma} \left[\frac{\partial}{\partial x_j} \left((\nu + \tilde{\nu}) \frac{\partial \tilde{\nu}}{\partial x_j} \right) + C_{b2} \left(\frac{\partial \tilde{\nu}}{\partial x_j} \right)^2 \right] - C_{w1} f_w \frac{\tilde{\nu}^2}{d^2} \qquad (4.19)$$

where:

$$\nu_t = \tilde{\nu} f_{\nu 1}, \quad f_{\nu 1} = \frac{\chi^3}{\chi^3 + C_{\nu 1}^3}$$

$$\tilde{S} = S + \frac{\tilde{\nu}}{\kappa^2 d^2} f_{\nu 2}, \quad f_{\nu 2} = 1 - \frac{\chi}{1 + \chi f_{\nu 1}}$$

$$\chi = \frac{\tilde{\nu}}{\nu}, \quad S = \sqrt{2 \Omega_{ij} \Omega_{ij}}, \quad \Omega_{ij} = \frac{1}{2} \left(\frac{\partial u_i}{\partial x_j} - \frac{\partial u_j}{\partial x_i} \right)$$

$$f_w = g \left(\frac{1 + C_{w3}^6}{g^6 + C_{w3}^6} \right)^{1/6}, \quad g = r + C_{w2} \left(r^6 - r \right), \quad r = \frac{\tilde{\nu}}{\tilde{S} \kappa^2 d^2}$$

The model coefficients are set as:

$$C_{b1} = 0.1355, \quad C_{b2} = 0.622, \quad \sigma = \frac{2}{3}, \quad C_{\nu 1} = 7.1$$

$$C_{w1} = \frac{C_{b1}}{\kappa^2} + \frac{(1 + C_{b2})}{\sigma}, \quad C_{w2} = 0.3, \quad C_{w3} = 2.0$$

4.1.3   Two-Equation Models

Most industrial turbulence models work is based on the linear eddy viscosity assumption (or Boussinesq assumption). The turbulent-eddy viscosity can be expressed as $k^m \varepsilon^n - k^p \varepsilon^q$ in different two-equation models. Theoretically, a two-equation model could be accurately derived from another [9].

4.1.3.1   The $k-\varepsilon$ Models

Standard k–ε *Model*

The standard $k-\varepsilon$ model and its improved versions are the most widely used two-equation models. They are developed based on the work of Zhou (1945), Davidov (1961), and Harlow and Nakayama (1986), and the most famous one, which is also the standard $k-\varepsilon$, was proposed by Jones and Launder in 1972 [10]. The $k-\varepsilon$ model is actually a semi-empirical model, in which the turbulent kinetic energy equation is obtained from accurate derivation and modeling.

However, it is hard to build an accurate equation for the dissipation term ε due to the lack of higher order correlation information. Based on the physics of turbulence, the following can be assumed:

production term:

$$C_{\varepsilon 1} \cdot \frac{\varepsilon}{k} \times \text{turbulent production}$$

gradient diffusive term:

$$\frac{\mu_t}{\sigma_\varepsilon} \cdot \frac{\partial \varepsilon}{\partial x_i}$$

dissipation term:

$$C_{\varepsilon 2} \cdot \frac{\varepsilon}{k} \times \text{dissipation of turbulent kinetic energy.}$$

Accordingly, the standard $k-\varepsilon$ model equation could be expressed as:

$$\frac{\partial(\bar\rho k)}{\partial t} + \frac{\partial(\bar\rho k \tilde u_j)}{\partial x_j} = \frac{\partial}{\partial x_j}\left[\left(\mu_l + \frac{\mu_t}{\sigma_k}\right)\frac{\partial k}{\partial x_j}\right] + P_k - \bar\rho\varepsilon - Y_M \tag{4.20}$$

$$\frac{\partial(\bar\rho\varepsilon)}{\partial t} + \frac{\partial(\bar\rho\varepsilon\tilde u_j)}{\partial x_j} = \frac{\partial}{\partial x_j}\left[\left(\mu_l + \frac{\mu_t}{\sigma_\varepsilon}\right)\frac{\partial\varepsilon}{\partial x_j}\right] + C_{\varepsilon 1}\frac{\varepsilon}{k}P_k - C_{\varepsilon 2}\bar\rho\frac{\varepsilon^2}{k} \tag{4.21}$$

To emphasize the compressible effect, the term Y_M is added to the standard form of the k equation in the above $k-\varepsilon$ model. Further information about the term Y_M will be given in the following discussion on the compressible modification.

The eddy viscosity coefficient above has a form:

$$\mu_t = \bar\rho C_\mu \frac{k^2}{\varepsilon}$$

The production of kinetic energy is:

$$P_k = \tau_{ij}\frac{\partial\tilde u_i}{\partial x_j}$$

The suggested model constants are listed here:

$$C_{\varepsilon 1} = 1.44, \quad C_{\varepsilon 2} = 1.92, \quad C_\mu = 0.09, \quad \sigma_k = 1.0, \quad \sigma_\varepsilon = 1.3$$

Notably, the $k-\varepsilon$ model is established for the fully developed turbulent field, while the effect of molecular viscosity is not well considered. Thus, the standard $k-\varepsilon$ could only work in principle for the fully developed turbulent field.

RNG k–ε Model

In the mid-1980s, Yakhot and Orszag [11] made a serious analysis of the turbulent field with the renormalization group (RNG) method. They obtained a theoretical RNG $k-\varepsilon$. Although no empirical assumption is used, it could afford similar model constants as the standard $k-\varepsilon$ model. As the molecular viscosity in the sub-layer is omitted in the derivation of the RNG method, the RNG model should not be expected to work in the sub-layer.

The transport equation of the RNG $k-\varepsilon$ model has a similar form to that of the standard $k-\varepsilon$ model, and is:

$$\frac{\partial(\bar{\rho}k)}{\partial t} + \frac{\partial(\bar{\rho}k\tilde{u}_j)}{\partial x_j} = \frac{\partial}{\partial x_j}\left[\alpha_k\mu_{\text{eff}}\frac{\partial k}{\partial x_j}\right] + P_k - \bar{\rho}\varepsilon - Y_M \tag{4.22}$$

$$\frac{\partial(\bar{\rho}\varepsilon)}{\partial t} + \frac{\partial(\bar{\rho}\varepsilon\tilde{u}_j)}{\partial x_j} = \frac{\partial}{\partial x_j}\left[\alpha_\varepsilon\mu_{\text{eff}}\frac{\partial \varepsilon}{\partial x_j}\right] + C_{1\varepsilon}\frac{\varepsilon}{k}P_k - C_{2\varepsilon}\bar{\rho}\frac{\varepsilon^2}{k} - R_\varepsilon \tag{4.23}$$

Notably, the term Y_M in the k equation is the compressible modification term. Compared with the standard $k-\varepsilon$ model, the diffuse term in the k and ε equations is different, while another term R_ε appears in the ε equation, which can be expressed as:

$$R_\varepsilon = \frac{C_\mu\bar{\rho}\eta^3(1-\eta/\eta_0)}{1+\beta\eta^3}\frac{\varepsilon^2}{k} \tag{4.24}$$

where:

$$\eta = S\frac{k}{\varepsilon}, \quad S = \sqrt{2S_{ij}S_{ij}}, \quad S_{ij} = \frac{1}{2}\left(\frac{\partial \tilde{u}_i}{\partial x_j} + \frac{\partial \tilde{u}_j}{\partial x_i}\right), \quad \eta_0 = 4.38, \beta = 0.012$$

The parameters in the above equation under high Reynolds number can be set to:

$$\mu_{\text{eff}} = \mu + \mu_{\text{t}}, \quad \mu_t = \bar{\rho}C_\mu\frac{k^2}{\varepsilon}$$

$$C_\mu = 0.0845, \quad a_k = a_\varepsilon = 1.39$$

The turbulence production term is:

$$P_k = \tau_{ij}\frac{\partial \tilde{u}_i}{\partial x_j},$$

with model constants:

$$C_{1\varepsilon} = 1.42, \quad C_{2\varepsilon} = 1.68$$

When low Reynolds number flow or near-wall flow is considered, the effective viscous coefficient can be given as:

$$d\left(\frac{\bar{\rho}^2 k}{\sqrt{\varepsilon\mu}}\right)\Big/ d\hat{v} = 1.72\frac{\hat{v}}{\sqrt{\hat{v}^2 - 1 + C_v}} \tag{4.25}$$

where $\hat{v} = \mu_{\text{eff}}/\mu, C_v \approx 100$

Application of the RNS model indicates that it could more effective than the standard $k-\varepsilon$ model in rapidly strained and swirling flow.

4.1.3.2   The $k-\omega$ Models

The modern $k-\omega$ model was developed by Wilcox [10] based on the work of Kolmogorov and Saffman, in which the length scale of turbulence is expressed by the dissipation ratio (ω). To overcome the sensitivity of the model to the initial conditions, two improved two-zone models were proposed (Menter, 1994 [12]), called the baseline model (BSL) and the shear-stress transport model (SST). These two zonal models have become popular industrial turbulence models in recent years. The Wilcox $k-\omega$ model and $k-\omega$ SST model will be given below.

Standard **k–ω** *Model*
The standard $k-\omega$ model [7] is also a kind of empirical model, with the turbulent kinetic energy and dissipation ratio equation introduced:

$$\frac{\partial(\bar{\rho}k)}{\partial t} + \frac{\partial(\bar{\rho}k\tilde{u}_j)}{\partial x_j} = \tau_{ij}\frac{\partial\tilde{u}_i}{\partial x_j} - \beta^*\bar{\rho}\omega k + \frac{\partial}{\partial x_j}\left[(\mu_1 + \sigma^*\mu_t)\frac{\partial k}{\partial x_j}\right] \tag{4.26}$$

$$\frac{\partial(\bar{\rho}\omega)}{\partial t} + \frac{\partial(\bar{\rho}\omega\tilde{u}_j)}{\partial x_j} = \alpha\frac{\omega}{k}\tau_{ij}\frac{\partial\tilde{u}_i}{\partial x_j} - \beta\bar{\rho}\omega^2 + \frac{\partial}{\partial x_j}\left[(\mu_1 + \sigma\mu_t)\frac{\partial\omega}{\partial x_j}\right] \tag{4.27}$$

The model constants are:

$$\alpha = 0.555, \quad \beta = 0.075, \quad \beta^* = 0.09, \quad \sigma = 0.5, \quad \sigma^* = 0.5$$

The turbulent viscosity coefficient is:

$$\mu_t = \frac{\bar{\rho}k}{\omega} \tag{4.28}$$

The relationship between the $k-\omega$ model and other turbulence models is:

$$\varepsilon = \beta^*\omega k, \quad l = k^{1/2}/\omega \tag{4.29}$$

SST **k–ω** *Model*
Menter (1994) [12] adopted a zonal concept to blend the Wilcox $k-\omega$ model used in the near wall region and the $k-\varepsilon$ model for the high Reynolds number flow in the outer region of the boundary layer. In the near wall region, the advantage of the $k-\omega$ model is that it can present the effect of the reverse pressure gradient and thus can model large scale separation problems. In the region away from the wall, the $k-\varepsilon$ model could help overcome the initial sensitivity problem of the standard $k-\omega$ model and so increase the stability of the turbulence model. Menter noticed that the ratio between the turbulence production and dissipation can be much larger than 1.0 in the case with a reverse pressure gradient. Thus, the shear stress would be overestimated when the term $C_\mu = 0.9$ is used. Based on the shear-stress transport assumption,

the turbulent viscosity coefficient (ν_t) is redefined so that it has similar features to the JK model [13] in order to consider the effect of the shear stress transport, and the prediction of the reverse pressure gradient problems is obviously improved [14,15].

The SST model can be obtained from the blend of transformed $k-\varepsilon$ model and $k-\omega$ model, with the expression:

$$\frac{\partial(\bar{\rho}k)}{\partial t} + \frac{\partial(\bar{\rho}k\tilde{u}_j)}{\partial x_j} = \widetilde{P}_k - \beta^*\bar{\rho}\omega k + \frac{\partial}{\partial x_j}\left[(\mu_l + \sigma_k\mu_t)\frac{\partial k}{\partial x_j}\right] \tag{4.30}$$

$$\frac{\partial(\bar{\rho}\omega)}{\partial t} + \frac{\partial(\bar{\rho}\omega\tilde{u}_j)}{\partial x_j} = P_\omega - \beta\bar{\rho}\omega^2 + \frac{\partial}{\partial x_j}\left[(\mu_l + \sigma_\omega\mu_t)\frac{\partial\omega}{\partial x_j}\right]$$
$$+ 2(1-F_1)\bar{\rho}\sigma_{\omega2}\frac{1}{\omega}\frac{\partial k}{\partial x_j}\frac{\partial\omega}{\partial x_j} \tag{4.31}$$

where:

$$\widetilde{P}_k = \min(P_k, 10\cdot\beta^*\bar{\rho}k\omega), \quad P_k = \mu_t\frac{\partial\tilde{u}_i}{\partial x_j}\left(\frac{\partial\tilde{u}_i}{\partial x_j} + \frac{\partial\tilde{u}_j}{\partial x_i}\right)$$

The viscous coefficient is:

$$\nu_t = \frac{a_1 k}{\max(a_1\omega, \Omega F_2)}, \quad \Omega = \sqrt{2S_{ij}S_{ij}}$$

Here, the blend functions have the following forms:

$$F_1 = \tanh\left(\eta_1^4\right), \quad \eta_1 = \min\left[\max\left(\frac{\sqrt{k}}{0.09\omega y}, \frac{500\nu}{\omega y^2}\right), \frac{4\bar{\rho}\sigma_{\omega2}k}{CD_{k\omega}y^2}\right] \tag{4.32}$$

$$F_2 = \tanh\left(\eta_2^2\right), \quad \eta_2 = \max\left(\frac{2\sqrt{k}}{0.09\omega y}, \frac{500\nu}{\omega y^2}\right) \tag{4.33}$$

$$CD_{k\omega} = \max\left(2\bar{\rho}\sigma_{\omega2}\frac{1}{\omega}\frac{\partial k}{\partial x_j}\frac{\partial\omega}{\partial x_j}, 10^{-10}\right) \tag{4.34}$$

Two points are emphasized by Menter (2009) [13] concerning the above equation. First, the turbulence production term P_k is limited with a new term $\widetilde{P}_k$, which is proposed to avoid the extra turbulence production near the stagnant point; Second, the small number in the second part of the $CD_{k\omega}$ function could be set as 10^{-10} rather than the original 10^{-20}.

Given the model constants θ_1 and θ_2 in the $k-\omega$ and $k-\varepsilon$ models, respectively, the blend model constants θ in the $k-\omega$ SST model could be written as:

$$\theta = F_1\theta_1 + (1-F_1)\theta_2 \tag{4.35}$$

The model constants in the $k-\omega$ model could be set as:

$$\sigma_{k1} = 0.85, \quad \sigma_{\omega1} = 0.5, \quad \beta_1 = 0.075, \quad a_1 = 0.31, \quad \beta^* = 0.09, \quad \kappa = 0.41,$$
$$\gamma_1 = \beta_1/\beta^* - \sigma_{\omega1}\kappa^2/\sqrt{\beta^*}$$

while the counterparts in the $k-\varepsilon$ model could be defined as:

$$\sigma_{k2} = 1.0, \quad \sigma_{\omega 2} = 0.856, \quad \beta_2 = 0.0828, \quad a_1 = 0.31, \quad \beta^* = 0.09, \quad \kappa = 0.41,$$
$$\gamma_2 = \beta_2/\beta^* - \sigma_{\omega 2}\kappa^2/\sqrt{\beta^*}$$

The SST model could effectively avoid too much turbulence production in the near wall region (compared with the dissipation), which would overestimate the shear stress. Applications [13] indicate that the SST model is more effective than the original $k-\omega$ model.

4.1.4   Turbulence Model Modification

Traditionally, turbulence models are derived for incompressible flow, and the Favre average was usually adopted to extend these models to moderate compressible flows in applications. However, for flows with a higher Mach number, this extension would encounter some other problems, which have received attention in recent years. Actually, compressible effect and turbulence model modification on some basic flows, such as wall turbulence, turbulent mixing layer, and shock turbulent interaction, began in the early 1990s, and some of the compressible modifications have been in use since that time. Sarkar [16], Zeman [17] and Wilcox have published their conclusions on the compressible effect in high speed flow investigations, and proposed their compressible modification accordingly.

The effective modifications proposed presently are on the pressure-dilatation, dilation-dissipation, shear layer structural modification and shock wave boundary layer interaction unsteadiness modification. The former two have been well adopted, while the later ones are relatively new and should be selected carefully. It should also be noted that most of the modifications were proposed to either the pure shear layer or the wall bounded turbulence, while misuse of the modification would import greater errors and make the simulation worse.

In supersonic or hypersonic flow with a Mach number (M) of < 5 (or even to $M < 8$), the compressible effect has no obvious influence on the wall turbulence if there is no extra pressure jump (induced by a shock wave for example); in subsonic flow with a wall temperature $T_w/T_e < 6$, the compressible effect will also have no obvious influence on the flow field. Based on the above, Morkovin proposed that the density variation should be considered in the case of moderate Mach numbers, while the density fluctuation could simply be omitted. This means that only the average density effect should be considered in the compressible flow without shock and not hypersonic flow, and the Favre average is enough to describe the turbulence. This is the main idea behind the early turbulence modeling for compressible flow.

Notably, the Morkovin hypothesis has obvious limitations in applications. First, the term $\rho'/\bar{\rho}$, which usually appears with the heater transfer and combustion, would not be small enough to be omitted; second, in the flow with shear layer the pressure fluctuation would be much larger. The density fluctuation $\rho'/\bar{\rho}$ in the shear layer with $M = 1$ is almost at the same level as the turbulent boundary layer with $M = 5$. It has been well recognized that the turbulence model based on the Morkovin hypothesis could not predict the development of the mixing layer thickness [7].

Most of the compressible modification is based on the extra terms introduced by the Favre-average in the Reynolds-averaged governing equations. The compressible continuity equation, momentum equation, and energy equation have the same form as the traditional

Reynolds-averaged equations. However, the turbulent kinetic equation is somewhat different and has to be treated more cautiously. The complete turbulent kinetic equation can be expressed as:

$$\frac{\partial}{\partial t}(\bar{\rho}k) + \frac{\partial}{\partial x_j}(\bar{\rho}\tilde{u}_j k) = \tau_{ij}\frac{\partial \tilde{u}_i}{x_j} - \bar{\rho}\varepsilon + \frac{\partial}{\partial x_j}\left[\overline{t_{ji}u_i''} - \overline{\rho u_j''(u_i''u_i''/2)} - \overline{p'u_j''}\right]$$

$$-\overline{u_i''\frac{\partial P}{\partial x_i}} + \overline{p'\frac{\partial u_i''}{\partial x_i}}$$

(4.36)

where the molecular diffusion and the turbulent transportation term could be modeled by:

$$\overline{t_{ji}u_j''} - \overline{\rho u_j''(u_i''u_i''/2)} = \left(\mu + \frac{\mu_t}{\sigma_k}\right)\frac{\partial k}{\partial x_j}$$

(4.37)

The pressure diffusion term $\overline{p'u_j''}$ has a very limited effect in a complex turbulence field. Due to the lack of relative information, it could be usually be abandoned. Thus, the remaining terms in the above equation awaiting modeling are the pressure work and pressure dilatation terms:

Pressure work:

$$-\overline{u_i''\frac{\partial P}{\partial x_i}}$$

Pressure dilatation:

$$\overline{p'\frac{\partial u_i''}{\partial x_i}}$$

Most compressible turbulence models are based on an incompressible flow, with the two terms usually omitted. To take the compressible effect into account, we have to model them appropriately. For the pressure-work term, Wilcox and Alber (1972) [10] proposed their closure models, which, however, were not as effective as expected. Moreover, they have also violated the Galilean invariance. By contrast, for the pressure dilatation term, Sarkar and Zeman separately obtained successful models based on the DNS simulation database.

Except the above modifications, the compressible effect on the turbulence structure and turbulent dynamics could also enhance the dissipation rate remarkably. To predict the compressible mixing layer accurately, some modification of the turbulent dissipation would be necessary, which is called the dilation-dissipation term.

4.1.4.1 Dilation-Dissipation

Based on some simple derivation from the dissipation rate definition, the dissipation term could be split into two parts:

$$\bar{\rho}\varepsilon = \bar{\nu}\overline{\rho\omega_i''\omega_i''} + \frac{4}{3}\bar{\nu}\overline{\rho u_{i,i}''u_{i,i}''} = \bar{\rho}\varepsilon_s + \bar{\rho}\varepsilon_d$$

(4.38)

The former part (ε_s) is the non-dilation term, while the later part (ε_d) denotes the dilation induced dissipation. The term ε_s is usually assumed to be the same as the original ε in the dissipation equation, while ε_d is the one we need to model.

Sarkar [18], Zeman [19] and El Baz and Launder [20] proposed their pressure diffusion and dilation-dissipation modification terms for the compressible mixing layer by considering the effect of turbulent Mach Number. Based on the work of Sarkar and Zeman, Wilcox proposed his dilation-dissipation modification, which could be used in the shear layer and turbulent boundary layer. In the assumption of Sarkar and Zeman, the dilation-dissipation should be a function of the turbulent Mach number M_t, which is defined as:

$$M_t = \frac{\sqrt{2k}}{a} \tag{4.39}$$

The term ε_s is considered to be independent of the compressible effect, and it could be given directly by the original dissipation equation. The modification lies in the term ε_d, which has the form:

$$\varepsilon_d = \xi^* F(M_t)\varepsilon_s \tag{4.40}$$

Here ξ^* is a model constant, and $F(M_t)$ is a function of the turbulent Mach number.

In the framework of the $k-\varepsilon$ model, the turbulent kinetic energy equation and dissipation equation are given below:

$$\bar{\rho}\frac{dk}{dt} = -\bar{\rho}(\varepsilon_s + \varepsilon_d) + \cdots$$
$$\bar{\rho}\frac{d\varepsilon_s}{dt} = -C_{\varepsilon 2}\bar{\rho}\varepsilon_s^2/k + \cdots \tag{4.41}$$

For the $k-\omega$ model, the modification could be added to the model constants β and β^*:

$$\beta^* = \beta_0^*[1 + \xi^* F(M_t)]$$
$$\beta = \beta_0 - \beta_0^* \xi^* F(M_t) \tag{4.42}$$

Although the Sarkar and Zeman modifications have a similar form, they are obtained from different physical aspects. Sarkar proposed that the dilation-dissipation increase with M_t, while Zeman insists that the appearance of shocklets is the main reason behind the extra dissipation, and there should be a limit Mach number below which the dilation-dissipation will be zero. The model constants of the Sarkar, Zeman, and Wilcox modifications are listed below:

Sarkar model:

$$\xi^* = 1, \quad F(M_t) = M_t^2 \tag{4.43}$$

Zeman model:

$$\xi^* = \frac{3}{4},$$
$$F(M_t) = \left\{1 - \exp\left[-\frac{1}{2}(r+1)(M_t - M_{t0})^2/\Lambda^2\right]\right\}H(M_t - M_{t0}) \tag{4.44}$$

Wilcox model:

$$\xi^* = \frac{3}{2}, \quad M_{t0} = \frac{1}{4}, \quad F(M_t) = \left[M_t^2 - M_{t0}^2\right]H(M_t - M_{t0}) \tag{4.45}$$

where H is a Heaviside function:

$$H(M_t - M_{t0}) = \begin{cases} 0 & M_t < M_{t0} \\ 1 & M_t \geq M_{t0} \end{cases} \tag{4.46}$$

Other parameters in the Zeman model should be considered separately for the turbulent boundary layer and free shear layer. Generally, the parameters in the free shear layer could be set as $\Lambda = 0.6$, $M_{t0} = 0.10\sqrt{2/(r+1)}$, while in the turbulent boundary layer $\Lambda = 0.66$, $M_{t0} = 0.25\sqrt{2/(r+1)}$ could be adopted.

Validation of the Wilcox modification confirmed that the unmodified $k-\omega$ model would over-predicate the development speed of the free shear layer, while all of the Sarkar, Zeman, and Wilcox modifications could provide reasonable results. However, when the modifications are conducted to the plane boundary layer between Mach 0 and 5, the skin friction coefficient would depart from the experimental results, while the original $k-\omega$ model could provide better results. This means that the Sarkar and Zeman modification are likely to induce more errors when they are used to the boundary flow.

4.1.4.2 Pressure Dilatation

It should be mentioned that the pressure dilatation term is at the same level as the dilation-dissipation term, and the two modifications are usually proposed together. Thus, the model constants should be adjusted considering each other.

Sarkar Model [18]

Sarkar supposed that the effect of the pressure-dilatation term could be remarkable. Based on theoretical analysis of the pressure fluctuation equation for the compressible flow and the DNS simulation results, an algebraic model for the pressure-dilatation modification was proposed:

$$\overline{p'\frac{\partial u_i''}{\partial x_i}} = -\alpha_3 P_k M_t^2 + \alpha_4 \bar{\rho}\varepsilon_s M_t^2 \tag{4.47}$$

where P_k is the production of the turbulent kinetic energy, and α_3 and α_4 are model constants.

To consider the pressure dilatation term and the above dilation-dissipation term together, Sarkar obtained the compromised model constants for the two modifications based on the DNS database:

$$\alpha_1 = 0.5, \quad \alpha_3 = 0.4, \quad \alpha_4 = 0.2$$

where α_1 is the model constant for the previous dilation-dissipation term ξ^*.

Zeman Model [17]

The Zeman model is obtained by introducing some assumptions in the pressure fluctuation expression in order to simplify and fit the model with DNS data:

$$\overline{p'\frac{\partial u_i''}{\partial x_i}} = g(M_t)\left(\frac{\partial \bar{\rho}}{\partial y}\right)^2 \frac{k}{\varepsilon}\frac{a^2}{\bar{\rho}}\overline{v'^2} \tag{4.48}$$

$$g(M_t) = 0.2\left[1 - \exp\left(-M_t^2/0.02\right)\right] \tag{4.49}$$

where v' is the transverse velocity fluctuation in the boundary layer.

4.1.5   Nonlinear Eddy Viscosity Model

As mentioned above, most turbulence models in industry applications are based on the Boussinesq approximation, for which the turbulent eddy viscosity coefficient has an isotropic feature. This departs from most practical turbulent fields, and limits its applications in simulations with a strong reverse pressure gradient. Usually, the second-order turbulence model can provide a more accurate Reynolds stress and prediction result in an application. However, it is much more expensive, considering five extra transportation equations, and the numerical schemes are demanding. In recent years, nonlinear turbulence models have attracted increasing attention. Compared with traditional nonlinear turbulence models, these nonlinear models could effectively predict the turbulent shear stress in different directions, which could help overcome the inherent limitation.

The nonlinear eddy viscosity model and algebraic stress model lies between the linear turbulence model and Reynolds stress model, which enables them to combine the advantages of the two. The nonlinear eddy viscosity model is a kind of two-equation model, but it abandons the linear Boussinesq eddy viscosity approximation between the Reynolds tress tensor and the averaged strain rate tensor. It has a similar two-equation form, but is able to reflect the anisotropism of the Reynolds stress, which could only be presented previously by the Reynolds stress model (containing the evolution equation of every term in the Reynolds stress tensor) [21, 22]. The two-equation nonlinear turbulence model has proved to be effective in the prediction of complex flow with pressure gradient, separation, impinge, curved streamlines, whirling flow, and so on. Its simplicity makes it a good choice in balancing simulation accuracy and computing cost [23].

In nonlinear eddy viscosity models, there is a kind algebraic stress model that keeps the strain–stress relationship in the two-order stress model, which makes it closer to the Reynolds stress model. A graph of the "cost/complexity" and "realism/dynamic range" given by Gatski [21] (also shown in Figure 4.1) presents the location of the nonlinear eddy viscosity models among different turbulence closure models. The main point of the nonlinear eddy viscosity model is to build a higher order correlation between the Reynolds stress tensor and strain rate tensor based on the linear eddy viscosity model. Then, the coefficients for the higher order tensor are fixed with the help of the DNS database. The low Reynolds number turbulence model has the advantage of dealing with the damp of eddy viscosity close to the wall, which makes it more suitable to present the anisotropy in the near wall region. Thus, most the nonlinear models are developed based on the low-Reynolds number $k-\varepsilon$ model.

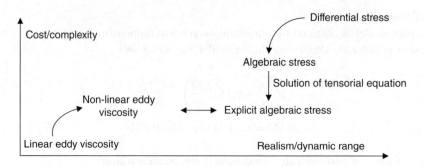

Figure 4.1   Hierarchy of single-point turbulent closure models.

Under the framework of the $k-\varepsilon$ model, the transport equations for a typical nonlinear eddy viscosity model could be:

$$\frac{D(\bar{\rho}k)}{Dt} = \frac{\partial}{\partial x_j}\left[\left(\mu+\frac{\mu_t}{\sigma_k}\right)\frac{\partial k}{\partial x_j}\right] + P_k - \bar{\rho}[\varepsilon^* + D] \tag{4.50}$$

$$\frac{D(\bar{\rho}\varepsilon^*)}{Dt} = \frac{\partial}{\partial x_j}\left[\left(\mu+\frac{\mu_t}{\sigma_\varepsilon}\right)\frac{\partial \varepsilon^*}{\partial x_j}\right] + [C_{\varepsilon1}P_k - C_{\varepsilon2}f_2\bar{\rho}\varepsilon^*]\frac{\varepsilon^*}{k} + S_1 + S_\varepsilon \tag{4.51}$$

where:

$$\mu_t = \bar{\rho}C_\mu f_\mu \frac{k^2}{\varepsilon^*} \tag{4.52}$$

where ε^* is the isotropic dissipation, and the relationship between it and the true dissipation can be expressed as:

$$\varepsilon^* = \varepsilon - D$$

Here the Launder & Sharma model [25] is used as the basic low-Reynolds number model, with the term $D = 2\nu\left(\partial\sqrt{k}/\partial x_j\right)^2$. This could guarantee the isotropic dissipation $\varepsilon^* \to 0$ when $y \to 0$, which enables us to integrate to the wall.

The damp function f_μ and f_2 are given as:

$$f_\mu = \exp\left[-3.4/\left(1+0.02R_t^2\right)\right] \tag{4.53}$$

$$f_2 = 1 - 0.3\exp\left(-R_t^2\right) \tag{4.54}$$

here $R_t = k^2/(\nu\varepsilon^*)$, with model constants:

$$\sigma_k = 1.0, \quad \sigma_\varepsilon = 1.3, \quad C_{\varepsilon1} = 1.44, \quad C_{\varepsilon2} = 1.92$$

The term S_l is added to avoid overestimating the turbulence length scale in the low-Reynolds number model when the reverse pressure gradient appears near the wall. It has the form:

$$S_l = \max\left[0.83\bar{\rho}\frac{\varepsilon^{*2}}{k}\left(\frac{l}{l_e}-1\right)\left(\frac{l}{l_e}\right)^2, 0\right] \tag{4.55}$$

where the length scale is given by: $l = k^{3/2}/\varepsilon^*$, $l_e = C_\mu^{-3/4}\kappa y$, $\kappa = 0.41$.

S_ε is a wall related term that depends on the adopted the nonlinear eddy viscosity model. The anisotropic Reynolds stress tensor **b** in compressible turbulence could be expressed as:

$$b_{ij} = \frac{\widetilde{u_i'u_j'}}{k} - \frac{2}{3}\delta_i^j, \quad \left(k = \frac{\widetilde{u_i'u_i'}}{2}\right)$$

The non-dimensional averaged strain tensor and vorticity tensor is:

$$S_{ij} = \frac{k}{2\varepsilon^*}\left(\tilde{u}_{i,j} + \tilde{u}_{j,i} - \frac{2}{3}\tilde{u}_{k,k}\delta_{ij}\right) \tag{4.56}$$

$$W_{ij} = \frac{k}{2\varepsilon^*}\left(\tilde{u}_{i,j} - \tilde{u}_{j,i}\right) \tag{4.57}$$

The nonlinear eddy viscosity model has almost a same expression on the Reynolds stress tensor as the Algebraic stress model. Thus, the power law of the Reynolds stress tensor could be used when building the nonlinear eddy viscosity model. Gatski (2000) [21] gave a universal isotropic tensor of the nonlinear eddy viscosity models based on a ten-term tensor base:

$$b_{ij} = \sum_{n=1}^{N} a_n T_{ij}^{(n)} \tag{4.58}$$

Here, T_{ij}^n is the tensor base, a_n denoted the model constants. The ten-term tensor base is [21] [22]:

$$T^{(1)} = S, \quad T^{(2)} = SW - WS, \quad T^{(3)} = S^2 - \frac{1}{3}\{S^2\}I, \quad T^{(4)} = W^2 - \frac{1}{3}\{W^2\}I,$$

$$T^{(5)} = WS^2 - S^2W, \quad T^{(6)} = W^2S + SW^2 - \frac{2}{3}\{SW^2\}I, \quad T^{(7)} = WSW^2 - W^2SW,$$

$$T^{(8)} = SWS^2 - S^2WS, \quad T^{(9)} = W^2S^2 + S^2W^2 - \frac{2}{3}\{S^2W^2\}I, \quad T^{(10)} = WS^2W^2 - W^2S^2W$$

In the early stages of nonlinear turbulence model development, the Reynolds stresses were all given by the averaged velocity gradient, and the tensor form was not used. However, it is straightforward to express the Reynolds stress using the above terms by proper transformation. Several typical nonlinear eddy viscosity models are given below, which contains the second and third order expressions of the averaged strain rate tensor and vorticity tensor.

4.1.5.1 WR (Wilcox & Rubesin) Second-Order Eddy Viscosity Model

The Wilcox–Rubesin (WR) second-order model [26] originates from the work of Saffman. It is the earliest second-order model based on the low-Reynolds number eddy viscosity model. Its expression on the tensor base is:

$$b = \alpha_1 S + \alpha_2 (SW - WS) \tag{4.59}$$

The model coefficients above are given as:

$$\alpha_1 = -2C_\mu f_\mu, \quad \alpha_2 = \frac{8/9f_\mu}{1/C_\mu + 2\{S^2\}}$$

where the model constant C_μ could be fixed as 0.09. The low-Reynolds number modification term in the near wall region is:

$$S_\varepsilon = 2\nu\mu_t \left(\frac{\partial^2 \tilde{u}_i}{\partial x_j \partial x_k}\right)^2 \tag{4.60}$$

Note that the model constant C_μ has no relation with the averaged strain rate tensor and vorticity tensor. This means it has no obvious advantages in comparison with the linear turbulence model.

4.1.5.2 SZL (Shih, Zhu & Lumley) Second-Order Eddy Viscosity Model

SZL second-order eddy viscosity model [27] includes the terms $T^{(3)}$ and $T^{(4)}$, and the coefficient C_μ is given as a function of averaged strain tensor but not a constant anymore. The isotropic tensor in this model is given by:

$$b = a_1 S + a_2 (SW - WS) + a_3 \left(S^2 - \frac{1}{3}\{S^2\}I\right) + a_4 \left(W^2 - \frac{1}{3}\{W^2\}I\right) \tag{4.61}$$

The model coefficients are fixed based on experimental data and the DNS database for the free shear flow and inertia layer. Applications to typical cases such as the rotational shear flow, backward step, and confined jet flow confirm that the SZL model has better a prediction ability than the traditional linear viscosity model. The coefficients for the high order tensor base are given by:

$$C_\mu = \frac{2/3}{1.25 + S + 0.9\Omega}, \quad \alpha_1 = -2C_\mu f_\mu, \quad \alpha_2 = \frac{15f_\mu}{1000 + S^3},$$

$$\alpha_3 = \frac{3f_\mu}{1000 + S^3}, \quad \alpha_4 = \frac{-19f_\mu}{1000 + S^3}$$

The term is $S = \sqrt{2S_{ij}S_{ij}}$, $\Omega = \sqrt{2W_{ij}W_{ij}}$

The Low-Reynolds number modification term in the near wall region is:

$$S_\varepsilon = \nu \mu_t \left(\frac{\partial^2 \tilde{u}_i}{\partial x_j \partial x_k} \right)^2 \tag{4.62}$$

The damp function is the same as in the WR model:

$$f_\mu = \exp\left[-3.4 / \left(1 + 0.02\, R_t^2 \right) \right] \tag{4.63}$$

Due to the lack of third-order tensor information, it is hard to make the coefficient for the second-order tensor base have a universal value, which means that it could not be used generally.

4.1.5.3 CLS (Craft, Launder, and Suga) Third-Order Eddy Viscosity Model

The isotropic term in the Craft–Launder–Suga (CLS) third-order eddy viscosity model [28] can be expressed as:

$$b = a_1 S + a_2 (SW - WS) + a_3 \left(S^2 - \frac{1}{3}\{S^2\}I \right) + a_4 \left(W^2 - \frac{1}{3}\{W^2\}I \right)$$
$$+ a_5 (W^2 S - SW^2) + c \left(\{S^2\} + \{W^2\} \right) S \tag{4.64}$$

Notably, the choice of tensor base is not exclusive, which means that the linear combination of the tensor base could also work as a new tensor base. Thus, the CLS below is not exactly given by the standard form.

Craft *et al.* proposed that the averaged strain tensor and vorticity tensor is an important parameter affecting the eddy viscosity coefficient. They have successfully described the nonlinear relationship between the term C_μ and averaged strain tensor and vorticity tensor. The term $(\{S^2\} + \{W^2\}) S$ is added to avoid extra turbulent kinetic energy. The coefficients in the CLS model can be given as:

$$C_\mu = \frac{0.3}{1 + 0.35 (\max[|S|, \Omega])^{1.5}} \left(1 - \exp\left[\frac{-0.36}{\exp(-0.75 \max[|S|, \Omega])} \right] \right) \tag{4.65}$$

where $|S|$ and Ω have the same definition as in the SZL model above.

The following coefficients and terms can be given:

linear term coefficient: $\alpha_1 = -2 f_\mu C_\mu$,

second-order coefficient: $\alpha_2 = 0.4 f_\mu C_\mu$, $\alpha_3 = -0.4 f_\mu C_\mu$, $\alpha_4 = -1.04 f_\mu C_\mu$,

third-order coefficient: $\alpha_5 = 80 f_\mu C_\mu^3$, $c = -40 f_\mu C_\mu^3$,

damp function: $f_\mu = 1 - \exp\left[-\sqrt{R_t/90} - (R_t/400)^2 \right]$, low-Reynolds number modification near the wall:

$$
S_\varepsilon =
\begin{cases}
0.0022 \dfrac{|S|\mu_t k^2}{\varepsilon^*} \left(\dfrac{\partial^2 \tilde{u}_i}{\partial x_j \partial x_k} \right)^2 & \mathrm{Re}_t \le 250 \\[4mm]
0 & \mathrm{Re}_t > 250
\end{cases}
\tag{4.66}
$$

here $R_t^2 = k^2/(\nu\varepsilon)$

The generalization and accuracy of the CLS model is better than that for most linear and second-order nonlinear turbulence models. It has good performance in simulations of obvious streamline curving, whirling flow, impinge jet, and unsteady separation flow, which makes it a practical third-order viscosity model.

4.1.6   Reynolds-Stress Model

It is generally considered that the reason behind the anisotropy of the turbulence is the departure of the principal axis direction of the Reynolds-stress tensor from that of the train tensor, which means the eddy viscosity coefficient μ_t should not be an isotropic scalar. Thus, the modeling of turbulence analogized from the viscosity Newton flow with an isotropic relationship and conception of turbulent eddy viscosity has actually violated the physical foundation. Additionally, the eddy viscosity assumption has omitted the pressure–strain effect, which makes it unable to present the isotropy of Reynolds-stress induced by the redistribution of turbulent energy on different principal axis directions, and the potential reverse energy transfer phenomenon could not be captured. To avoid these drawbacks, we need to abandon the conception of the eddy viscosity coefficient (μ_t) and directly establish the Reynolds stress transport equations, and model the high-order correlation terms before we solve the equations. The accurate Reynold-stress transport equation and relative theories were established by Zhou Peiyuan in the 1940s. The development of computer techniques and numerical methods after the 1960s promoted development of the Reynold-stress model and several kinds of second order enclosure model appeared.

Starting from the original N-S equations and Reynolds averaging, we can obtain the Reynolds-stress transport equations as below:

$$
\frac{\partial}{\partial t}\left(\rho \overline{u_i' u_j'} \right) + \frac{\partial}{\partial x_k}\left(\rho U_k \overline{u_i' u_j'} \right) = D_{ij} + \varphi_{ij} + G_{ij} - \varepsilon_{ij}
\tag{4.67}
$$

The two items in the left-hand side are the time variety ratio and advection term, while D_{ij}, φ_{ij}, G_{ij}, and ε_{ij} denote the diffusion term, pressure-strain term, production term, and dissipation term, respectively. The exact expressions of these terms are given below:

$$
D_{ij} = -\frac{\partial}{\partial x_k}\left(\rho \overline{u_i' u_j' u_k'} + \overline{p' u_j'}\delta_{jk} + \overline{p' u_j'}\delta_{jk} - \mu \frac{\partial}{\partial x_k}\overline{u_i' u_j'} \right)
\tag{4.68}
$$

$$\varphi_{ij} = p\overline{\left(\frac{\partial u_i'}{\partial x_j} + \frac{\partial u_j'}{\partial x_i}\right)} \tag{4.69}$$

$$G_{ij} = -\rho\left(\overline{u_i' u_k'}\frac{\partial U_j}{\partial x_k} + \overline{u_j' u_k'}\frac{\partial U_i}{\partial x_k}\right) \tag{4.70}$$

$$\varepsilon_{ij} = 2\mu\overline{\frac{\partial u_i'}{\partial x_k}\frac{\partial u_j'}{\partial x_k}} \tag{4.71}$$

The diffusion term D_{ij} is given in a conservative form, which does not change the magnitude of the Reynolds stress in the whole system, but only redistributes the Reynolds stress inner the system, making it tend to be homogeneous. The production term G_{ij} provides the interaction between the Reynolds stress and average velocity gradient, which works as a source of the Reynolds stress. The dissipation term ε_{ij} presents the dissipation effect of the molecular viscosity on the turbulent fluctuation, which always tends to reduce the Reynolds stress. The pressure–strain term φ_{ij} shows the correlation between the pressure and turbulent strain ratio. It can be observed that except for the term G_{ij} the other three terms all contain the second or third correlation, which needs additional assumptions to enclose these terms before we can solve the Reynolds-stress equations.

Diffusion Term D$_{ij}$

The last term in expression D_{ij} presents the molecular diffusion, which could be omitted in the high Reynolds number flow. The first three terms reflect the effect of third-order velocity fluctuation correlation and pressure fluctuation on the diffusion. To simplify the problem, the effect of pressure fluctuation could be allowed to emerge in the third-order velocity fluctuation correlation, and we can adopt the gradient diffusion model of Launder:

$$D_{ij} = \frac{\partial}{\partial x_k}\left(C_s\frac{k}{\varepsilon}\overline{u_k' u_l'}\frac{\partial \overline{u_i' u_j'}}{\partial x_l}\right) \tag{4.72}$$

Here the constant C_s could be given as 0.21.

Dissipation Term ε$_{ij}$

The dissipation of turbulence mainly occurs in the small scale eddies. Theoretical and experimental studies have proved that the small scale eddies tend to be isotropic. Thus, we could omit the isotropy on the dissipation term, which means that the shear stress induced dissipation is close to zero, while the viscosity only imports the normal stress and turbulent dissipation. Thus, the tensor of dissipation term could be simplified with a scalar of dissipation rate:

$$\varepsilon_{ij} = 2\mu\overline{\frac{\partial u_i'}{\partial x_k}\frac{\partial u_j'}{\partial x_k}} = \frac{2}{3}\mu\overline{\left(\frac{\partial u_l'}{\partial x_k}\right)^2}\delta_{ij} = \frac{2}{3}\rho\varepsilon\delta_{ij} \tag{4.73}$$

Because the modeling approach lacks concrete proof, it has obvious shortfalls in some situations.

Pressure–Strain Term

The term φ_{ij} plays an important part in the transport of Reynolds stress, especially in a flow field with quick pressure variance. To model φ_{ij}, the equation for pressure fluctuation is first solved, and then the critical factors influencing φ_{ij} are investigated and modeled separately.

The terms φ_{ij} could be expressed as:

$$\varphi_{ij} = \overline{p\left(\frac{\partial u_i'}{\partial x_j} + \frac{\partial u_j'}{\partial x_i}\right)} = \varphi_{ij1} + \varphi_{ij2} + \varphi_{ijw} \tag{4.74}$$

Here, φ_{ijw} is the Green function weighted surface integration of wall pressure fluctuation and strain fluctuation, which could be called a wall reflection term or just a wall term. This term only makes a notable contribution very close to the wall, and can usually be omitted.

In the modeling of φ_{ij1}, Rotta proposed a scheme based on its isotropy function:

$$\varphi_{ij1} = -C_1 \frac{\varepsilon}{k} \rho \left(\overline{u_i' u_j'} - \frac{2}{3}\delta_{ij}k\right) \tag{4.75}$$

The two terms inside the parentheses stand for the anisotropy parts of the Reynolds stress. We can prove that the term φ_{ij1} could really play a role in enhancing the isotropy of Reynolds stress. The model constant C_1 can be set between 1.5 and 3.

Similarly, Naot *et al.* proposed a modeling scheme for φ_{ij2}:

$$\varphi_{ij2} = -C_2 \left(G_{ij} - \frac{1}{3}\delta_{ij}G\right) \tag{4.76}$$

Here, G_{ij} and G denote the production of Reynolds stress and turbulent kinetic separately, which involved the interaction between the mean velocity gradient and turbulent fluctuation. The constant C_2 can be given in a range of $C_2 \leq 0.6$.

Based on the above discussion, we can obtain a standard Reynolds-stress deferential equation:

$$\frac{\partial}{\partial t}\left(\rho \overline{u_i' u_j'}\right) + \frac{\partial}{\partial x_k}\left(\rho U_k \overline{u_i' u_j'}\right) = \frac{\partial}{\partial x_k}\left[C_s \rho \frac{k}{\varepsilon} \overline{u_k' u_l'} \frac{\partial}{\partial x_l}\left(\overline{u_i' u_j'}\right) - C_1 \frac{\varepsilon}{k}\rho\left(\overline{u_i' u_j'} - \frac{2}{3}\delta_{ij}k\right)\right.$$
$$\left. - C_2\left(G_{ij} - \frac{2}{3}\delta_{ij}G\right) - \frac{2}{3}\delta_{ij}\rho\varepsilon + G_{ij}\right] \tag{4.77}$$

The Reynolds stress model is a second-order symmetry tensor, which possesses six different independent components. Thus, the above function actually presents six independent differential equations. The turbulence parameter k could be obtained as the sum of the three normal Reynolds stress components, while ε needs to be solved from a separate transfer equation. Because there is no conception of eddy viscosity coefficient, the equation for ε is somewhat different from the one in the $k-\varepsilon$ model in the basic form:

$$\frac{\partial \varepsilon}{\partial t} + \frac{\partial}{\partial x_j}\left(\rho U_j \varepsilon\right) = \frac{\partial}{\partial x_j}\left(C_s \rho \frac{k}{\varepsilon} \overline{u_i' u_j'} \frac{\partial \varepsilon}{\partial x_i}\right) + C_{\varepsilon 1}\frac{k}{\varepsilon}G - C_{\varepsilon 2}\frac{\varepsilon^2}{k} \tag{4.78}$$

The constants in the model are usually given as:

$$C_s = 0.15, \quad C_{\varepsilon 1} = 1.34, \quad C_{\varepsilon 2} = 1.8$$

The above ε equation and Reynolds stress transport equations composed the seven-equation differential Reynolds stress model. Notably, in most applications, the advantage of the Reynolds-stress transport model is not obvious, and its improvement of prediction accuracy is limited. This may be not enough to outweigh its high cost in terms of computation and complexity [29]; in addition, the Reynolds-stress transport model is still under development.

4.1.7 Comments on the Models

4.1.7.1 Linear Eddy Viscosity Model

A review by Wilcox in 2001 [30] has presented the evaluation results on the prediction ability of algebraic models, one-equation models, and two-equation models based on a series of experimental results. It is supposed that if a turbulence model is expected to work in a complex turbulent field it should first work for some basic flow. If not, its application in a more complex field would be doubtful. For this purpose, three kinds of basic turbulent flow were selected: the free shear flow (containing the wake flow, mixing layer, plane jet flow, round jet flow, and transverse jet flow), attached turbulent boundary layer, and separation flow.

In the evaluation of the zero-equation models (CS model, BL model, and half-equation JK model), it is found that the prediction of the JK model is rather poor in the attached boundary layer with adverse pressure gradient, while the CS model is relatively better in such a flow; For the separated flow, although most zero-equation models could not work, the JK model could give reasonable results.

The evaluation of one-equation models suggests that the SA model has good prediction accuracy in the above tests except for jet flows, which means that the SA model could be a good choice in industrial applications such as airfoil flow. For the attached boundary flow, the BB model gives a much smaller skin friction, while the SA and BL models provide similar results. In the separated flow, the BB model produced a much larger separation region, while SA model could provide an acceptable separation and attachment location.

As for the two-equation models (Wilcox $k-\omega$ model, Robinson $k-\xi$ model, Launder–Sharma $k-\varepsilon$ model, and RNG $k-\varepsilon$ model), it was found that for the free shear flow the $k-\omega$ model and $k-\xi$ model have higher precision, which produces an average error of about 5% and 7%, respectively, while the $k-\varepsilon$ model and RNG $k-\varepsilon$ model produce larger errors of up to 18% and 29%. For the attached boundary flow, the prediction of the $k-\omega$ model is also found to be much better (with an error of 3.5%) than that of the $k-\varepsilon$ model (with an error of about 29%). For the separated turbulent flow, the separation point obtained from $k-\omega$ is also seen to be much closer to the experimental results (with an error of about 4%) compared with the $k-\varepsilon$ model.

Based on the above analysis, the author concluded that:

1. A zero-equation model could be used to predict the attached turbulent boundary flow, while the JK model could be used for applications with moderate separation. However, both of them could not be used in free shear flows.

2. The SA model has good performance in free shear flows, except jet flows, while the BB model usually has poor performance for wall-bounded turbulence.
3. The $k-\omega$ model works very well in the free shear flow, attached turbulent flow, and moderately separated flow, while the $k-\varepsilon$ model is relatively poor in the corresponding predictions.

4.1.7.2 Nonlinear Eddy Viscosity Model

Loyau [24] investigated the effect of different nonlinear eddy viscosity models on the turbulent kinetic energy production and model coefficient C_μ. It was found that the term P_k obtained from the WR model and linear eddy viscosity model is identical to the one in the two-dimensional shear flow. Because of the dependence of C_μ on the strain and vorticity, the production term in the SZL and CLS models is somewhat different. For all the models, $C_\mu = 0.09$ when the strain is at an equilibrium state. However, all the coefficients C_μ drop quickly when the strain exceeds the equilibrium value, except in the WR model. This will make the eddy viscosity decrease obviously, which means that the boundary is more likely to be separated under a reverse pressure gradient. Thus, the turbulence will become more sensitive to shock waves, which could help to improve the simulation results. Because the coefficient C_μ in the WR model is given as a constant, it will make no obvious difference in this aspect.

H. Loyau (1998) [24] and X.D. Yang and H.Y. Ma [23] (2003) evaluated the performance of different nonlinear turbulence models in comparison with the traditional linear turbulence models, based on two standard transonic experiments: transonic two-dimensional (2D) bump (Delery 1981 [31]) and transonic axi-symmetric bump (Bachalo and Johnson 1986 [32]). It was found that the linear $k-\varepsilon$ model and nonlinear WR model could not present the plateau associated with the shock-induced separation well, while the other nonlinear turbulence models could produce reasonable results. The third-order models performed better than the second order models.

4.2   Theories and Equations of Large Eddy Simulation

Large eddy simulation (LES) is a numerical method for turbulence simulation between direct numerical simulation (DNS) and Reynolds average numerical simulation (RANS). With the rapid development in computer hardware, large eddy simulation has attracted a lot of research and applications.

4.2.1   Philosophy behind LES

Turbulence contains a series of large and small turbulent eddies, which have a broad range of scales. To simulate the turbulent flows, we always expect the grid to be small enough to resolve the motion of the minimum vortex. However, concerning current computer capacity, the scale of the finest grid that we can use is still larger than that of the smallest eddies.

In turbulent flows, the transport of momentum, mass, energy, and other physical quantities is mainly due to the large-scale eddies. The large-scale eddies are closely related to the problem solving. Affected by the geometry and boundary conditions, the structures of the large-scale eddies differ from each other. However, the small-scale eddies are barely affected by the geometry and boundary conditions. The small-scale eddies tend to be isotropic and their dynamic behavior is more universal. Therefore, we no longer attempt a simulation of all-scale eddies, but use the N-S equations to calculate only the turbulent eddies that are larger than the grid scale. The impact of small-scale eddies on the large-scale eddies can be reflected through the turbulence modeling in the large-scale eddies instantaneous N-S equations, resulting in the current large-eddy simulation method (large eddy simulation, referred to as LES).

To carry out a large-eddy simulation, two important issues must be tackled. First, a filter function should be established. The function is used to filter eddies with scales smaller than the width of the filter function, and the effect of the filtered eddies is included by introducing an additional stress term in the equations. The additional stress is comparative to the Reynolds stress term in RANS, and is called the subgrid-scale (SGS) stress. The second issue is to establish the mathematical model of the stress term, which is called the SGS model.

4.2.2 LES Governing Equations

In LES, the instantaneous variables are divided by a filter into the large scale component $\bar{\phi}$ and the small scale component ϕ'. The term $\bar{\phi}$ is represented in the equations and can be solved directly, but ϕ' should be represented by a model.

Suppose $\phi(\bar{x},t)$ is an instantaneous flow variable at $x=\bar{x}$, and the large-scale component $\bar{\phi}(\bar{x},t)$ can be expressed by the weighted integral as follows:

$$\bar{\phi}(\bar{x},t) = \int G(|\bar{x}-\bar{x}'|)\phi(\bar{x}',t)\,\mathrm{d}V' \tag{4.79}$$

There are three popular filters: Box filters, Fourier-truncation filter, and Gauss filters. Of these filters, the Gauss filter shows the best performance. However, it brings in a large calculation. Therefore, the Box filter and Fourier-truncation filter are often used. The Box filter is expressed as follows:

$$G(|\bar{x}-\bar{x}'|) = \begin{cases} \dfrac{1}{\Delta x_1 \Delta x_2 \Delta x_3}, & |\bar{x}-\bar{x}'| \le \dfrac{\Delta x_i}{2} \quad i=1,2,3 \\ 0, & |\bar{x}-\bar{x}'| > \dfrac{\Delta x_i}{2} \quad i=1,2,3 \end{cases} \tag{4.80}$$

In this equation, x_i is the coordinate of a grid node, Δx_i is the scale of the filter in the i direction, and the large scale component $\bar{\phi}$ is the volume average in the rectangular cell with the center of x_i.

The width of the filter is not necessarily linked to the grid used in the numerical calculation. In principle, the scale of the grid in simulations should be smaller than the scale of the filter. To simplify the calculation, the scale of the two can be made the same. In the equation, the averaged component is the filtered variables, and it is the average in the time domain, not in the spatial domain. The term $G(|\bar{x}-\bar{x}'|)$ decides the eddy scale, which divides the large eddy and the small vortex. In other words, $\bar{\phi}$ only retains the variability of ϕ whose scale is greater than the width of $G(|\bar{x}-\bar{x}'|)$.

With the filter and Favre average, the LES governing equations are as follows:

$$\frac{\partial \bar{\rho}}{\partial t} + \frac{\partial (\bar{\rho}\tilde{u}_j)}{\partial x_j} = 0 \tag{4.81}$$

$$\frac{\partial(\bar{\rho}\widetilde{u}_i)}{\partial t} + \frac{\partial(\bar{\rho}\widetilde{u}_i\widetilde{u}_j + \bar{p}\delta_{ij})}{\partial x_j} = \frac{\partial\left(\bar{\tau}_{ij} + \tau_{ij}^{SGS}\right)}{\partial x_j} \tag{4.82}$$

$$\frac{\partial(\bar{\rho}\widetilde{E})}{\partial t} + \frac{\partial(\bar{\rho}\widetilde{H}\widetilde{u}_j)}{\partial x_j} = \frac{\partial\left[\widetilde{u}_i\left(\bar{\tau}_{ij} + \tau_{ij}^{SGS}\right) - \left(\bar{q}_j + Q_j^{SGS}\right)\right]}{\partial x_j} \tag{4.83}$$

In the equations, the overbar "$-$" denotes the spatial filter and "$\sim$" represents the spatial filter along with the Favre average. The term:

$$\tau_{ij}^{SGS} = \overline{\rho u_i u_j} - \bar{\rho}\widetilde{u}_i\widetilde{u}_j \tag{4.84}$$

is defined as the subgrid-scale stress; it reflects the effect of small eddies on the equations.

It could be observed that the filtered N-S governing equations are similar to the RANS equations. The difference is that the filtered variables in the equations are instantaneous values rather than averaged values. Meanwhile, the turbulence stress expressions are different. Due to the linear characteristics of the continuous equation, the continuity equation is the same between the filtered and averaged equations. To resolve the equations, the unsolved SGS should be expressed by a subgrid-scale model.

4.2.3 Subgrid-Scale Model

The SGS model plays a very important role in the LES. The earliest and most basic model was proposed by Smagorinsky, and several scholars have subsequently developed the model. τ_{ij}^{SGS} is often analogous with the viscous stress τ_{ij} in the unfiltered N-S equation, and thus it consists of anisotropic and isotropic components, $\tau_{ij}^{SGS} = \tau_{ij,d}^{SGS} + \tau_{kk}^{SGS}$, where τ_{ij}^{SGS} is the anisotropic component and $\tau_{ij,d}^{SGS}$ is proportional to the resolved strain rate component:

$$\widetilde{S}_{ij} = \frac{1}{2}\left(\partial\widetilde{u}_i/\partial x_j + \partial\widetilde{u}_j/\partial x_i\right).$$

$\tau_{ij,d}^{SGS}$ is modelled as follows:

$$\tau_{ij,d}^{SGS} = -2\bar{\rho}\nu_t\left[\widetilde{S}_{ij} - \frac{1}{3}\widetilde{S}_{kk}\delta_{ij}\right] \tag{4.85}$$

To enclose the SGS, the eddy viscosity (ν_t) and SGS kinetic energy (k^{SGS}) need to be determined.

4.2.3.1 Zero-Equation SGS Model

Smagorinsky Model
Based on the Smagorinsky's SGS model, the SGS stress is described as follows:

$$\tau_{ij}^{SGS} - \frac{1}{3}\tau_{kk}^{SGS}\delta_{ij} = -2\rho\nu_t\bar{S}_{ij} \tag{4.86}$$

In the equation, ν_t is the SGS eddy viscosity:

$$\nu_t = (C_s\Delta)^2|\bar{S}| \tag{4.87}$$

where:

$$|\bar{S}| = \sqrt{2\bar{S}_{ij}\bar{S}_{ij}}, \quad \Delta = \left(\Delta_x\Delta_y\Delta_z\right)^{1/3} \tag{4.88}$$

$$\bar{S}_{ij} = \frac{1}{2}\left(\frac{\partial \bar{u}_i}{\partial x_j} + \frac{\partial \bar{u}_j}{\partial x_i}\right) \tag{4.89}$$

where Δ_i is the grid scale in the i direction and C_s is the Smagorinsky constant. Theoretically, C_s can be obtained from the Kolmogorov constant (C_K):

$$C_s = \frac{1}{\pi}\left(\frac{3}{2}C_K\right)^{3/4}$$

Practical applications show that the dissipation in the near wall region is too high, usually resulting in an overestimated decay of the large-scale fluctuation in this region. Consequently, a smaller C_s should be adopted to decrease the influence of SGS stress dissipation. The Van Dries model suggests that:

$$C_s = C_{s0}\left(1-e^{-y^+/A^+}\right) \tag{4.90}$$

In the equation, $y^+ = yu_\tau/\nu_w$, which is the dimensionless distance to the wall, A^+ is a semi-empirical constant with a value of 25.0, and C_{s0} is the Van Driest constant, which is equal to 0.1.

Dynamic Smagorinsky Model

A dynamic Smagorinsky model was developed in 1992 by Germano et al. [33]. This model is proposed to solve the problem that the constant coefficient in the Smagorinsky model induces too high a dissipation in shear turbulent flows. The basic idea is that high-wavenumber information contained in large eddy simulation is used to infer the impact of subgrid scale motions on the resolved large eddies to determine the model coefficients dynamically. Therefore, the constant in the Smagorinsky model can adjust itself based on the local turbulent kinetic properties to reflect the instantaneous and local dynamic properties—even to automatically capture the intermittent turbulence and reverse energy transfer.

Germano (1992) introduced a test filter, whose filter scale Δ_2 is greater than sub-scale Δ_1. The variable φ is filtered twice by Δ_1 and Δ_2, respectively, which is denoted by the subscript. If the filter is linear:

$$(\varphi)_{fg} = (\varphi)_g \tag{4.91}$$

Based on the N-S equations, a Germano correlation between $(\tau_{ij})_{fg}$ obtained from two consecutive filtering and $[(\tau_{ij})_f]_g$ obtained by two separate filterings can be derived:

$$\left[(\tau_{ij})_f\right]_g = (\tau_{ij})_{fg} - L_{ij} \tag{4.92}$$

In the equation:

$$L_{ij} = (u_i)_{fg}(u_j)_{fg} - \left[(u_i)_f(u_j)_f\right]_g$$

is an enclosure variable.

The resolved velocity filtered by Δ_1 is denoted by superscript "–", The resolved velocity filtered by Δ_2 is denoted by superscript "^", and their Smagorinsky subgrid stress is as follows, respectively:

$$(\tau_{ij})_f - \frac{1}{3}(\tau_{kk})_f \delta_{ij} = 2(\nu_t)_f \bar{S}_{ij} = 2C_{D1}\Delta_1^2|\bar{S}|\bar{S}_{ij} \tag{4.93}$$

$$(\tau_{ij})_g - \frac{1}{3}(\tau_{kk})_g \delta_{ij} = 2(\nu_t)_g \hat{S}_{ij} = 2C_{D2}\Delta_2^2|\hat{S}|\hat{S}_{ij} \tag{4.94}$$

In the equation C_{D1} and C_{D2} are the dynamic coefficients instead of Smagorinsky coefficients. Assuming filter scales A and B are within the inertial region, $C_{D1} = C_{D2}$.

Because $(\tau_{ij})_{fg} = (\tau_{ij})_g$, substituting Equations 4.93 and 4.94 into the Germano equation gives:

$$L_{ij} - \frac{1}{3}L_{kk}\delta_{ij} = 2C_D\left(\Delta_2^2|\hat{S}|\hat{S}_{ij} - \Delta_1^2\widehat{|\bar{S}|\bar{S}_{ij}}\right) \tag{4.95}$$

Given:

$$2\left(\Delta_2^2|\hat{S}|\hat{S}_{ij} - \Delta_1^2\widehat{|\bar{S}|\bar{S}_{ij}}\right) = M_{ij}$$

we have:

$$L_{ij} - \frac{1}{3}L_{kk}\delta_{ij} = C_D M_{ij} \tag{4.96}$$

This is an overdetermined equation, and C_D cannot be solved directly. Lilly used the minimum error method by minimizing the squared difference of the both sides of the equation:

$$\frac{\partial}{\partial C_D}\left\{L_{ij} - \frac{1}{3}L_{kk}\delta_{ij} - C_D M_{ij}\right\}^2 = 0 \tag{4.97}$$

$$C_D = \frac{M_{ij}L_{ij}}{M_{ij}M_{ij}} \tag{4.98}$$

The minimum error method above could lead to a negative value or a tiny denominator, resulting in simulation divergence. To overcome such difficulties in calculations, the average coefficient method can be used to obtain the ensemble average of the numerator and denominator on the right-hand side of Equation 4.98.

4.2.3.2 One-Equation SGS Model

One-Equation SGS with Constant Coefficients
If the SGS turbulent kinetic energy generation and dissipation can maintain a local balance, the Smagorinsky eddy viscosity model [34] can be applied directly. However, the local balance assumption is not always valid in many engineering applications. The drawback of the Smagorinsky model is its failure in reflecting the upstream history effect. When turbulent transport is strong, some of the energy transported by the subgrid-scale and resolved scale are equally important—then Smagorinsky eddy viscosity models and dynamic eddy viscosity models developed by Germano [33] may lead to large errors. Based on this understanding, Schumann [35] established a transport equation with SGS turbulent kinetic energy as a turbulence characteristic parameter. Yoshizawa *et al.* [36] and Chakravarthy *et al.* [37] proposed the modified form of the equation.

$$\frac{\partial(\bar{\rho}k^{SGS})}{\partial t} + \frac{\partial(\bar{\rho}k^{SGS}\tilde{u}_j)}{\partial x_j} = \frac{\partial}{\partial x_j}\left[\left(\bar{\rho}\frac{\nu_e}{\mathrm{Pr}_t}\right)\frac{\partial k^{SGS}}{\partial x_j}\right] + P_k^{SGS} - D^{SGS} \qquad (4.99)$$

where, $\nu_e = \nu + \nu_t$, and ν and ν_t are, respectively, the molecular viscosity and SGS viscosity coefficient. P_k^{SGS} and D^{SGS} are generation and dissipation terms of the SGS kinetic energy:

$$P_k^{SGS} = -\tau_{ij}^{SGS}\left(\partial\tilde{u}_i/\partial x_j\right) \qquad (4.100)$$

$$D^{SGS} = \frac{\partial}{\partial x_i}\left(\tilde{u}_j\tau_{ij}^{SGS}\right) \qquad (4.101)$$

Yoshizawa *et al.* [36] and Chakravarthy *et al.* [37] have given the following equations:

$$\nu_t \approx C_\nu\sqrt{k^{SGS}}\bar{\Delta} \qquad (4.102)$$

$$D^{SGS} \approx C_\varepsilon\bar{\rho}\left(k^{SGS}\right)^{3/2}/\bar{\Delta} \qquad (4.103)$$

There are two parameters to be determined, C_ν and C_ε. When the local is in partial equilibrium, the effective Smagorinsky coefficient in the Yoshizawa and Horiuti model [36] is:

$$(C_s)^2 = \sqrt{2}C_\nu(C_\nu/C_\varepsilon)^{\frac{1}{2}} \qquad (4.104)$$

$C_\nu = 0.05$ and $C_\varepsilon = 1.0$ in the Yoshizawa and Horiuti model, C_s is 0.126. The Smagorinsky coefficient is not a constant, with its value set as 0.2 in the uniform turbulence and as 0.065 in the shear turbulence. To ensure C_s in the appropriate interval, its value is usually taken as a compromise:

$$C_\nu = 0.02075, C_\varepsilon = 1.0$$

Chakravarthy *et al.* [37] proposed a combination of factors:

$$C_v = 0.067, C_\varepsilon = 0.916$$

One-Equation Dynamic Subgrid Model

We can know from the preceding discussion that the fixed coefficient one-equation SGS model also has the problem that it is difficult to give a proper model coefficient. In 1995, Kim and Menon [38] used SGS kinetic equations to develop the localized dynamic k^{SGS} model (LDKM), which dynamically determines the model coefficients. The idea is similar to the zero-equation dynamic SGS model, that is, through the introduction of two filter information in the local structure of turbulence SGS stress, the model coefficients are adjusted in the calculation process according to the local flow field information. The SGS eddy viscosity and SGS dissipation term are modeled as follows:

$$\nu_t \approx C_v \sqrt{k^{\mathrm{SGS}}} \, \bar{\Delta} \tag{4.105}$$

$$D^{\mathrm{SGS}} \approx C_\varepsilon \bar{\rho} \left(k^{\mathrm{SGS}} \right)^{3/2} / \bar{\Delta} \tag{4.106}$$

At this point we have the SGS stress model:

$$\tau_{ij}^{\mathrm{SGS}} = \bar{\rho} \left(\widetilde{u_i u_j} - \tilde{u}_i \tilde{u}_j \right) = -2 \bar{\rho} C_v \left(k^{\mathrm{SGS}} \right)^{\frac{1}{2}} \bar{\Delta} \left[\widetilde{S}_{ij} - \frac{1}{3} \widetilde{S}_{kk} \delta_{ij} \right] + \frac{2}{3} \bar{\rho} k^{\mathrm{SGS}} \delta_{ij} \tag{4.107}$$

To determine the relevant model parameters, the test filter $\hat{\mathrm{G}}$ is introduced. Its width is greater than the SGS filter scale. The test filter SGS stress is obtained by applying the test filter to the momentum equation:

$$
\begin{aligned}
\tau_{ij}^{\mathrm{SGS, Test}} &= \widehat{\bar{\rho} \tilde{u}_i \tilde{u}_j} - \frac{\widehat{\bar{\rho} \tilde{u}_i} \, \widehat{\bar{\rho} \tilde{u}_j}}{\hat{\bar{\rho}}} \\[2mm]
&\approx -2 C_v \hat{\bar{\rho}} \hat{\Delta} \left(\frac{\widehat{\bar{\rho} \widetilde{u_k u_k}}}{2 \hat{\bar{\rho}}} - \frac{\widehat{\bar{\rho} \tilde{u}_k} \, \widehat{\bar{\rho} \tilde{u}_k}}{2 \hat{\bar{\rho}} \, \hat{\bar{\rho}}} \right)^{\frac{1}{2}} \left(\widehat{\tilde{S}}_{ij} - \frac{1}{3} \widehat{S}_{kk} \delta_{ij} \right) \\[2mm]
&\quad + \frac{1}{3} \left(\widehat{\bar{\rho} \widetilde{u_k u_k}} - \frac{\widehat{\bar{\rho} \tilde{u}_k} \, \widehat{\bar{\rho} \tilde{u}_k}}{\hat{\bar{\rho}}} \right) \delta_{ij}
\end{aligned}
\tag{4.108}
$$

where:

$$\widetilde{S}_{ij} = \frac{1}{2} \left(\frac{\partial \tilde{u}_i}{\partial x_j} + \frac{\partial \tilde{u}_j}{\partial x_i} \right), \quad \widehat{\tilde{S}}_{ij} = \frac{1}{2} \left\{ \frac{\partial}{\partial x_i} \left(\frac{\widehat{\bar{\rho} \tilde{u}_j}}{\hat{\bar{\rho}}} \right) - \frac{\partial}{\partial x_j} \left(\frac{\widehat{\bar{\rho} \tilde{u}_j}}{\hat{\bar{\rho}}} \right) \right\} \tag{4.109}$$

The SGS stress tensor model error is:

$$E_{ij} \approx \widetilde{\overline{\rho u_i u_j}} - \frac{\overline{\rho \widetilde{u}_i} \, \overline{\rho \widetilde{u}_j}}{\hat{\bar{\rho}}} + 2\hat{\bar{\rho}} C_v \hat{\Delta} \left(\frac{\widetilde{\overline{\rho u_k u_k}}}{2\hat{\bar{\rho}}} - \frac{\overline{\rho \widetilde{u}_k} \, \overline{\rho \widetilde{u}_k}}{2\hat{\bar{\rho}} \, \hat{\bar{\rho}}} \right)^{\frac{1}{2}} \left(\widehat{\widetilde{S}_{ij}} - \frac{1}{3} \widehat{\widetilde{S}_{kk}} \delta_{ij} \right)$$

$$- \frac{1}{3} \left(\widetilde{\overline{\rho u_k u_k}} - \frac{\overline{\rho \widetilde{u}_k} \, \overline{\rho \widetilde{u}_k}}{\hat{\bar{\rho}}} \right) \delta_{ij} \tag{4.110}$$

It can be abbreviated as:

$$E_{ij} = L_{ij} + 2C_v D_{ij} \tag{4.111}$$

where L_{ij} and $2C_v D_{ij}$ are the exact expression and mode form of test stress:

$$L_{ij} = \widetilde{\overline{\rho u_i u_j}} - \frac{\overline{\rho \widetilde{u}_i} \, \overline{\rho \widetilde{u}_j}}{\hat{\bar{\rho}}} - \frac{1}{3} \left(\widetilde{\overline{\rho u_k u_k}} - \frac{\overline{\rho \widetilde{u}_k} \, \overline{\rho \widetilde{u}_k}}{\hat{\bar{\rho}}} \right) \delta_{ij} \tag{4.112}$$

$$D_{ij} = \hat{\Delta} \hat{\bar{\rho}} \left(\frac{\widetilde{\overline{\rho u_k u_k}}}{2\hat{\bar{\rho}}} - \frac{\overline{\rho \widetilde{u}_k} \, \overline{\rho \widetilde{u}_k}}{2\hat{\bar{\rho}} \, 2\hat{\bar{\rho}}} \right)^{\frac{1}{2}} \left(\widehat{\widetilde{S}_{ij}} - \frac{1}{3} \widehat{\widetilde{S}_{kk}} \delta_{ij} \right) \tag{4.113}$$

To acquire the smallest root mean square of model error, it should satisfy the following equation:

$$\frac{\partial E_{ij} E_{ij}}{\partial C_v} = 4 D_{ij} L_{ij} + 8 C_v D_{ij} L_{ij} = 0 \tag{4.114}$$

Thus we have:

$$C_v = - \frac{L_{ij} D_{ij}}{2 D_{ij} D_{ij}} \tag{4.115}$$

In 1997, Nelson constructed the SGS dissipation model.
The original dissipation model is:

$$\varepsilon \approx \widetilde{\mu} \left[\frac{\widetilde{\partial u_j} \, \partial u_j}{\partial x_i \, \partial x_i} - \frac{\partial \widetilde{u}_j}{\partial x_i} \frac{\partial \widetilde{u}_j}{\partial x_i} \right] = \frac{C_\varepsilon \bar{\rho} (k^{\mathrm{SGS}})^{\frac{3}{2}}}{\bar{\Delta}} \tag{4.116}$$

The test dissipation model is given as:

$$\varepsilon \approx \hat{\mu} \left[\frac{\widehat{\partial \widetilde{u}_j} \, \partial \widetilde{u}_j}{\partial x_i \, \partial x_i} - \frac{\partial}{\partial x_i} \left(\frac{\widehat{\overline{\rho \widetilde{u}_j}}}{\hat{\bar{\rho}}} \right) \frac{\partial}{\partial x_i} \left(\frac{\widehat{\overline{\rho \widetilde{u}_j}}}{\hat{\bar{\rho}}} \right) \right] \approx \frac{C_\varepsilon \hat{\bar{\rho}}}{\hat{\Delta}} \left(\frac{\widehat{\overline{\rho u_k u_k}}}{2\hat{\bar{\rho}}} - \frac{\overline{\rho \widetilde{u}_k} \, \overline{\rho \widetilde{u}_k}}{2\hat{\bar{\rho}} \, \hat{\bar{\rho}}} \right)^{\frac{3}{2}} \tag{4.117}$$

Similarly, the following equation is obtained:

$$
C_\varepsilon = \frac{\hat{\mu}\hat{\Delta}}{\hat{\bar{\rho}}} \frac{\left[\widehat{\frac{\partial \widetilde{u_j}}{\partial x_i} \frac{\partial \widetilde{u_j}}{\partial x_i}} - \frac{\partial}{\partial x_i}\left(\frac{\widehat{\bar{\rho}\widetilde{u_j}}}{\hat{\bar{\rho}}} \right) \frac{\partial}{\partial x_i}\left(\frac{\widehat{\bar{\rho}\widetilde{u_j}}}{\hat{\bar{\rho}}} \right) \right]}{\left(\frac{\widehat{\bar{\rho}\widetilde{u_k u_k}}}{2\hat{\bar{\rho}}} - \frac{\widehat{\bar{\rho}\widetilde{u_k}}}{2\hat{\bar{\rho}}} \frac{\widehat{\bar{\rho}\widetilde{u_k}}}{\hat{\bar{\rho}}} \right)^{\frac{3}{2}}}
\tag{4.118}
$$

4.2.4 Hybrid RANS/LES Methods

For turbulent flows, LES is generally expected to produce more accurate results than RANS models [39]. However, for high Reynolds number flows and wall-bounded turbulence problems, the use of LES is unrealistic in industrial applications due to the huge computational cost. Especially for wall turbulence, to capture the evolution of coherent structures in the boundary layer, we need a very fine mesh.

The hybrid RANS/LES idea was first proposed by Speziale [40]. Batten *et al.* [41] proposed LNS (limited numerical scales) based on this idea. However, this model switch between LES and RANS is actually very casual, and a more reasonable conversion between LES and RANS is a key issue for the hybrid RANS/LES model. Since the mid-1990s there has been various hybrid RANS/LES approaches, which can be grouped into four categories: damping RANS models, detached eddy simulation (DES) and its improved versions, the weighted averaged LES/RANS model, and a hybrid approach consistent with the turbulent energy spectrum. Currently, blending of the principles in the above methods has also evolved into various hybrid models.

Damping RANS Models

Speziale [40] proposed a modified turbulence model, which is called the resolution-dependent damping model. Turbulent stress can be adjusted by multiplying a correction function in the flow field, namely:

$$
\tau_{ij}^{\text{model}} = f_\Delta(\Delta/l_k)\tau_{ij}^{\text{RANS}}
\tag{4.119}
$$

where $f_\Delta(\Delta/l_k)$ is called the contribution function, which is used to damp the model, and the Kolmogorov scale estimation formula proposed by Speziale is $l_k \approx \nu^{3/4}/\varepsilon^{1/4}$. The contribution function given in the literature [40] is:

$$
f_\Delta\left(\frac{\Delta}{l_k}\right) = \left(1 - e^{-\beta\frac{\Delta}{l_k}}\right)^n
\tag{4.120}
$$

where n is used to determine the extent of function steepness and β is used to determine the resolution of the model where a contribution can be neglected. We can select $n = 1$, $\beta = 0.001$.

The term Δ is local grid scale and l_k is the Kolmogorov scale. In the region where the grid scale is small enough to be comparative with the Kolmogorov scale, calculation of the model is close to DNS. In the region with a coarser grid, the model resumes as a general turbulence model. This method consists of two elements: RANS models and contribution function. Any RANS model can be used. Speziale recommend using the algebraic Reynolds stress model, while simulations in the literature [42,43] adopted $k-\omega$ and $k-\varepsilon$ and acquired similar results.

DES and Its Improved Versions
In 1997, Spalart developed the detached eddy DNS simulation method [44] (known now as DES97). LES and RANS using the Spalart–Allmaras (SA) model [45] were combined based on the characteristics of large and small eddies. RANS is adopted in the zone where dissipation is the main feature, and LES is adopted in the zone where large eddy transportation is the main feature. Models are automatically switched by comparing the local grid scale with RANS calculated turbulence mixing length, namely, replacing the wall distance in the SA model with the length scale $\tilde{d}$, which is associated with the grid spacing and is defined as follows:

$$\tilde{d} = \min(d, C_{\text{DES}}\Delta), \quad \Delta = \max(\Delta x, \Delta y, \Delta z) \tag{4.121}$$

Obviously, the model is the SA turbulence model when $d \ll \Delta$, and it is automatically converted into the LES model when $d \gg \Delta$. This approach is known as the DES97 limiter.

When the grid layout is not appropriate (e.g., the grid scale in each direction near the wall is nearly the same and small enough), DES97 limiter will misjudge the cells and treat them with LES. In this case, velocity fluctuation in the wall boundary layer is not well resolved. The DES97 will reduce eddy viscosity and cause a mismatch of modeled Reynolds stress, which is called modeled stress depletion (MSD). Given that Menter *et al.* [13] used blending functions to appropriately achieve a switch from the wall boundary layer to the mainstream, Spalart *et al.* [46] used the same idea and adopted "reserved RANS mode" or "delayed LES function" to develop the DEES method (delayed DES). The SA one-equation model does not contain scale information such as $k\omega/y$ (Menter SST), but contains parameter r which is analogous to the ratio of the turbulence mixing length and the distance to the wall. For DDES, in order to be consistent with the SA-definition and become more robust in the potential flow region, r is modified as follows:

$$r_{\text{d}} = \frac{\nu_t + \nu}{\sqrt{U_{i,j}U_{i,j}}\kappa^2 d^2} \tag{4.122}$$

where:

ν_t is the dynamic eddy viscosity,
ν is the molecular viscosity,
$U_{i,j}$ is the velocity gradient,
κ is the Karman constant,
d is the distance to the wall,
the subscript "d" means "delayed".

Spalart [46] constructed a transfer function of r_{d}:

$$f_{\text{d}} = 1 - \tanh\left([8r_{\text{d}}]^3\right) \tag{4.123}$$

The transfer function in the LES region ($r_d \ll 1$) is 1 and is 0 in the RANS region (and is not sensitive to the case where r_d is slightly more than one in the near wall region); f_d is used to modify DES97 limiters to define a new DES length scale:

$$\tilde{d} = d - f_d \max(0, d - C_{DES}\Delta) \tag{4.124}$$

This is Spalart's DDES method. Spalart applied the method to the plate boundary layer and backward steps [46], and solved the MSD problem.

Weighted Average LES/RANS Models

Regarding the spurious separation problem caused by MSD, Fan *et al.* in 2002 proposed the weighted average RANS and LES method [47]. A blending function of the wall distance is used to combine the RANS equations and turbulence models with the LES equations and turbulence models. This method used Menter's SST k-ω turbulence model [12] construction ideas, which can ensure adoption of RANS model near the wall. This can avoid the use of LES in the near wall region. A hybrid RANS/LES approach defines the turbulent viscosity coefficient as the weighted average of the eddy viscosity turbulence model and SGS models, namely:

$$\mu_t = F\mu_t^{RANS} + (1-F)\mu_t^{LES} \tag{4.125}$$

The blending function F depends on the flow characteristics and grid scale at each point and varies from 0 to 1, achieving an automatic switch between the RANS and LES models.

The weighted average LES/RANS approach has two key issues: one is the choice of model, including the choice of turbulence model and the SGS model; the second is the construction of a blending function. To date, researchers have kept improving hybrid RANS/LES for the problem of separated flows especially under high Reynolds number adverse pressure gradient. First, more sophisticated turbulence models are used in the turbulent dissipation region, including $k-\omega$ BSL/SST, improved $k-\varepsilon$, $k-\zeta$ (enstrophy) [48] models. In terms of SGS model, the standard Smagorinsky model, one-equation Yoshizawa model, and multi-scale models were used for enclosure of the LES. Characteristics of the blending function directly affect the calculation of the boundary layer, and thus proper construction of the blending function is very important. Researchers construct various blending functions from different theories; Fan *et al.* proceeded in a manner completely analogous to the BSL/SST construction process and proposed the following blending function [47]:

$$F = \tanh\left(\eta^4\right)\eta = \max\left(\frac{\sqrt{k}}{0.09\omega d}, \frac{500\nu}{\omega d^2}\right) \tag{4.126}$$

The blending function is completely similar to F_1 in the Menter-SST model.

Based on Fan's research, Baurle *et al.* further considered the impact of grid scale on the LES [49]. They adopted the idea of a limited numerical scale (LNS), and presumed that:

$$F = \max\left[\tanh\left(\eta^4\right), F_{LNS}\right]$$

where:

$$F_{\text{LNS}} = AINT \left[\min \left(\frac{\mu^{\text{SGS}}}{\mu^{\text{RANS}}}, 1.0 \right) \right]$$

This approach avoids the problem that a coarse grid cannot be used for large-eddy simulation. Fan's blending functions impose a large eddy simulation approach in the region away from the wall. After the correction of Baurle, LES can be transferred to RANS on the coarse grid in the region away from the wall. Xiao [48] and Nichols et al. [50] further proposed improved blending functions.

In more recent studies, Edwards, Choi et al. [51] used the following method to obtain a blending viscosity coefficient. The blending function is established based on the ratio between the wall distance and Taylor micro-scale:

$$\Gamma = \frac{1}{2} \left(1 - \tanh \left[5 \left(\frac{\kappa}{\sqrt{C_\mu}} \eta^2 - 1 \right) - \phi \right] \right) \tag{4.127}$$

where:

$$\eta = \frac{d}{\alpha \chi}.$$

The Taylor scale is defined as:

$$\chi = \sqrt{\nu / (C_\mu \omega)}$$

d is the wall distance. The position where $\kappa \eta^2 = \sqrt{C_\mu}$ is called balanced position. The term ϕ is used to implement the blending function. The parameter α can be used to control the thickness of the RANS region in the blending function. It can be estimated according to local conditions and the wall boundary condition. Notably, the parameter α is not the same at each point in the flow field, which makes the use of the blending function inconvenient. Gieseking improved the blending function in subsequent studies [52,53], and the compression ramp flows were calculated, leading to good results.

Although there has been a lot of research on the modification and improvement of blending functions, how to exactly construct the blending function remains an unsolved problem. No conclusion has been established for issues such as: whether the position where the blending function begins to transfer should be in the logarithmic layer or wake layer of the boundary layer, and what fitting function should be adopted.

Turbulent Energy Spectrum Consistent Hybrid Approach

DES97, the DDES method, and the weighted average method described above basically adopt blending functions to combine RANS eddy viscosity with LES eddy viscosity. These functions adopt different shape functions. However, the current studies do not give construction principles of shape function. Arunajatesan et al. [54,55] developed a class of turbulent energy spectrum consistent hybrid RANS/LES simulation methods. They suggested that the local turbulent energy spectrum should be consistent, and turbulence scales assessed by RANS and LES only represent the truncated energy spectrum. Such turbulent energy spectrum information can be obtained by solving the turbulent kinetic energy. Then we can obtain the unsolved information of the turbulent kinetic energy by integrating all scales of turbulence. Therefore, the whole flow field can be solved using a unified approach, and combination of RANS and LES resolved information can be realized. After testing, Arunajatesan *et al.* suggested that the approach can provide a smooth transition process from the high eddy viscosity in the RANS region to the low viscosity in the LES region.

The turbulent energy spectrum consistent mixing model assumes that turbulent energy statistics at any point in the flow field can meet the turbulent energy spectrum. The turbulent energy spectrum can be formulated by energy k_e including wave number and Kolmogorov scale (η) as:

$$\hat{E}\left(\hat{k}\right) = C_e \hat{k}_e^{-5/3} \left(\frac{\hat{k}}{\hat{k}_e}\right)^4 \left[1 + \left(\frac{\hat{k}}{\hat{k}_e}\right)^2\right]^{\frac{-17}{6}} \exp\left(-\frac{3}{2}a\hat{k}^{4/3}\right) \tag{4.128}$$

where:

k is the wave number,
k_e is the energy including wave number,
$a = 1.5$,
$\hat{k} = k\eta$, and ν is the dynamic viscosity coefficient,
C_e is a coefficient used to adjust the dissipation spectrum to be consistent with the local turbulent dissipation rate.

Arunajatesan *et al.* [54] first solved the LES turbulent kinetic energy equation to obtain K^{SGS}, and K^{SGS} is correlated with the turbulent kinetic energy spectrum as follows:

$$K^{SGS} = \int_{k_\Delta}^{\infty} E(k)\,dk \tag{4.129}$$

where k_Δ is the wave number corresponding to the grid. Then, Arunajatesan *et al.* solved the RANS turbulent kinetic energy equation to obtain K^{RANS}, which is correlated with the turbulent kinetic energy spectrum as follows:

$$K^{RANS} = \int_{k_\eta}^{\infty} E(k)\,dk \tag{4.130}$$

where k_η is Kolmogorov wave number.

With the turbulent kinetic energy provided, the above two equations can be solved through iterative solution to obtain the energy, including wave number, and then we can obtain other scales ranges integrals based on the turbulent energy spectrum.

This turbulence energy spectrum consistent mixed model developed by Arunajatesan *et al.* is complete in theory, but it is subject to hypotheses made in the model, such as the spectral distribution of turbulent energy, correlation between Re_λ and k^{RANS}, and correlation between k_e and Re_λ, restraining the application of this approach. The linear interpolation can introduce certain errors. Furthermore, the determination of the local grid scale also causes errors in the integration results. For example, the use of $\max(\Delta x, \Delta y, \Delta z)$ or $(\Delta x \times \Delta y \times \Delta z)^{1/3}$ makes a significant difference for different grid distributions. Arunajatesan found that the adoption of $(\Delta x \times \Delta y \times \Delta z)^{1/3}$ for the grid in a simulation of cylindrical jet can capture more accurately instability development at the jet exit.

4.3 Two-Phase Turbulence Model

We have already discussed the single-phase fluid turbulence model in detail. To describe complex physical and chemical processes in a liquid rocket engine, including atomization, flow, and combustion, we should establish a multiphase turbulence model that can simultaneously reflect the interaction between the droplets and gas. Currently, there are two ways to describe multiphase turbulent flows, namely, the Eulerian–Lagrangian method and the Eulerian–Eulerian method. The former takes the continuous phase as continuous medium that is described in Eulerian coordinates. The method treats the particles as discrete system described by a tracking method in the Lagrangian coordinate system. The latter method not only takes the continuous phase as continuous medium but also takes the discrete phase as pseudo-fluid or semi-continuous medium. Two phases penetrate mutually and are described in an Eulerian coordinate system. We pay more attention to the two-fluid model. Pseudo-fluid characteristics such as particle viscosity, thermal conductivity, and diffusion rate are introduced for the discrete phase in this model. Particle turbulent kinetic energy and particle Reynolds stress concepts are introduced by analogy. Thus, we can develop various two-fluid and multi-fluid models from single-phase turbulence models.

4.3.1 Hinze–Tchen Algebraic Model for Particle Turbulence

The Hinze–Tchen algebraic particle turbulence model [56] (Ap model) is similar to the mixing length model for the fluid turbulence. Based on the assumption that the particle fluctuation is induced by the flow fluctuation, the relationship between the particles and the gas turbulent viscosity coefficient is established. It is a simple model to describe two-phase turbulence. Tchen [56] studied the fluctuation of a particle caused by the gas fluctuation in the alternating flow. He used Taylor's statistical theory to describe the gas turbulent fluctuation. Hinze made a further derivation and obtained the ratio between particle turbulent viscosity and the gas turbulent viscosity, namely, the ratio between particle turbulent diffusion coefficient and the gas turbulent diffusion coefficient:

$$\frac{\nu_P}{\nu_T} = \frac{D_P}{D_T} = \left(\frac{k_P}{k}\right)^2 = \left(1 + \frac{\tau_{r1}}{\tau_T}\right)^{-1} \tag{4.131}$$

where ν_P and ν_T are particle and gas turbulent viscosity coefficients, respectively, and D_P and D_T are particle and gas turbulent diffusion coefficients respectively; $\tau_{r1} = \rho_s d_P^2/18\mu$ is the relaxation time that is caused by relative fluctuation between particle and gas as stokes resistance;

$\tau_T = k/\varepsilon$ is the gas turbulent fluctuation characteristic time. Here, particle viscosity is established to reflect the particles fluctuation by analogy with fluid, which has no direct relation with particle collisions.

The Ap model is simple and intuitive, and has been widely used in two-phase fluids. Because this model is based on the theory that the particle motion follows the local flow fluctuation, the particle fluctuation calculated by this model is always less than the gas fluctuation, and the larger the particle, the smaller the fluctuation, the slower the diffusion. However, some experimental results show that the gas fluctuation is weaker than particle fluctuation and fluctuation of large particles is stronger than small particles in many cases. Thus, the Ap model assumes that the particle fluctuation only depends on local fluid turbulent fluctuation and ignores the convection and diffusion of the particle turbulent kinetic energy (upstream impact or historical effect) and the generation of the average kinetic energy.

4.3.2   Two-Phase Turbulence Model k-ε-k$_p$ and k-ε-A$_p$

Similar to the single-phase turbulence model, Boussinesq's assumption and scalar viscosity coefficient are introduced in the two-phase Reynolds stress, resulting in the **k-ε-k$_p$** model, which describes the two-phase turbulent fluctuation. The closed equations are list below [57]:

$$\overline{v_i v_j} = \frac{2}{3}k\delta_{ij} - \nu_T\left(\frac{\partial V_i}{\partial x_j} + \frac{\partial V_j}{\partial x_i}\right) \tag{4.132}$$

$$\overline{v_{pi} v_{pj}} = \frac{2}{3}k_p\delta_{ij} - \nu_p\left(\frac{\partial V_{pi}}{\partial x_j} + \frac{\partial V_{pj}}{\partial x_i}\right) \tag{4.133}$$

$$\overline{n_p v_{pi}} = -\frac{\nu_p}{\sigma_p}\frac{\partial n_p}{\partial x_i} \tag{4.134}$$

$$\overline{n_p v_{pj}} = -\frac{\nu_p}{\sigma_p}\frac{\partial n_p}{\partial x_j} \tag{4.135}$$

$$\frac{\partial}{\partial t}(\rho k) + \frac{\partial}{\partial x_j}(\rho V_i k) = \frac{\partial}{\partial x_j}\left(\frac{\mu_e}{\sigma_k}\frac{\partial k}{\partial x_j}\right) + G + G_p - \rho\varepsilon \tag{4.136}$$

$$\frac{\partial}{\partial t}(\rho\varepsilon) + \frac{\partial}{\partial x_j}(\rho V_i\varepsilon) = \frac{\partial}{\partial x_j}\left(\frac{\mu_e}{\sigma_\varepsilon}\frac{\partial\varepsilon}{\partial x_j}\right) + \frac{\varepsilon}{k}[c_{\varepsilon 1}(G + G_p) - c_{\varepsilon 2}\rho\varepsilon] \tag{4.137}$$

$$\frac{\partial}{\partial t}(n_p k_p) + \frac{\partial}{\partial x_j}(n_p V_{pi} k_p) = \frac{\partial}{\partial x_j}\left(\frac{n_p\nu_p}{\sigma_p}\frac{\partial k_p}{\partial x_j}\right) + P_p + P_g - n_p\varepsilon_p \tag{4.138}$$

$$G = \mu_t\left(\frac{\partial V_i}{\partial x_j} + \frac{\partial V_j}{\partial x_i}\right)\frac{\partial V_i}{\partial x_j}, G_p = \sum_p\frac{2m_p n_p}{\tau_{rp}}\left(c_p^k\sqrt{kk_p} - k\right) \tag{4.139}$$

$$P_p = n_p\nu_p\left(\frac{\partial V_{pi}}{\partial x_j} + \frac{\partial V_{pj}}{\partial x_i}\right)\frac{\partial V_{pi}}{\partial x_j} \tag{4.140}$$

$$P_g = \frac{2m_p n_p}{\tau_{rp}}\left(c_p^k\sqrt{kk_p} - k_p\right) \tag{4.141}$$

$$\varepsilon_p = -\frac{1}{\tau_{rp}} \left[2 \left(c_p^k \sqrt{k_p k} - k_p \right) + \frac{1}{n_p} \overline{n_p v_{pi}} \left(V_i - V_{pi} \right) \right] \qquad (4.142)$$

$$\mu_e = \mu + \mu_T, \nu_T = c_\mu \frac{k^2}{\varepsilon}, \mu_T = \rho \nu_T, \nu_p = c_{\mu p} \frac{k_p^{\ 2}}{|\varepsilon_p|} \qquad (4.143)$$

The k-ε-k_p model can better reveal the two-phase turbulent convection, diffusion, generation, and dissipation of turbulence and the interaction between the two phases. In addition, combining the A_p model with Boussinesq's assumption, k equation and ε equation, we can obtain the k-ε-A_p equation.

In addition, researchers also established a two-phase fluid model based on the probability density function and particle trajectory model based on particle phase Lagrangian description [58] and so on. Multiphase turbulent large eddy simulation (LES) and direct numerical simulation (DNS) research work has gradually expanded, and drawn many useful conclusions [59–62].

References

[1] Barina J E, Huang P G, Coakley T. Turbulence Modeling Validation, Testing, and Development. NASA Technical Memorandum 110446, 1997.

[2] Zhang Z S, Cui G X, Xu C X. *Turbulence theory and simulation*. Press of Tsinghua University, 2005 (in Chinese).

[3] Smith A M O, Cebeci T. Numerical Solution of the Turbulent Boundary-Layer Equations. Douglas Aircraft Division Report DAC.33735, 1967.

[4] Baldwin B S, Lomax H. Thin-Layer Approximation and Algebraic Model for Separated Turbulent Flows. AIAA Paper 78–257, 1978.

[5] Johnson D A, King L S A. Mathematically simple turbulence closure model for attached and separated turbulent boundary layers. *AIAA Journal*, 23 (11):1684–1692, 1985.

[6] Yan C. *Computational fluid dynamics method and application*. Press of Beijing University of Aeronautics and Astronautics, 2006 (in Chinese).

[7] Baldwin B S, Barth T J. A One-Equation Turbulence Transport Model for High Reynolds Number Wall-Bounded Flows. AIAA Paper 91–0610, 1991.

[8] Spalart P R, Allmaras S R. A One-Equation Turbulence Model for Aerodynamic Flows. AIAA Paper 92–439, 1992.

[9] Marvin J G, Huang G P. Turbulence Modeling - Progress and Future Outlook. NASA Technical Memorandum 110414, 1996.

[10] Wilcox D C. *Turbulence Modeling for CFD*. La Canada. California: DCW Industries, Inc., 1998.

[11] Yakhot V, Orszag S A. Renormalised group analysis of turbulence: I basic theory. J Sci Comput, 1986 (1):3–5.

[12] Menter F R. Two-Equation Eddy-Viscosity Turbulence Models for Engineering Applications. *AIAA Journal*, 1994, **32** (8):1598–1605.

[13] Menter F R. Review of the shear-stress transport turbulence model experience from an industrial perspective. *International Journal of Computational Fluid Dynamics*, 2009, **23** (4):305–316.

[14] Menter F R, Kuntz M, Langtry R. Ten years of experience with the SST turbulence model. *Turbulence, Heat and Mass Transfer*, 2003, **4**:625–632.

[15] Godin P, Zingg D W, Nelson T. High-lift aerodynamic computations with one- and two-equation turbulence models. *AIAA Journal*, 1997, **35** (2):237–243.

[16] Sarkar S, Balakrishnan L. Application of a Reynolds-Stress Turbulence Model to the Compressible Shear Layer. ICASE Report 90–18, NASA CR 182002, 1990.

[17] Zeman O. A new model for supersonic turbulent boundary layer. AIAA Paper 93–0897, 1993.

[18] Sarkar S. The pressure-dilation correlation in compressible flows. *Physics of Fluids A*, 1992, **4**:2674–2682.

[19] Zeman O. Dilatation dissipation: The concept and application in modeling compressible mixing layers. *Physics of Fluids A*, 1990, **2**:178–188.

[20] Baz A M E, Launder B E. Second moment modeling of compressible mixing layer. Eng. *Turbulence Modeling Exp.*, 1993, **2**:63–72.

[21] Gatski T B, Johgen T. Nonlinear eddy viscosity and algebraic stress models for solving complex turbulent flow. *Progress in Aerospace Sciences*, 2000, **36**:655–682.

[22] Lin BoYing, Chen YiLiang. Nonlinear eddy viscosity coefficient and algebraic stress model to solve complex turbulence. *Advances Mechanics*, 2005, **35** (2):260–282 (in Chinese).

[23] Yang X D, Ma H Y. The nonlinear turbulence models in the interaction between shock wave and boundary layer. *Acta Mechanica Sinica*, 2003, **35** (1):57–63 (in Chinese).

[24] Loyau H, Batten P, Leschziner M A. Modeling shock/boundary-layer interaction with nonlinear eddy-viscosity closures. Flow, Turbulence and Combustion, 1998, (60):257–281.

[25] Launder B E, Sharma B I. Application of the energy-dissipation model of turbulence to the calculation of flow near a spinning disc. *Letters in heat and mass transfer*, 1974, **1** (2):131–137.

[26] Wilcox D C, Rubesin M W. Progress in turbulence modeling for complex flow field including effects of compressibility. NASA.TP1517, 1980.

[27] Shih T, Zhu J, Lumley J L. A realizable Reynolds stress algebraic equation model. Tm 105993 ICOMP-92-27, CMOTT-92-14, NASA. 1992.

[28] Craft T J, Launder B E, Suga K. Development and application of a cubic eddy-viscosity model of turbulence. *J. Heat Fluid Flow*, 1996, **17**:8.

[29] Zingg D W, Godin P. A perspective on turbulence models for aerodynamic flows. *International Journal of Computational Fluid Dynamics*, 2009, **23** (4): 327–335.

[30] Wilcox D C. Turbulence Modeling: An Overview. DCW Industries, Inc., La Canada, CA. AIAA Paper 0724, 2001.

[31] Delery J. Investigation of strong turbulent boundary-layer interaction in 2D flows with emphasis on turbulence phenomena. AIAA. Paper 81–1245, 1981.

[32] Bachalo W D, Johnson D A. Transonic turbulent boundary-layer separation generated on an axisymmetric flow model. *AIAA Journal.*, 1986, **24** 437–443.

[33] Germano M. Turbulence: the filtering approach. *J.l Fluid Mechanics*, 1992, **238**:325.

[34] Smagorinsky J. General circulation experiments with the primitive equations I, the basic experiment. *Monthly Weather Review*, 1963, **91** (3):99–164.

[35] Schumann U. Subgrid scale model for finite difference simulations of turbulent flows in plane channels and annuli. *J. Computational Physics*, 1975, **18**:376–404.

[36] Yoshizawa A, Horiuti K. Statistically derived subgrid scale kinetic energy model for Large-eddy simulation of turbulent flows. *Journal of the Physical Society of Japan*, 1985, **54**:2834–2839.

[37] Chakravarthy V, Menon S. Large eddy simulations of turbulent premixed flames in the flamelet regime. *Combustion Science and Technology*, 2001, **162**:175–222.

[38] Kim W W, Menon S. A new dynamic one-equation subgrid-scale model for large-eddy simulations. AIAA Paper 95–0356, 1995.

[39] Georgiadis N J, Rizzetta D P, Fureby C. Large-eddy simulation: Current capabilities, recommended practices and future research. *AIAA Journal*, 2010, **48** (8):1772–1784.

[40] Speziale C G. Turbulence modeling for time dependent RANS and VLES: a review. *AIAA Journal*, 1998, **36** (2):173–184.

[41] Batten P, Goldberg U, Chakravarthy S. Sub-grid turbulence modeling for unsteady flow with acoustic resonance. AIAA Paper 2000–0473, 2000.

[42] Zhang H L, Bachman C, Fasel H F. Application of a new methodology for simulation of complex turbulent flows. AIAA Paper 2000–2535, 2000.

[43] Fasel H F, Seidel J, Wernz S. A methodology for simulation of complex turbulent flows. *Journal of Fluid Eng.*, 2002, **124**:933–975.

[44] Spalart P R. Comments on the feasibility of LES for wings and on a hybrid RANS/LES approach. 1st Air force office of scientific research international conference on DNS/LES, Ruston, Louisiana, 4–8 August, 1997.

[45] Spalart P R, Allmaras S R. A one-equation turbulence model for aerodynamic flows. *La Rech Aerospace*, 1994, **1**:5–21.

[46] Spalart P R, Deck S, Shur M L, Squires K D. A new version of detached-eddy simulation, resistant to ambiguous grid densities. *Theory Computation Fluid Dynamics*, 2006, **20** (1):181–195.

[47] Fan T C, Xiao X D, Edwards J R, Hassan H A, Baurle R A. Hybrid LES/RANS simulation of a shock wave/boundary layer interaction. AIAA Paper 2002–0431, 2002.

[48] Xiao X D, Edwards J R, Hassan H A, Baurle R A. Inflow boundary conditions for LES/RANS simulations with applications to shock wave/boundary layer interactions. AIAA Paper 2003–79, 2003.

[49] Baurle R A, Tam C J, Edwards J R, Hassan H A. Hybrid simulation approach for cavity flows: blending, algorithm and boundary treatment issues. *AIAA Journal*, 2003, **41** (8):1463–1480.

[50] Nichols R H, Nelson C C. Application of hybrid RANS/LES turbulence model. AIAA Paper 2003–0083, 2003.

[51] Edwards J R, Choi J I, Boles J A. Hybrid LES/RANS simulation of a Mach 5 compression-corner interaction. AIAA Paper 2008–718, 2008.

[52] Gieseking D A, Choi J I, Edwards J R, Hassan H A. Simulation of shock/boundary layer interactions using improved LES/RANS method. AIAA Paper 2010–111. 2010.

[53] Gieseking D A, Edwards J R. Simulation of a Mach 3 compression-ramp interaction using LES/RANS models. AIAA Paper 2011–762, 2011.

[54] Arunajatesan S, Sinha N. Unified Unsteady RANS-LES Simulations of Cavity Flow fields. AIAA Paper 2001–0516, 2001.

[55] Arunajatesan S A, Dash S M. Progress Towards Hybrid RANS-LES Modeling For High-Speed Jet Flows. AIAA Paper 2002–0428, 2002.

[56] Tchen C M. Mean value and correlation problem connected with the motion of small particles in a turbulent field. PhD Thesis, Delft university, Martinus Nijhoff, 1947.

[57] Xu X C, Zhou L X. *Combustion Technology Handbook*. Press of Chemical Industry, 2007 (in Chinese).

[58] Gosman A D, Ioannides E. Aspects of computer simulation of liquid-fuelled combustors. AIAA Paper 81–0323, 1981.

[59] Squires K D, Eaton J K. Preferential concentration of particles by turbulence. *Physics of Fluids A*. 1991, **3**:1169.

[60] Miller R S, Bellan J. Direct numerical simulation and subgrid analysis of a transitional droplet laden mixing layer. *Physics of Fluids*, 2000, **12** (3):650–671.

[61] Portela L M, Oliemans R V A. Eulerianian–Lagrangian DNS/LES of particle–turbulence interactions in wall-bounded flows. *International Journal for Numerical Methods in Fluids*. 2003, **43**:1045–1065.

[62] Okong N, Leboissetier A, Bellan J. Detailed characteristics of drop-laden mixing layers: Large eddy simulation predictions compared to direct numerical simulation. *Physics of Fluids*, 2008, **20**:103305.

5

Turbulent Combustion Model

The flow within liquid rocket engines is essentially turbulent, where the combustion process is a typical turbulent combustion and there exist strong interactions between turbulence and combustion. Turbulence influences the combustion in two respects: (i) by influencing the mixing process between the fuel and the oxidizer by turbulent transport; (ii) by influencing the chemical process by inducing fluctuations of temperature and species. Contrariwise, combustion influences turbulence mainly by changing the local Reynolds number. On one hand, the heat released may increase the molecular viscosity by increasing the temperature, which decreases the local Reynolds number and laminarizes the turbulence. On the other hand, combustion may dilate and accelerate the fluid, which increases the local Reynolds number and enhances the turbulence. Therefore, it is necessary to take the interactions between turbulence and combustion into account so that the combustion flow inside the liquid rocket engine can be simulated with good accuracy.

5.1 Average of Chemical Reaction Term

A single-step chemical reaction can be expressed by the following stoichiometric formulas:

$$\sum_{i=1}^{N} v_i' M_i' \rightarrow \sum_{i=1}^{N} v_i'' M_i'' \tag{5.1}$$

where: v_i' and v_i'' are the stoichiometric coefficients of the reactants and products, respectively; M_i' and M_i'' are the chemical symbols (chemical species) of the reactants and products, respectively, and N is the number of chemical species.

Internal Combustion Processes of Liquid Rocket Engines: Modeling and Numerical Simulations,
First Edition. Zhen-Guo Wang.
© 2016 National Defense Industry Press. Published 2016 by John Wiley & Sons Singapore Pte Ltd.

The reaction rate of species i (R_i) is defined as the mass consumed or produced by reactions per unit volume and per unit time, i.e.:

$$R_i = -\left(\frac{dC_{M_i}}{dt}\right)_{chem} \tag{5.2}$$

The classical law of mass action gives:

$$R_i = k_i \prod_{i=1}^{N} C_{M_i}^{v_i} \tag{5.3}$$

where C_{M_i} is usually the absolute mass concentration and k_i is a proportionality constant, known as the reaction rate constant. For a given chemical reaction, the value of k_i is independent of concentration (C_{M_i}) and is only a nonlinear function of temperature.

Consider a simple chemical reaction process:

$$A_1 + A_2 \underset{b}{\overset{f}{\rightleftharpoons}} A_3 + A_4 \tag{5.4}$$

The forward reaction rate given by the law of mass action can be expressed as:

$$R_f = k_f[A_1][A_2] = k_f \rho^2 Y_1 Y_2 \tag{5.5}$$

where k_f is the forward reaction rate constant, which can be expressed using the Arrhenius law:

$$k_f = BT^\alpha \ \exp\left(-\frac{E_a}{R_u T}\right) \tag{5.6}$$

where E_a is the activation energy, a measure of the energy needed for the reaction to take place; R_u is the universal gas constant; BT^α is the collision frequency in which the exponential term α is the Boltzmann factor and B is called the pre-exponential factor or frequency factor. The values of B, α, and E_a are related to the property of elementary reactions. For a given chemical reaction, these parameters are independent of concentration and temperature.

Therefore:

$$R_f = BT^\alpha \rho^2 Y_1 Y_2 \ \exp\left(-\frac{E_a}{R_u T}\right) \tag{5.7}$$

For laminar gas phase reactions, the reaction source term can be calculated directly using the above formula. When considering turbulence, due to the effect of turbulence, the species concentrations and temperature as well as the chemical reaction rates fluctuate with time. Therefore, numerical simulation of turbulent combustion is faced with not only all the problems of turbulent flow, including how to deal with the transport equation of a fluctuating scalar, but also the issues of modeling a time-averaged chemical reaction rate $\bar{R}_f$. This is a unique problem of turbulent combustion. The rate $\bar{R}_f$ is a strong nonlinear function of fluctuating temperature and species concentrations. Since the chemical reaction source terms such as the production (consumption) rate of species or the release rate of energy are strongly nonlinear functions of species concentrations and temperature, the time-averaged reaction rate in turbulent combustion is not equal to the reaction rate expressed by the time-averaged species concentrations and temperature, which is the most fundamental problem of turbulent combustion modeling.

To find the expression of $\bar{R}_f$ it is necessary to carry out a Reynolds decomposition of Equation 5.7, where the instantaneous value of an independent variable is expressed as the sum of a time-averaged value and a fluctuation, e.g.:

$$T = \bar{T} + T'$$
$$Y_1 = \bar{Y}_1 + Y_1'$$
$$Y_2 = \bar{Y}_2 + Y_2' \tag{5.8}$$
$$\rho = \bar{\rho} + \rho'$$

Then Reynolds decomposition of Equation 5.7 is:

$$R_f = B(\bar{\rho} + \rho')^2 (\bar{T} + T')^\alpha (\bar{Y}_1 + Y_1')(\bar{Y}_2 + Y_2') \exp\left(-\frac{T_f}{\bar{T} + T'}\right) \tag{5.9}$$

When considering a multi-component calculation, the expression of R_f will be more complex, and the corresponding expression of $\bar{R}_f$ will also be very complicated.

How to model the average chemical reaction source term $\bar{R}_f$ is the main research area of turbulent combustion modeling. At present, the following models are usually used: (i) fast chemistry reaction model, such as the traditional presumed PDF (probability density function)—fast chemistry diffusion combustion model, as well as the eddy break-up (EBU) premixed combustion model, which is able to consider a simple finite rate reaction; (ii) flamelet model; (iii) probability density function transport equation method (transported PDF); and (iv) linear eddy model (LEM).

5.2   Presumed PDF—Fast Chemistry Model for Diffusion Flame

To use the presumed PDF model to close the source terms of turbulence–chemistry interactions one needs to find a number of scalars to describe the chemically thermodynamic parameters of the combustion system, establish transport equations of these scalars, and assume their probability density functions. Thus, one can completely determine the time-averaged properties of all scalars in the turbulent combustion process through the probability integral. If one uses a fast chemistry reaction model, which assumes that once the reactants are mixed the chemical reactions are completed in an instant, one can obtain all the statistical properties of the thermodynamic parameters through the statistical properties of conserved scalars without the need to directly calculate the time-averaged chemical reaction rate.

5.2.1   Concepts and Assumptions

Simple Chemical Reaction Model

The chemical reaction can be expressed as a single-step, an irreversible one, in which fuel and oxidizer react at the stoichiometric ratio of s:

$$1\,\text{kg fuel} + s\,\text{kg oxidizer} \rightarrow (1+s)\,\text{kg product}$$

that is, fuel + oxidant $\rightarrow$ product:

$$R_f + R_{ox}/s \rightarrow -R_{pr}/(1+s)$$

Conserved Scalar and Mixture Fraction

The governing equations of fuel and oxidizer mass fractions are:

$$\frac{\partial(\rho Y_{fu})}{\partial t} + \frac{\partial}{\partial x_j}\left(\rho u_j Y_{fu}\right) = \frac{\partial}{\partial x_j}\left(\Gamma_{eff,fu}\frac{\partial Y_{fu}}{\partial x_j}\right) + \bar{R}_{fu} \tag{5.10}$$

$$\frac{\partial(\rho Y_{ox})}{\partial t} + \frac{\partial}{\partial x_j}\left(\rho u_j Y_{ox}\right) = \frac{\partial}{\partial x_j}\left(\Gamma_{eff,ox}\frac{\partial Y_{ox}}{\partial x_j}\right) + \bar{R}_{ox} \tag{5.11}$$

where $\Gamma_{eff} = \Gamma_t + \Gamma_1$; the diffusion coefficients of all species, Γ_{fu}, Γ_{ox}, and Γ_{pr} are equal to each other and equal to the diffusion coefficient of total enthalpy (Γ_b), which means we have a unit Lewis number, Le = 1.

Dividing Equation 5.11 by s and then subtracting Equation 5.10, one obtains the following equation:

$$\frac{\partial}{\partial t}\left[\rho\left(Y_{fu} - \frac{Y_{ox}}{s}\right)\right] + \frac{\partial}{\partial x_j}\left[\rho u_j\left(Y_{fu} - \frac{Y_{ox}}{s}\right)\right] = \frac{\partial}{\partial x_j}\left[\Gamma_{eff}\frac{\partial}{\partial x_j}\left(Y_{fu} - \frac{Y_{ox}}{s}\right)\right] + \bar{R}_{fu} - \frac{\bar{R}_{ox}}{s}$$

Introducing the Zeldovich conversion, one can define:

$$X = Y_{fu} - \frac{Y_{ox}}{s} \tag{5.12}$$

$$R_x = \bar{R}_{fu} - \frac{\bar{R}_{ox}}{s} \tag{5.13}$$

According to simple chemical reaction assumptions, there should be:

$$\Gamma_{eff,fu} = \Gamma_{eff,ox} = \Gamma_{eff} \quad R_x = 0 \tag{5.14}$$

Then one obtains:

$$\frac{\partial \rho X}{\partial t} + \frac{\partial}{\partial X_j}\left(\rho u_j X\right) = \frac{\partial}{\partial X_j}\left(\Gamma_{eff}\frac{\partial X}{\partial X_j}\right) \tag{5.15}$$

It can be seen that X is a conserved scalar because its conservation equation does not include source terms.

For the diffusion-controlled reacting flow, one can introduce a mixture fraction:

$$Z \equiv \frac{X - X_2}{X_1 - X_2}$$

where X_1 and X_2 represent the value on the fuel and oxidizer side, respectively, that is:

$$X_1 = Y_{F_1} = 1, \quad X_2 = -\frac{Y_{ox,2}}{s} = -\frac{1}{s}$$

$$Z_1 = 1, \quad Z_2 = 0$$

Generally speaking, the mixture fraction "Z" represents the mixed degree of the two species in space. In addition, Z is also a conserved scalar, the conservation equation for which can be obtained from Equation 5.15:

$$\frac{\partial}{\partial t}(\rho Z) + \frac{\partial}{\partial X_j}(\rho u_j Z) = \frac{\partial}{\partial X_j}\left(D\rho \frac{\partial Z}{\partial X_j}\right) \tag{5.16}$$

It can be found that the conservation equation of Z is also sources free.

We usually need to introduce the fluctuating mean square $\overline{Z'^2}$ of the mixture fraction Z when determining the chemically thermodynamic state. For convenience, $\overline{Z'^2}$ is usually denoted as g and the governing equation of g is:

$$\frac{\partial(\rho g)}{\partial t} + \frac{\partial(\rho u_j g)}{\partial X_j} = \frac{\partial}{\partial X_j}\left(\frac{\mu_{eff}}{\sigma_g}\frac{\partial g}{\partial X_j}\right) + C_{g1}G_g - C_{g2}\rho g \frac{\varepsilon}{\kappa} \tag{5.17}$$

where G_g is the production rate of g per unit volume, and:

$$G_g = \mu_t \left(\frac{\partial Z}{\partial X_i}\right)^2 \tag{5.18}$$

where C_{g1}, C_{g2}, and σ_g are constants, and are usually set to 2.8, 2.0, and 0.9, respectively.

Fast Chemistry Hypothesis

Fast chemistry refers to a class of reactions where the chemical reaction rates greatly exceed the mixing rate. That is, the reaction between the fuel and oxidizer would be complete immediately after they are mixed, so that they could not coexist at any point in space at the same time (but the time-averaged values in the turbulent conditions can coexist in the same position). Finite rate reactions are those with finite reaction rates. The following mainly focuses on the modeling of fast chemistry reactions.

5.2.2 $\kappa - \varepsilon - Z - g$ Equations

Turbulent fluctuations play an important role in turbulent diffusion flames, therefore the establishment of a turbulent diffusion combustion model must consider turbulent fluctuations. Spalding proposed the $\kappa - \omega - g$ model to calculate turbulent diffusion flames in 1971, which later evolved into the $\kappa - \varepsilon - Z - g$ model. The main points are as below:

1. use the $\kappa - \varepsilon$ model to simulate the turbulent transport effects;
2. adopt the fast chemistry assumption;
3. establish governing equation of $g = \overline{Z'^2}$;
4. solve equations of $\bar{Z}$ and $\overline{Z'^2}$, assume the probability density function $P(Z)$ of Z, and express it as a function of $\bar{Z}$ and $\overline{Z'^2}$;
5. use the mixture fraction equation to avoid direct modeling of chemical reaction source terms.

Based on the Favre average, one can establish governing equations:

$$\frac{\partial \bar{\rho}}{\partial t} + \frac{\partial}{\partial X_j}\left(\bar{\rho}\widetilde{u}_j\right) = 0 \tag{5.19}$$

$$\frac{\partial}{\partial t}\left(\bar{\rho}\widetilde{u}_i\right) + \frac{\partial}{\partial X_j}\left(\bar{\rho}\widetilde{u}_j\widetilde{u}_i\right) = \frac{\partial}{\partial X_j}\left(\mu_e \frac{\partial \widetilde{u}_i}{\partial X_j}\right) + S_{u_i} \tag{5.20}$$

$$\frac{\partial}{\partial t}\left(\bar{\rho}k\right) + \frac{\partial}{\partial X_j}\left(\bar{\rho}\widetilde{u}_j k\right) = \frac{\partial}{\partial X_j}\left(\frac{\mu_e}{\sigma_R}\frac{\partial k}{\partial X_j}\right) + G_k - \bar{\rho}\varepsilon \tag{5.21}$$

$$\frac{\partial}{\partial t}\left(\bar{\rho}\varepsilon\right) + \frac{\partial}{\partial X_j}\left(\bar{\rho}\widetilde{u}_j\varepsilon\right) = \frac{\partial}{\partial X_j}\left(\frac{\mu_e}{\sigma_\varepsilon}\frac{\partial \varepsilon}{\partial X_j}\right) + \frac{\varepsilon}{k}\left(C_1 G_k - C_2 \bar{\rho}\varepsilon\right) \tag{5.22}$$

$$\frac{\partial}{\partial t}\left(\bar{\rho}\widetilde{Z}\right) + \frac{\partial}{\partial X_j}\left(\bar{\rho}\widetilde{u}_j\widetilde{Z}\right) = \frac{\partial}{\partial X_j}\left(\frac{\mu_e}{\sigma_Z}\frac{\partial \widetilde{Z}}{\partial X_j}\right) \tag{5.23}$$

$$\frac{\partial}{\partial t}\left(\bar{\rho}g\right) + \frac{\partial}{\partial X_j}\left(\bar{\rho}\widetilde{u}_j g\right) = \frac{\partial}{\partial X_j}\left(\frac{\mu_e}{\sigma_g}\frac{\partial g}{\partial X_j}\right) + C_{g1}\mu_t \left(\frac{\partial \widetilde{Z}}{\partial X_j}\right)^2 - C_{g2}\left(\frac{\bar{\rho}g\varepsilon}{k}\right) \tag{5.24}$$

5.2.3 Probability Density Distribution Function

For the random mixture fraction Z which fluctuates between 0 and 1, the probability of its appearance in the interval $Z \sim Z + dZ$ is defined as $P(Z)\,dZ$, where $P(Z)$ is known as the probability density function. Clearly, it satisfies:

$$\int_0^1 P(Z)\,dZ = 1 \tag{5.25}$$

The mean and fluctuation mean square of Z are:

$$\bar{Z} = \int_0^1 ZP(Z)\,dZ \tag{5.26}$$

$$\overline{Z'^2} = \overline{(Z-\bar{Z})^2} = \overline{Z^2} - (\bar{Z})^2 = \int_0^1 (Z-\bar{Z})^2 P(Z)\,dZ = \int_0^1 Z^2 P(Z)\,dZ - (\bar{Z})^2 \tag{5.27}$$

Similarly, for any scalar function $\varphi(Z)$, the mean and fluctuation mean square of φ are:

$$\bar{\varphi}(x) = \int_0^1 \varphi(Z)P(Z,x)\,dZ \tag{5.28}$$

$$\overline{\varphi'^2} = \int_0^1 \varphi^2(Z)P(Z,x)\,dZ - (\bar{\varphi})^2 \tag{5.29}$$

Therefore, the time-averaged mass fractions of the fuel and oxygen are:

$$\bar{Y}_F = \int_0^1 Y_F(Z)P(Z)\,dZ = \int_0^1 \frac{Z-Z_F}{1-Z_F}P(Z)\,dZ \tag{5.30}$$

$$\bar{Y}_{ox} = \int_0^1 Y_{ox}(Z)P(Z)\,dZ = \int_0^1 \left(1-\frac{Z}{Z_F}\right)P(Z)\,dZ \tag{5.31}$$

So far, the relationship between the time-averaged mass fraction and random mixture fraction has been established, but a concrete expression of the probability density function still needs to be provided.

5.2.4 Presumed PDF

Spalding proposed a simple PDF function in 1970, assuming that Z may only take two values, Z_+ and Z_-. Furthermore, it is assumed that the time fraction of Z equal to Z_- is α and that of Z equal to Z_+ is $(1-\alpha)$. In this case, PDF is a kind of "battlements type" distribution, so $P(Z)$ has two peaks at $Z=Z_-$ and $Z=Z_+$, and when $Z \neq Z_-$ or $Z \neq Z_+$, $P(Z)$ equals 0. It is expressed as:

$$P(Z) = \alpha\delta(Z_-) + (1-\alpha)\delta(Z_+) \tag{5.32}$$

To determine $P(Z)$, it is necessary to obtain Z_+, Z_-, and α. Since:

$$\bar{Z} = \int_0^1 ZP(Z)\,dZ = \int_0^1 Z[\alpha\delta(Z_-) + (1-\alpha)\delta(Z_+)]\,dZ$$

$$\bar{Z} = \alpha Z_- + (1-\alpha)Z_+ \tag{5.33}$$

$$\overline{Z'^2} = \int_0^1 (Z-\bar{Z})^2 P(Z)\,dZ = \int_0^1 (Z-\bar{Z})^2 [\alpha\delta(Z_-) + (1-\alpha)\delta(Z_+)]\,dZ$$

Then:

$$\overline{Z'^2} = g = \alpha(Z_- - \bar{Z})^2 + (1-\alpha)(Z_+ - \bar{Z})^2 \tag{5.34}$$

When $\alpha = 0.5$ (the opportunities of taking the upper limit and the lower limit are equal), one obtains:

$$\bar{Z} = \frac{Z_- + Z_+}{2} \tag{5.35}$$

$$g = \frac{(Z_- - \bar{Z})^2 + (Z_+ - \bar{Z})^2}{2} \tag{5.36}$$

Thus:

$$Z_- = \bar{Z} - g^{1/2} \tag{5.37}$$

$$Z_+ = \bar{Z} + g^{1/2} \tag{5.38}$$

Notably, to ensure the reasonableness of the value of Z, Z_- and Z_+ must meet the following conditions according to the definition:

$$Z_- \geq 0 \tag{5.39}$$

$$Z_+ \leq 1 \tag{5.40}$$

In general, for this "battlements type" distribution of $P(Z)$, the specific steps to determine α, Z_-, and Z_+ are as follows:

1. Assume that $\alpha = 0.5$, then solve Z_- and Z_+, and judge whether the conditions of $Z_- \geq 0$ and $Z_+ \leq 1$ are met. If not, further processing is needed.
2. If $Z_+ > 1$, let $Z_+ = 1$; using Equations 5.33 and 5.34 one can obtain:

$$\alpha = \left[\frac{1+g}{(1-\bar{Z})^2}\right]^{-1} \tag{5.41}$$

$$Z_- = \frac{\bar{Z} - g}{1 - \bar{Z}} \tag{5.42}$$

3. If $Z_- < 0$, let $Z_- = 0$; using the same Equations 5.33 and 5.34, one can obtain:

$$\alpha = \left(\frac{1+\bar{Z}^2}{g}\right)^{-1} \tag{5.43}$$

$$Z_+ = \bar{Z} + \frac{g}{\bar{Z}} \tag{5.44}$$

Thus, one obtains the values of α, Z_-, and Z_+ for different situations; therefore, the "battlements type" $P(Z)$ is determined. Using the rapid reaction hypothesis and the linear relationship among conserved quantities, one can obtain the corresponding $\phi(Z_-)$ and $\phi(Z_+)$, where ϕ can be the fuel or oxidizer mass fraction. According to Equation 5.27, one can solve the mean and fluctuation mean square of all kinds of chemically thermodynamic parameters:

$$\bar{\phi} = \alpha\phi(Z_-) + (1-\alpha)\phi(Z_+) \tag{5.45}$$

$$\overline{\phi'^2} = \alpha\left[\phi(Z_-) - \bar{\phi}\right]^2 + (1-\alpha)\left[\phi(Z_+) - \bar{\phi}\right]^2 \tag{5.46}$$

5.2.5 Truncated Gaussian PDF

The valid range of the Gaussian distribution is $(-\infty, \infty)$, but $Z \in [0,1]$, so the PDF has no physical meaning when $Z < 0$ or $Z > 1$. Thus, two δ functions can be assumed at $Z = 0$ and $Z = 1$. One then obtains the truncated Gaussian probability distribution:

$$P(Z) = \frac{1}{\sqrt{2\pi}\sigma}\exp\left[-\frac{1}{2}\left(\frac{Z-\mu}{\sigma}\right)^2\right]\cdot[D(Z) - D(Z-1)] + A\delta(0) + B\delta(1) \tag{5.47}$$

where μ is the value of Z that leads to the greatest probability, σ is the variance, and $D(Z)$ is the Heaviside step function, i.e.:

$$D(Z) = \begin{cases} 0 & Z \le 0 \\ 1 & Z > 0 \end{cases} \tag{5.48}$$

$\delta(Z)$ is the Dirac function, defined as:

$$\delta(Z) = \begin{cases} 1 & Z = 0 \\ 0 & Z \neq 0 \end{cases} \tag{5.49}$$

Therefore, the time-averaged mixture fraction $\bar{Z}$ is:

$$\bar{Z} = \int_0^1 Z\left\{\frac{1}{\sqrt{2\pi}\sigma}\exp\left[-\frac{1}{2}\left(\frac{Z-\mu}{\sigma}\right)^2\right]\cdot[D(Z) - D(Z-1)] + A\delta(0) + B\delta(1)\right\}dZ \tag{5.50}$$

where:

$$A = \int_{-\infty}^0 P_0(Z)\,dZ \tag{5.51}$$

$$B = \int_1^\infty P_0(Z)\,dZ \tag{5.52}$$

$$P_0(Z) = \frac{1}{\sqrt{2\pi}\sigma}\exp\left[-\frac{1}{2}\left(\frac{Z-\mu}{\sigma}\right)^2\right]$$ (5.53)

The fluctuation mean square of the mixture fraction is:

$$g = \int_0^1 (Z-\bar{Z})^2 \left\{ \frac{1}{\sqrt{2\pi}\sigma}\exp\left[-\frac{1}{2}\left(\frac{Z-\mu}{\sigma}\right)^2\right] [D(Z)-D(Z-1)] + A\delta(0) + B\delta(1) \right\} dZ$$ (5.54)

Then the iterative method can be used to determine the four constants σ, μ, A, and B, so that $P(Z)$ can be determined accordingly.

If a function of Z is known as $\phi(Z)$, its mean and fluctuation mean square can be determined according to Equations 5.28 and 5.29, respectively.

5.3   Finite Rate EBU—Arrhenius Model for Premixed Flames

A flame is said to be premixed if the fuel and oxidizer are uniformly mixed before entering the combustion zone.

Spalding proposed the EBU (eddy-break-up) model for premixed flames in 1971. Basic points concerning this model are:

1. the turbulent combustion zone is considered as a mixture of burned and unburned gas packets;
2. chemical reactions occur at the interface of these two packets;
3. assume that the rate of chemical reactions depends on the rate at which the unburned gas packets broke into smaller ones via turbulence;
4. assume that the breakup rate is proportional to the decay rate of turbulent kinetic energy.

The EBU reaction rate is expressed as:

$$\bar{R}_{\text{fu,T}} = C_R \bar{\rho} \frac{\varepsilon}{k}(g_f)^{1/2}$$ (5.55)

where C_R is a constant, generally taken as 1.07, and g_f is the fluctuation mean square of the local fuel mass fraction, i.e.:

$$g_f = \overline{Y'^2_{\text{fu}}}$$ (5.56)

The term g_f can be expressed as an algebraic expression associated with $\bar{Y}_{\text{fu}}$ or its gradients, such as:

$$g_f = C\left(\tilde{Y}_{\text{fu}}\right)^2$$ (5.57)

or:

$$g_f = l^2\left(\frac{\partial \tilde{Y}_{\text{fu}}}{\partial X_j}\right)^2$$ (5.58)

where l is the length of turbulent mixing.

More generally, it can be obtained by solving a differential equation:

$$\frac{D(\bar{\rho}g_f)}{Dt} = \frac{\partial}{\partial x_j}\left(\left(\frac{\mu_e}{Sc} + \frac{\mu_t}{\sigma_t}\right)\frac{\partial g_f}{\partial x_j}\right) + c_{g1}\mu_t\left(\frac{\partial\tilde{Y}_{fu}}{\partial x_j}\right)^2 - c_{g2}\bar{\rho}g_f\frac{\varepsilon}{k} \tag{5.59}$$

where σ_t, c_{g1}, and c_{g2} are constants, usually taken as 0.7, 2.8, and 1.79, respectively.

A turbulent combustion system may involve some regions where the average velocity gradients are large but the mixture temperature is low, so that there is no intense chemical reaction. Obviously, Equation 5.55 cannot give a reasonable rate of reaction in these regions. To overcome this disadvantage, another Arrhenius-type burning rate formula expressed in terms of average parameters can be introduced:

$$\bar{R}_{fu,A} = BT^\alpha\bar{Y}_1\bar{Y}_2\,\exp\left(-\frac{E}{k\bar{T}}\right) \tag{5.60}$$

We take the actual reaction rate $\bar{R}_{fu}$ as the smaller one of $\bar{R}_{fu,A}$ and $\bar{R}_{fu,T}$, i.e.:

$$\bar{R}_{fu} = -\min(\bar{R}_{fu,A}, \bar{R}_{fu,T}) \tag{5.61}$$

The average concentration can be used instead the fluctuating concentration in Equation 5.55 [1] and, thus, the expression of $\bar{R}_{fu,T}$ becomes:

$$\bar{R}_{fu,T} = A\rho\frac{\varepsilon}{k}\min\left[\bar{Y}_{fu}, \frac{\bar{Y}_{ox}}{s}, \frac{Y_{pr}}{B(1+s)}\right] \tag{5.62}$$

The EBU model highlights the control effect of turbulent mixing (or flow state) on the reaction rate, which is reasonable; however, it fails to consider the effects of molecular transport and detailed chemical kinetics, which is inadequate. As a result, this model is only suitable for turbulent combustion processes with high Reynolds numbers.

5.4 Moment-Equation Model

The moment-equation approach for turbulent combustion is similar to the closure model for turbulent flow. The nonlinear exponential term in the reaction rate is expanded to series, so that the fluctuations of the nonlinear term can be expressed as an infinite series of the temperature fluctuations. However, this series is convergent only if $E/R\bar{T} < 1$ and $T'/\bar{T} \ll 1$, which are not satisfied in many practical combustion systems, resulting in severe errors in the series-expanding approach. When the above two conditions are satisfied, i.e., the series is convergent, the closure for the time-averaged chemical reaction rate actually becomes the closure for the second-order moments $\overline{Y_f'Y_o'}$, $\overline{Y_f'T'}$, $\overline{T'Y_o'}$, and $\overline{T'^2}$. Usually, transport equations for these moments can be deduced and gradient-type closure is adopted. Six equations for these second-order scalar fluctuation moments are needed, so this approach is also called the transport-equation model for moments. Certainly, the introduction of these equations greatly increases the computational costs.

5.4.1 Time-Averaged Chemical Reaction Rate

The nonlinear exponential term in the reaction rate can be expanded as:

$$\exp\left(-\frac{E}{RT}\right) = \exp\left[-\frac{E}{R(\bar{T}+T')}\right] = \exp\left[-\frac{E}{R\bar{T}}\left(1+\frac{T'}{\bar{T}}\right)^{-1}\right] \tag{5.63}$$

For $T'/\bar{T} \ll 1$, one obtains:

$$\exp\left(-\frac{E}{RT}\right) \approx \exp\left(-\frac{E}{R\bar{T}}\right)\exp\left(-\frac{E}{R\bar{T}^2}T'\right) \tag{5.64}$$

If, and only if:

$$\frac{E}{R\bar{T}}\frac{T'}{\bar{T}}$$

is relatively small, the following approximation holds:

$$\exp\left(-\frac{E}{RT}\right) \approx \exp\left(-\frac{E}{R\bar{T}}\right)\left[1+\frac{E}{R\bar{T}}\frac{T'}{\bar{T}}+\frac{1}{2}\left(\frac{E}{R\bar{T}}\frac{T'}{\bar{T}}\right)^2\right] \tag{5.65}$$

Then the time-averaged chemical reaction rate can be expressed as:

$$\bar{w}_s = \overline{B\rho^2 Y_1 Y_2 \exp\left(-\frac{E}{RT}\right)}$$

$$= \overline{B\rho^2 (\bar{Y}_1 + Y_1')(\bar{Y}_2 + Y_2')\exp\left(-\frac{E}{R\bar{T}}\right)\left[1+\frac{E}{R\bar{T}}\frac{T'}{\bar{T}}+\frac{1}{2}\left(\frac{E}{R\bar{T}}\frac{T'}{\bar{T}}\right)^2\right]} \tag{5.66}$$

or:

$$\bar{w}_s = B\rho^2 \bar{Y}_1 \bar{Y}_2 \exp\left(-\frac{E}{R\bar{T}}\right)\left[1+\frac{\overline{Y_1'Y_2'}}{\bar{Y}_1\bar{Y}_2}+\frac{E}{R\bar{T}}\left(\frac{\overline{T'Y_1'}}{\bar{T}\bar{Y}_1}+\frac{\overline{T'Y_2'}}{\bar{T}\bar{Y}_2}\right)+\frac{1}{2}\left(\frac{E}{R\bar{T}}\right)^2\frac{\overline{T'^2}}{\bar{T}}\right] \tag{5.67}$$

5.4.2 Closure for the Moments

To close Equation 5.67, one may solve transport equations for $\overline{Y_1'Y_2'}$, $\overline{T'Y_1'}$, $\overline{T'Y_2'}$, and $\overline{T'^2}$, respectively. Take $\overline{Y_1'Y_2'}$ as an example:

$$\frac{\partial}{\partial t}\left(\rho\overline{Y_1'Y_2'}\right)+\frac{\partial}{\partial x_j}\left(\rho v_j\overline{Y_1'Y_2'}\right)$$

$$= \frac{\partial}{\partial x_j}\left(\frac{\mu_e}{\sigma_Y}\frac{\partial\overline{Y_1'Y_2'}}{\partial x_j}\right)-c_1\mu_T\left(\frac{\partial\bar{Y}_1}{\partial x_j}\right)\left(\frac{\partial\bar{Y}_2}{\partial x_j}\right)-c_2\frac{\varepsilon}{k}\rho\overline{Y_1'Y_2'} \tag{5.68}$$

To save storage and computational costs while retaining the main characteristics of the model, the transport-equation model can be simplified to an algebraic model, similar to the treatment in a turbulent algebraic stress model, i.e., eliminating the convection and diffusion terms in the equations:

$$\overline{Y_1' Y_2'} = c_Y \frac{k^3}{\varepsilon^2} \frac{\partial \bar{Y}_1}{\partial x_j} \frac{\partial \bar{Y}_2}{\partial x_j} \tag{5.69}$$

$$\overline{T' Y'}_1 = c_{Y_1} \frac{k^3}{\varepsilon^2} \frac{\partial \bar{T}}{\partial x_j} \frac{\partial \bar{Y}_1}{\partial x_j} \tag{5.70}$$

$$\overline{T' Y'}_2 = c_{Y_2} \frac{k^3}{\varepsilon^2} \frac{\partial \bar{T}}{\partial x_j} \frac{\partial \bar{Y}_2}{\partial x_j} \tag{5.71}$$

$$\overline{T'^2} = c_T \frac{k^3}{\varepsilon^2} \left(\frac{\partial \bar{T}}{\partial x_j} \right)^2 \tag{5.72}$$

The algebraic expressions Equations 5.69–5.72 have reasonable physical meaning. That is, the fluctuations of species, temperature, and their cross term is proportional to the product of the turbulent scale and the average gradient. Similar to the mixing-length model, the root mean square of the fluctuating velocity is proportional to the product of the turbulent scale and the average velocity gradient since $k^3 / \varepsilon^2 = l^2$. Notably, this algebraic model can only be adopted in shear flows, it cannot be used under certain conditions, where the average gradients of species and temperature are zero while their fluctuations are nonzero.

5.5   Flamelet Model for Turbulent Combustion

Peters [2] proposed the flamelet model for turbulent combustion based on the study of laminar diffusion flames. The basic physical view embedded in the flamelet model is that turbulent combustion flowfields consist of plentiful flamelets and surrounding non-reactive turbulence. These flamelets are thin reactive-diffusive layers. If these layers are thin compared to the size of a Kolmogorov eddy, the turbulent combustion regime will present as "wrinkled" laminar flamelets. Though the application of the flamelet model is limited to specific combustion regimes, these regimes are frequently encountered in practice. In the "wrinkled" laminar flamelet regime, the inner structure of the flamelets is not influenced by the turbulent eddies but only stretched and distorted by the turbulence, since the thickness of the flamelets is even smaller than the size of a Kolmogorov eddy. Under this physical condition, the inner layer of the flame and the influence of turbulence can be separately considered. Intuitively, the flamelet concept focuses on the transport of the flame surfaces rather than the reactive scalars usually considered in a PDF approach. The location of the flame surface is defined as an iso-surface of a non-reacting scalar quantity, for which a suitable transport equation can be derived. For non-premixed combustion, the mixture fraction Z is that scalar quantity. For premixed combustion, however, there is not such a scalar quantity. Wirth and Peters [3] proposed a scalar G to describe the location of the premixed flame, which originates from the level set method [4]. Because the flamelet is very thin, it can be approximately viewed as one-dimensional. Accordingly, the inner structure of the flamelet can be described by the distribution of reactive scalars in the direction perpendicular to the flame surface. These scalars are attached to the flamelets and transported together with the flamelets. Usually, the one-dimensional distributions of these scalars are described by a group of flamelet equations. Flamelet equations have been derived for both premixed and non-premixed combustion. These equations solve the conserved scalars, then the transport of reactive scalars in the direction perpendicular to the flame surface can be obtained by introducing other parameters.

5.5.1 Diffusion Flamelet Model

Numerical simulation of turbulent combustion using the laminar diffusion flamelet model includes three steps: (i) Derive and solve the corresponding laminar flamelet equations, generating a flamelet database. (ii) Calculate the turbulent flowfield and obtain the distribution of variables. That is, solve the parameters (mixture fraction (Z) and scalar dissipation rate (χ)) on which the scalars in the flamelet database (species concentration and temperature) depend. Then obtain the corresponding variables by looking up and interpolating in the database, which avoids solving the transport equation for each variable and reduces the computational cost considerably. (iii) Based on the presumed PDF, couple the flamelet database and turbulent flowfield to obtain the mean of each variable.

5.5.1.1 Generation of Flamelet Database

The laminar flamelet database can be constructed based on the laminar counter-diffusion flame. The flamelet equations can be located in the mixture fraction space via coordinate transformation under certain flame stretch conditions. Peters derived a group of flamelet equations by using the assumption of unity Lewis number and neglecting the variation of pressure and radiation heat loss. Further simplification leads to the quasi-steady equations:

$$\rho \frac{\chi}{2} \frac{\partial^2 Y_i}{\partial Z^2} + \omega_i = 0 \tag{5.73}$$

$$\rho \frac{\chi}{2} \frac{\partial^2 T}{\partial Z^2} - \sum_{i=1}^{n} \frac{h_i}{c_p} \omega_i = 0 \tag{5.74}$$

where:

ρ is density,
Y_i denotes the mass fraction of species i,
χ is the scalar dissipation rate,
ω_i is the reaction source, which obeys the Arrhenius law,
Z is the mixture fraction, a conserved scalar, which can be expressed as:

$$Z = \frac{Z_i - Z_{i,o}}{Z_{i,f} - Z_{i,o}} \tag{5.75}$$

where Z_i is the mass fraction of element I, $Z_{i,o}$ is the mass fraction of element i in the oxidizer, and $Z_{i,f}$ is the mass fraction of element i in the fuel.

Equations 5.73 and 5.74 constitute the flamelet model of diffusion combustion in the mixture fraction space. First, one can solve Equations 5.73 and 5.74 to obtain Y_i and T, which are functions of mixture fraction Z and scalar dissipation rate χ. Then, the results are saved to generate the flamelet database.

Coupling of Turbulent Flowfield and Flamelet Database

Solving Equations 5.16 and 5.17, one can obtain the spatio-temporal distribution of the Favre-averaged mixture fraction $(\widetilde{Z})$ and its fluctuation $(\widetilde{Z'^2})$.

According to the values of $\widetilde{Z}$ and χ calculated from turbulent flowfield, scalars $Y_i(Z)$ and $T(Z)$ can be obtained via interpolation in the flamelet database. Furthermore, the averaged values for $Y_i(Z)$ and $T(Z)$ can be obtained by integrating them for the mixture fraction between 0 and 1 based on the presumed probability density function $P(Z)$. The integration can be expressed as:

$$Y_i = \int_0^1 Y_i(Z)P(Z)\,\mathrm{d}Z \tag{5.76}$$

$$T = \int_0^1 T(Z)P(Z)\,\mathrm{d}Z \tag{5.77}$$

where $P(Z)$ is the presumed PDF, for which the β distribution is usually adopted:

$$P(Z) = \frac{Z^{\alpha-1}(1-Z)^{\beta-1}}{\int Z^{\alpha-1}(1-Z)^{\beta-1}\,\mathrm{d}Z} \tag{5.78}$$

where:

$$\alpha = \widetilde{Z}\frac{\widetilde{Z}(1-\widetilde{Z})}{\widetilde{Z'^2}} - 1 \tag{5.79}$$

$$\beta = (1-\widetilde{Z})\frac{\widetilde{Z}(1-\widetilde{Z})}{\widetilde{Z'^2}} - 1 \tag{5.80}$$

Combined with $\widetilde{Z}$ and $\widetilde{Z'^2}$ calculated from turbulent flowfields, $P(Z)$ can be determined.

5.5.2   Premixed Flamelet Model

As premixed flames widely exist in practical combustion apparatus, this situation can be described by the flamelet assumption: the flame thickness (δ_F) is very small and is similar to the Kolmogorov scale (η), the combustion characteristic time (τ_c) is very short and is similar to the flow characteristic time (τ). As a result, the flame maintains a laminar flame structure; thus, it can be considered that the flame propagates as a very thin flame front.

The distance function G is introduced to describe the position of the premixed combustion flamelets. The model equation describing the propagation of the flame which depends on the convective transport and normal combustion is called the G equation. Then the flame propagation can be modeled as the propagating scalar G and its conservative form can be written as:

$$\frac{\partial \rho G}{\partial t} + \nabla \cdot \rho u G = -\rho_0 S_L |\nabla G| \tag{5.81}$$

where:

G is $G(x, t)$, a process variable that defines the flame position;
u denotes the velocity vector;
ρ_0 denotes the density of reactants at reference temperature;
S_L denotes the local laminar flame propagation velocity.

It is defined that $G < 0$ in the unburned area, $G > 0$ in the burnt area, and $G = 0$ in the thin flamelets. On the flame surface, there exists a balance between the fluid flow velocity and laminar flame propagation speed (S_L) so that the flame is stable.

Using the Favre-averaged method, G can decomposed into $\widetilde{G}$ and $\widetilde{G'^2}$, which, respectively, represent the average distance away from the flamelet and the flamelet thickness; the equations for them are as follows:

$$\frac{\partial \left(\bar{\rho}\widetilde{G}\right)}{\partial t} + \nabla \cdot \left(\bar{\rho}\widetilde{u}\widetilde{G}\right) = \bar{\rho}S_T|\nabla\widetilde{G}| - \bar{\rho}D_t\kappa|\nabla\widetilde{G}| \tag{5.82}$$

$$\sqrt{\widetilde{G'^2}} = 1.78 l_t \tag{5.83}$$

where κ denotes the mean curvature of the flamelet, D_t denotes the turbulent diffusion coefficient, and S_T denotes the turbulent flame speed, which is determined by the following relational expression:

$$S_T = S_L \left[1 - \frac{0.39\, l_t}{2\ l_F} + \sqrt{\left(\frac{0.39\, l_t}{2\ l_F}\right)^2 + 4\left(\frac{0.39\, l_t}{2\ l_F}\right)\frac{v'}{S_L}} \right] \tag{5.84}$$

where:

l_F and S_L denote, respectively, the laminar flame thickness and the laminar flame speed;
$l_t = 0.37\dfrac{v'^3}{\varepsilon}$ denotes the integral length scale, where $v' = \sqrt{\frac{2}{3}k}$;
k and ε denote, respectively, the turbulent kinetic energy and its dissipation rate.

Since $\widetilde{G}$ is a distance function, it may become irregular after some iteration. That is, two $\widetilde{G}$ lines may merge, resulting in greatly increased gradients of $\widetilde{G}$, which can cause divergence of the calculation, so it is necessary to re-initialize the value of $\widetilde{G}$ at each time step to satisfy $|\nabla \widetilde{G}| = 1$ when $\widetilde{G} \neq G_0$.

When the value of G at position x and time t is clarified, one can construct a Favre-averaged Gauss probability density function:

$$\widetilde{P}(G,x,t) = \frac{1}{\sqrt{2\pi \widetilde{G'^2}}} \exp\left[-\frac{\left(G-\widetilde{G}\right)^2}{2\widetilde{G'^2}}\right] \tag{5.85}$$

From the flamelet database, one can obtain the transient temperature (T), density (ρ), and species concentration (Y_i). Using the simplified PDF form, one can obtain their Favre-averaged values:

$$\widetilde{T}(x,t) = \int_{-\infty}^{+\infty} T(G,t)\widetilde{P}(G,x,t)\,dG \tag{5.86}$$

$$\widetilde{\rho}(x,t) = \left\{\int_{-\infty}^{+\infty} \left[\rho(G,t)\right]^{-1}\widetilde{P}(G,x,t)\,dG\right\}^{-1} \tag{5.87}$$

$$\widetilde{Y}(x,t) = \int_{-\infty}^{+\infty} Y_i(G,t)\widetilde{P}(G,x,t)\,dG \tag{5.88}$$

The first step of numerical calculation is to solve the mass, momentum, and energy equations and turbulence model equations. Then one can solve the conservation equations of the mean and variation of G. Based on the pre-generated laminar flamelet database, one can find the mean of variables using the simplified PDF method.

5.6   Transported PDF Method for Turbulent Combustion

The transported PDF method treats the turbulent flowfields from a completely random point of view. The transport equations of the joint PDF of the velocity and scalars are solved to get all of the single point statistical information of the turbulent flowfields. The PDF method stems from the pioneering work of Dopazo [5], Pope [6] and so on. In the PDF method, the closures of scalar fluctuation moments, vector fluctuation moments, scalar-vector fluctuation moments, and the nonlinear chemical reaction sources are based completely on the joint probability density function, without the need for modeling. This method has obvious advantages in combustion processes, where we need to consider finite rate reaction and detailed reaction kinetics.

5.6.1   Transport Equations of the Probability Density Function

Based on theories of probability and statistics, an accurate transport equation of the joint probability density function can be established for turbulent combustion. By defining the mass fraction of the flow field $Y(x, t)$ as a scalar field, using the probability density function, the ensemble average of the mass fraction scalar field $Y(x, t)$ can be calculated from the initial average. Given initial condition $Y(x,t) = Y_0$, the calculated Y will have its own probability density function if Y_0 satisfies a certain probability density function. The ensemble average of any function, such as $G[Y(x_1,t), Y(x_2,t), \cdots]$, can be obtained by multiplying it with the probability density function and integrating over the whole Y space. The probability density function of a sample in the ensemble is called the sample probability density function (P_{fg}). Assume that the function $Y^*(x, t)$ is a sample of the random field $Y(x, t)$, then:

$$P_{fg}\left(\hat{Y}; x, t\right) = \delta\left[Y^*(x,t) - \hat{Y}\right] \tag{5.89}$$

Obviously, $P_{fg}(\hat{Y}; x, t)$ is a function of $\hat{Y}$ and $Y^*(x, t)$.

P_{fg} has all the properties of the probability density function. According to the properties of a δ function, one can obtain the normalized condition:

$$\int P_{fg}\left(\hat{Y}; x, t\right) d\hat{Y} = 1 \tag{5.90}$$

Thus:

$$Y^*(x,t) = \int \hat{Y}^* P_{fg}\left(\hat{Y}; x, t\right) d\hat{Y} \tag{5.91}$$

The samples of a random field can also be written in the form of the probability density function using $P_{fg}(\hat{Y}; x, t)$. The ensemble average of $P_{fg}(\hat{Y}; x, t)$ denotes the usual probability density function, and is expressed as:

$$P\left(\hat{Y}; x, t\right) = \left\langle P_{fg}\left(\hat{Y}; x, t\right)\right\rangle \tag{5.92}$$

where $\langle\,\rangle$ denotes the ensemble average. $P(\hat{Y}; x, t)$ denotes the probability that the mass fraction $Y(x, t)$ will be within $\hat{Y} \pm d\hat{Y}$ at time t and x.

Next, we will derive the PDF equation in the simplest case where the fluid density and mass diffusion coefficients are constant. The conservation equations of species are:

$$\frac{\partial Y}{\partial t} + v \cdot \nabla Y = D\nabla^2 Y + \frac{\dot{\omega}}{\rho} \tag{5.93}$$

Taking the derivative of Equation 5.89 on t yields:

$$\frac{\partial P_{fg}}{\partial t} = \frac{\partial \delta}{\partial t} = \frac{\partial \delta}{\partial Y}\frac{\partial Y}{\partial t} = -\frac{\partial P_{fg}}{\partial \hat{Y}}\frac{\partial Y}{\partial t} \tag{5.94}$$

Substituting the $\partial Y/\partial t$ of Equation 5.93 into the above formula yields:

$$\frac{\partial P_{fg}}{\partial t} = -\frac{\partial P_{fg}}{\partial \hat{Y}}\left(-v \cdot \nabla Y + D\nabla^2 Y + \frac{\dot{\omega}}{\rho}\right) \tag{5.95}$$

The second term on the right-hand side of the equation can be written as:

$$D\frac{\partial^2 Y}{\partial x_k \partial x_k}\frac{\partial P_{fg}}{\partial \hat{Y}} = D\frac{\partial}{\partial x_k}\left(\frac{\partial Y}{\partial x_k}\frac{\partial P_{fg}}{\partial \hat{Y}}\right) - D\frac{\partial Y}{\partial x_k}\frac{\partial^2 P_{fg}}{\partial x_k \partial \hat{Y}} \qquad (5.96)$$

Therefore:

$$\frac{\partial P_{fg}}{\partial t} + \boldsymbol{v}\cdot\nabla P_{fg} - D\nabla^2 P_{fg} + \frac{\partial}{\partial \hat{Y}}\left(P_{fg}\cdot\frac{\dot{\omega}}{\rho}\right) + D\frac{\partial Y}{\partial x_k}\frac{\partial Y}{\partial x_k}\frac{\partial^2 P_{fg}}{\partial \hat{Y}\partial \hat{Y}} = 0 \qquad (5.97)$$

Assuming that the system has three kinds of chemical species, the sample probability density function is the product of the three δ functions:

$$P_{fg} = \left(\hat{Y}_1, \hat{Y}_2, \hat{Y}_3; x, t\right) = \prod_{a=1}^{3}\delta\left[Y_a(x,t) - \hat{Y}_a\right] \qquad (5.98)$$

Taking the derivative on t:

$$\frac{\partial P_{fg}}{\partial t} = -\sum_{a=1}^{3}\frac{\partial P_{fg}}{\partial \hat{Y}_a}\cdot\frac{\partial \hat{Y}_a}{\partial t} \qquad (5.99)$$

Substituting it into the conservation equations of Y_a:

$$\frac{\partial P_{fg}}{\partial t} = \boldsymbol{v}\cdot\sum_{a=1}^{3}\frac{\partial P_{fg}}{\partial \hat{Y}_a}\nabla Y_a - D\sum_{a=1}^{3}\nabla^2 Y_a\frac{\partial P_{fg}}{\partial \hat{Y}_a} - \sum_{a=1}^{3}\frac{\partial}{\partial \hat{Y}_a}\left(\frac{\dot{\omega}}{\rho}P_{fg}\right) \qquad (5.100)$$

Simple algebraic transformation then yields:

$$\frac{\partial P_{fg}}{\partial t} + \boldsymbol{v}\cdot\nabla P_{fg} - D\sum_{a=1}^{3}\nabla^2 P_{fg} + \sum_{a=1}^{3}\frac{\partial}{\partial \hat{Y}_a}\left(\frac{\dot{\omega}}{\rho}P_{fg}\right) + D\sum_{a=1}^{3}\sum_{\gamma=1}^{3}\nabla Y_\gamma\cdot\nabla Y_a\frac{\partial^2 P_{fg}}{\partial \hat{Y}_\gamma\partial \hat{Y}_a} = 0 \qquad (5.101)$$

This is the partial differential equation of sample probability density, which means that any infinitely differentiable function $\varphi(\hat{Y})$ multiplying the terms on the right-hand side of Equation 5.101 and integrating in the $\hat{Y}$ space leads to zero. Taking the ensemble average of Equation 5.101 yields:

$$\frac{\partial P}{\partial t*} + \langle\boldsymbol{v}\cdot\nabla P_{fg}\rangle + \sum_{a=1}^{3}\frac{\partial}{\partial \hat{Y}_a}\left(\frac{\dot{\omega}}{\rho}P\right) = D\sum_{a=1}^{3}\nabla^2 P - D\sum_{a=1}^{3}\sum_{\gamma=1}^{3}\frac{\partial^2}{\partial \hat{Y}_\gamma\partial \hat{Y}_a}\langle\nabla Y_a\cdot\nabla Y_\gamma P_{fg}\rangle \qquad (5.102)$$

It is seen that the dependent variable P in the equation is a function of five independent variables, Y_1, Y_2, Y_3, x, t, while the original dependent variable Y is just a function of x, t. This is the cost for getting rid of the closure problem of the production terms. The $\dot{\omega}$ in the equation can be considered as a known parameter, because it is only a function of $\hat{Y}$ and no integral or differential problems in physical space is involved in its definition, this is true for arbitrary complex chemical reaction kinetics. It is the main advantage of the transported probability density function method.

The general combustion problem contains N reactants, three components of velocity, density, and enthalpy, and at the same time it is asymmetric in space and time. In this general combustion case, the simple PDF will be a function of $N + 9$ variables; the computational cost is thus very great. In the derivation of Equation 5.102, it is assumed that the convection velocity field is known. That is, the effects of chemical reactions on turbulent flow are neglected.

5.6.2   The Closure Problem of Turbulence PDF Equation

In Equation 5.102, the convective term $\langle v \cdot \nabla P_{\text{fg}} \rangle$ and the molecular diffusion term $-D(\partial/\partial Y^*)\langle \nabla^2 Y P_{fg} \rangle$ are unclosed. The convection term consists of two parts, one is caused by the average velocity and the other by the turbulent fluctuating velocity. The convection term of fluctuating velocity usually adopts the gradient transport model, i.e.:

$$\langle v \cdot \nabla P_{\text{fg}} \rangle = v \cdot \nabla P - \nabla \cdot K \nabla P \qquad (5.103)$$

where K is the eddy diffusion tensor.

In turbulence research, modeling of the molecular diffusion term is one of the most difficult problems. If it is assumed to be independent of $\nabla^2 Y$ and P_{fg} statistics, the meaning that it represents the microscale mixing will be lost and therefore the results have no realistic significance. In the case of moderate Reynolds numbers, this term is smaller than the turbulent transport term, so it can be ignored. The molecular diffusion term in Equation 5.102 is:

$$-D\frac{\partial^2}{\partial \hat{Y}^2}\left(\frac{\partial Y}{\partial x_i}\frac{\partial Y}{\partial x_i}P_{\text{fg}}\right)$$

Obviously, the factor $(\partial Y/\partial x_i)(\partial Y/\partial x_i)$ is a dissipation function of the scalar field $Y(x, t)$ and it is always greater than zero, similar to P_{fg}.

At present, the closed model for the aforementioned PDF transport equation has been developed, in which the model used to simulate the molecular diffusion process is called the small-scale mixing model, while the corresponding model used to simulate the transport of PDF in velocity space is called the random velocity model. At present, there are basically three types of small scale mixing models: the deterministic model, particle interaction model, and the model constructed by mapping closures. The common random velocity models are simplified Langevin model (SLM) and general Langevin model (GLM). Research has progressed greatly in these aspects recently. Interested readers are referred to the literature [7].

5.6.3   Transport Equation for the Single-Point Joint PDF with Density-Weighted Average

In most combustion problems, the density fluctuation is large and thus cannot be ignored. Owing to the influences of compressible turbulence, the density in combustion problems is generally a random variable. To consider the effects of density fluctuation, the density-weighted average is usually required. Therefore, the sample density function in ensemble is redefined as:

$$\widetilde{P}_{\text{fg}}\left(\hat{Y},\hat{\rho};x,t\right) = \delta\left[Y(x,t)-\hat{Y}\right]\delta[\rho(x,t)-\hat{\rho}] \tag{5.104}$$

The probability density function with density-weighted average is defined as:

$$\widetilde{P}\left(\hat{Y};x,t\right) = \frac{1}{\langle\rho(x,t)\rangle}\int\hat{\rho}\langle P_{\text{fg}}\left(\hat{Y},\hat{\rho};x,t\right)\rangle\mathrm{d}\rho \tag{5.105}$$

According to this definition, we can derive the transport equation for the joint PDF with density weighted average in the system consisting of N scalars:

$$\begin{aligned}
&\bar{\rho}\frac{\partial\widetilde{P}}{\partial t}+\widetilde{\rho u}_i\frac{\partial\widetilde{P}}{\partial x_i}-\frac{\partial}{\partial x_j}\left[\frac{u}{\rho}\frac{\partial\widetilde{P}}{\partial x_j}\right]+\sum_{k=1}^{N}\frac{\partial}{\partial\hat{\phi}}\left[\bar{\rho}\widetilde{P}S(\hat{\phi})\right]\\
&=-\frac{\partial}{\partial x_j}\left\{\bar{\rho}\left(u''_j\widetilde{P}_{\text{fg}}\right)\right\}-\sum_{k=1}^{N}\sum_{l=1}^{N}\frac{\partial^2}{\partial\hat{\phi}_k\partial\hat{\phi}_l}\left\langle\frac{\mu}{\sigma}\frac{\partial\phi_k}{\partial x_j}\frac{\partial\phi_l}{\partial x_j}\widetilde{P}_{\text{fg}}\right\rangle
\end{aligned} \tag{5.106}$$

where:

$$\widetilde{P}_{\text{fg}}=\prod_{k=1}^{N}\delta(\hat{\varphi}_k-\varphi_k)$$

and S denotes the source terms in the transport equation of ϕ.

The terms on the left-hand side of Equation 5.106 are closed, without the need of modeling. However, the two terms on the right-hand side need to be modeled and their modeling is more difficult than the closure of second-order statistical moments. Pope [6] pointed out that using only the local value in $\hat{\phi}$ phase space to model the last molecular mixing term cannot lead to satisfactory results, and the use of an integral of phase space is recommended.

5.6.4   Solution Algorithm for the Transport Equation of Probability Density Function

At present, the Lagrange method (or Monte Carlo method or particle method) is generally adopted to solve the transport equation of the probability density function.

The basic idea of the particle method is described as below:

1. Discretize the joint probability density function into an ensemble consisting of many samples (or particles); the state change of each particle is described by a set of stochastic differential equations, starting from the initial condition at the time $t = t_0$, given a very small time step size Δt, solving the stochastic differential equations will give the state parameters of each particle at the time $t = t_0 + \Delta t$.
2. At the same time, the entire flow field is divided into many units; calculate the various average values of the grid nodes represented by this unit according to the state parameter value of all particles within each unit, then substitute them into the stochastic differential equations to continue the calculation for the next time step.

Repeat steps 1 and 2 until the average values change no longer with time and then the statistically steady solution is obtained.

However, there are different methods to solve the turbulent combustion described by the probability density function. A class of hybrid algorithm is to combine the Monte Carlo method for solving the PDF equation and the traditional finite difference or finite volume method for solving equations of statistical moments to solve the flow field.

The emergence of an early hybrid algorithm is due to the incompleteness of the PDF equation, which does not contain all information of flow field, so that one must supplement the missing physical quantities by statistical moment models, thereby closing the model (e.g. hybrid algorithm PDF2DS).

After establishment of the velocity-dissipation rate-scalar joint PDF model, the hybrid algorithm is basically used to reduce the numerical error and improve the computational efficiency, which is mainly because the single particle method has the following disadvantages:

1. The average pressure term and the relevant gradient term are obtained through the Poisson equation, which is very difficult to solve.
2. The average flowfields in particle stochastic differential equations are provided by the particle statistical average value of the unit grid, which may lead to serious deviations;
3. To reduce the statistical error and ensure computational accuracy, the number of particles for simulation cannot be too small, while too many particles will cause unaffordable computational costs.

Early hybrid algorithms, such as the RANS/PDF hybrid algorithm proposed by Anand [8], Haworth and Tahry [9] and the PDF2DS hybrid algorithm [10] proposed by Correa and Pope, cannot meet the compatibility at the model level.

The main reason for this is that the turbulence model solved by the finite volume method in these two hybrid algorithms is hardly compatible with the stochastic velocity model of PDF transport equation.

The new hybrid approach PDF2DFV proposed by Muradoglu and Pope [11] in 1999 fundamentally solves the compatibility problem at the model level.

The most important feature of this method is that the average equations solved by FV are the first-order moment equations obtained by integrating the modeled PDF transport equation in the phase space, where the unclosed items are given by the PDF statistical moments—no other models are introduced. This satisfies the compatibility at the model level.

In addition, the Monte Carlo method is used to solve the PDF of the fluctuating velocity so that the average velocity field only needs to be solved once, greatly reducing the deviation. In addition, just a few particles can meet the accuracy requirements.

This hybrid algorithm shows good performance in the calculation of bluff body jet flames and lifted jet flames.

All the present hybrid algorithms are based on PDF2DFV but try to reduce the numerical error by changing the numerical iteration methods and introducing modified algorithms. However, the current modified algorithms are not yet mature but need further improvement.

5.7 Large Eddy Simulation of Turbulent Combustion

The basic idea of large-eddy simulation is to directly calculate the large-scale fluctuations and model the small-scale fluctuations. After filtering, the turbulent velocity can be decomposed into the sum of large-eddy portion ($\bar{u}_i$) and small-eddy portion (u''):

$$u_i = \bar{u}_i + u'' \tag{5.107}$$

The $\bar{u}_i$ can be calculated by large eddy simulation and is called resolvable-scale fluctuations; u'' is called unsolvable-scale fluctuations or sub-grid scale fluctuations. The influences of unsolvable-scale eddies on large eddy motions need to be expressed by the eddy viscosity relations via sub-grid scale models. In turbulent reacting flows, only if the fuel and oxidant are mixed at molecular level can chemical reactions occur, and the molecular mixing mainly occurs in small eddies, therefore it is necessary to develop sub-grid scale combustion models. We will first derive the governing equations of large eddy simulation for turbulent combustion below, and then briefly introduce the currently available sub-grid scale models for large eddy simulation of turbulent combustion.

5.7.1 Governing Equations of Large Eddy Simulation for Turbulent Combustion

The definition of Favre filtering for a variable is:

$$\tilde{f} = \frac{\overline{\rho f}}{\bar{\rho}} \tag{5.108}$$

where, "—" denotes the filtering in physical space. Consequently, the general variable $\overline{\rho f}$ can be given according to the following integration:

$$\overline{\rho f}(x_i, t) = \int_D \rho f(x_i', t) G(x_i - x_i', \Delta) \, dx_i' \tag{5.109}$$

where G denotes the filtering function and the integrating domain D is the whole flowfield. Generally there is:

$$\int_D G(x_i - x_i', \Delta) \, dx_i' = 1 \tag{5.110}$$

Using the box filtering function, G can be expressed as:

$$G = \begin{cases} \dfrac{1}{\Delta^3} & \dfrac{-\Delta}{2} \le x_i - x_i' \le \dfrac{\Delta}{2} \\ 0 & |x_i - x_i'| > \Delta/2 \end{cases} \tag{5.111}$$

Differing from the usual Favre average:

$$\widetilde{\widetilde{f}} \ne \widetilde{f} \text{ and } \overline{f''} \ne 0$$

for Favre filtering of variable f. After Favre filtering, the governing equations of turbulent reacting flow can be written as:

$$\frac{\partial \bar{\rho}}{\partial t} + \frac{\partial \bar{\rho}\widetilde{u}_i}{\partial x_i} = 0 \tag{5.112}$$

$$\frac{\partial \bar{\rho}\widetilde{u}_i}{\partial t} + \frac{\partial}{\partial x_i}\left(\bar{\rho}\widetilde{u}_i\widetilde{u}_j + \bar{P}\delta_{ij} - \overline{\tau}_{ij} + \tau_{ij}^{\text{sgs}}\right) = 0 \tag{5.113}$$

$$\frac{\partial \bar{\rho}\widetilde{E}}{\partial t} + \frac{\partial}{\partial x_i}\left[\left(\bar{\rho}\widetilde{E} + \bar{P}\right)\widetilde{u}_i + \overline{q}_i - \widetilde{u}_i\overline{\tau}_{ij} + H_i^{\text{sgs}} + \delta_i^{\text{sgs}}\right] = 0 \tag{5.114}$$

$$\frac{\partial \bar{\rho}\widetilde{y}_m}{\partial t} + \frac{\partial}{\partial x_i}\left(\bar{\rho}\widetilde{y}_m\widetilde{u}_i - \bar{\rho}D_m\frac{\partial\widetilde{y}_m}{\partial x_i} + \phi_{i,m}^{\text{sgs}} + \theta_{i,m}^{\text{sgs}}\right) = \overline{\dot{\omega}}_m \quad m = 1, N \tag{5.115}$$

In the above formula:

μ denotes the laminar viscous coefficient; the thermal conductivity $K = C_p\mu/\text{Pr}$, where Pr denotes the Prandtl number;
D_m denotes the molecular diffusion coefficient of component m;
C_p denotes the isobaric specific heat capacity.
The pressure is given by:

$$\bar{P} = \bar{\rho}\widetilde{T}\sum_{m=1}^{N}\widetilde{y}_m R/\omega_m;$$

R is the universal gas constant;
ω_m denotes the reaction source term of component m;
N is the total number of components.

The viscous stress tensor and heat flux tensor, respectively, are:

$$\overline{\tau}_{ij} = \mu\left(\frac{\partial\widetilde{u}_i}{\partial x_j} + \frac{\partial\widetilde{u}_j}{\partial x_i}\right), \quad \overline{q}_i = -K\frac{\partial\widetilde{T}}{\partial x_i} \tag{5.116}$$

The filtered total energy per unit volume is:

$$\bar{\rho}\widetilde{E} = \bar{\rho}\widetilde{e} + \frac{1}{2}\bar{\rho}\widetilde{u}_l\widetilde{u}_l + \frac{1}{2}\bar{\rho}(\widetilde{u_lu_l} - \widetilde{u}_l\widetilde{u}_l) \tag{5.117}$$

and the filtered internal energy is:

$$\widetilde{e} = \sum_{m=1}^{N}\widetilde{y}_m h_m - \bar{P}/\bar{\rho} \tag{5.118}$$

The reaction enthalpy of components is:

$$h_m = \Delta h_{\mathrm{f},m}^0 + \int_{T_0}^{T} C_{p,m}(\widetilde{T})\mathrm{d}\widetilde{T} \tag{5.119}$$

where $\Delta h_{\mathrm{f},m}^0$ denotes standard heat of formation at temperature T_0, and $C_{p,m}$ denotes the specific heat of component m at constant pressure. According to the Fick diffusion law, the diffusion rate can be approximately written as:

$$\bar{V}_{im} = (-D_m/\widetilde{Y}_m)(\partial\widetilde{y}_m/\partial x_i) \tag{5.120}$$

For mixed gas:

$$\sum_{m=1}^{N}\widetilde{Y}_m = 1, \sum_{m=1}^{N}V_{im} = 0$$

The unclosed sub-grid terms in governing equations, respectively, are: sub-grid stress tensor τ_{ij}^{sgs}, sub-grid heat flux H_i^{sgs}, the deformation work δ_i^{sgs} of unsolvable-scale viscous force, the sub-grid convective flux $\phi_{i,m}^{\mathrm{sgs}}$ and diffusion mass flux $\theta_{i,m}^{\mathrm{sgs}}$, and the filtered chemical reaction rate $\bar{\omega}_m$ of component m. They can be expressed as follows:

$$\begin{aligned}
\tau_{ij}^{\mathrm{sgs}} &= \bar{\rho}(\widetilde{u_iu_j} - \widetilde{u}_i\widetilde{u}_j) \\
H_i^{\mathrm{sgs}} &= \bar{\rho}(\widetilde{Eu_i} - \widetilde{E}\widetilde{u}_i) + (\overline{Pu_i} - \bar{P}\widetilde{u}_i) \\
\delta_i^{\mathrm{sgs}} &= \overline{u_i\tau_{ij}} - \widetilde{u}_i\widetilde{\tau}_{ij} \\
\phi_{i,m}^{\mathrm{sgs}} &= \bar{\rho}(\widetilde{u_iy_m} - \widetilde{u}_i\widetilde{y}_m) \\
\theta_{i,m}^{\mathrm{sgs}} &= \bar{\rho}(\widetilde{V_{i,m}y_m} - \widetilde{V}_{i,m}\widetilde{y}_m)
\end{aligned} \tag{5.121}$$

To close the governing equations, we use the characteristic length scale–local mesh size ($\bar{\Delta}$) and sub-grid turbulent kinetic energy (k^{sgs}) to determine the sub-grid stress tensor (τ_{ij}^{sgs}). Then the modeled sub-grid stress tensor is:

$$\tau_{ij}^{\text{sgs}} = -2\bar{\rho}\upsilon_t\left(\widetilde{S}_{ij} - \frac{1}{3}\widetilde{S}_{kk}\delta_{ij}\right) + \frac{2}{3}\bar{\rho}k^{\text{sgs}}\delta_{ij} \qquad (5.122)$$

where the resolvable scale strain rate tensor is:

$$\widetilde{S}_{ij} = \left(\partial\widetilde{u}_i/\partial x_j + \partial\widetilde{u}_j/\partial x_i\right)/2$$

and the sub-grid turbulent kinetic energy:

$$k^{\text{sgs}} = \left(\widetilde{u_k^2} - \widetilde{u}_k^2\right)/2$$

can be obtained from the following transport equation:

$$\frac{\partial\bar{\rho}k^{\text{sgs}}}{\partial t} + \frac{\partial}{\partial x_i}(\bar{\rho}\widetilde{u}_k k^{\text{sgs}}) = P^{\text{sgs}} - D^{\text{sps}} + \frac{\partial}{\partial x_i}\left(\bar{\rho}\frac{\upsilon_t}{\text{Pr}_t}\frac{\partial k^{\text{sgs}}}{\partial x_i}\right) \qquad (5.123)$$

where Pr_t denotes the turbulent Prandtl number (a value 0.9 may be used); the relationship between the sub-grid turbulence intensity and k^{sgs} is:

$$u'_{\text{sgs}} = \sqrt{\frac{2}{3}k^{\text{sgs}}} \qquad (5.124)$$

P^{sgs} and D^{sps} are, respectively, the generation term and consumption term in the sub-grid turbulent kinetic energy equation:

$$P^{\text{sgs}} = -\tau_{ij}^{\text{sgs}}\frac{\partial\widetilde{u}_i}{\partial x_j}, D^{\text{sps}} = \frac{C_\varepsilon\bar{\rho}(k^{\text{sgs}})^{3/2}}{\bar{\Delta}} \qquad (5.125)$$

The sub-grid turbulent viscosity coefficient $\upsilon_t = C_\text{v}(k^{\text{sgs}})^{1/2}\bar{\Delta}$; C_v and C_ε are constants and can be determined by dynamic model or simply taken as 0.2 and 0.916, respectively. The sub-grid heat flux (H_i^{sgs}) can be given according to the conventional gradient-diffusion model:

$$H_i^{\text{sgs}} = -\bar{\rho}\frac{\upsilon_t}{\text{Pr}_t}\frac{\partial\widetilde{H}}{\partial x_i} \qquad (5.126)$$

where, $\widetilde{H}$ denotes the total enthalpy of resolvable-scale mixed gas per unit mass, $\widetilde{H} = \widetilde{E} + \bar{P}/\bar{\rho}$. The sub-grid cross terms between components and velocity can also be modeled as:

$$\phi_{i,m}^{\text{sgs}} = -\bar{\rho}\frac{\upsilon_t}{\text{Sc}_t}\frac{\partial\widetilde{y}_m}{\partial x_i} \qquad (5.127)$$

where the turbulent Schmidt number $(\text{Sc}_t = \text{Pr}_t\text{Le}_m)$, where Le_m denotes the Lewis number of component m. The sub-grid diffusive mass flux $(\theta_{i,m}^{\text{sgs}})$ and unsolvable viscous deformation work (δ_i^{sgs}) are usually very small and thus can be ignored.

When considering the compressible effects, one may redefine ν_t and k^{sgs}:

$$\nu_t = C_R (D\Delta)^2 \left(2\widetilde{S}_{ij}\widetilde{S}_{ij} \right)^{1/2}, k^{\text{sgs}} = C_1 (D\Delta)^2 \left(2\widetilde{S}_{ij}\widetilde{S}_{ij} \right) \tag{5.128}$$

where the two model coefficients $C_R \approx 0.01$ and $C_1 \approx 0.007$. The near-wall Van Driest damping function D is:

$$D = 1 - \exp\left[1 - \frac{(y^+)^3}{26^3} \right] \tag{5.129}$$

where $y^+ = y u_\tau / \nu$. The sub-grid heat flux is modeled as:

$$H_i^{\text{sgs}} = -\rho \frac{\nu_t}{\text{Pr}_t} \left(\frac{\partial \widetilde{h}}{\partial x_i} + u_i \frac{\partial u_i}{\partial x_j} + \frac{1}{2} \frac{\partial k^{\text{sgs}}}{\partial x_j} \right) \tag{5.130}$$

5.7.2   Sub-Grid Scale Combustion Models

At present, the sub-grid combustion models that are available for large eddy simulation of turbulent combustion [12, 13] are: combustion models similar to those of RANS, such as sub-grid EBU combustion model, diffusion flamelet model, G-equation flamelet model for premixed combustion, probability density function model (filtered mass density function model), and linear eddy model, and so on. These models will be introduced one by one in the following sections.

5.7.2.1   Sub-Grid EBU Model

Since the chemical reaction rate depends on the mixing of fuel and oxygen, the reaction rate depends on the mixing rate. The sub-grid EBU model can be used to close the filtered reaction rate, $\overline{\omega}_m$. It can be assumed that the time required for molecular mixing is equal to that required for a sub-grid eddy to be completely dissipated; thus, the sub-grid fluid mixing time is proportional to the sub-grid turbulent kinetic energy (k^{sgs}) divided by its dissipation rate (ε^{sgs}):

$$\tau_{\text{mix}} \sim \frac{k^{\text{sgs}}}{\varepsilon^{\text{sgs}}} \sim \frac{C_{\text{EBU}}\overline{\Delta}}{\sqrt{2k^{\text{sgs}}}} \tag{5.131}$$

Here, setting the model coefficient $C_{\text{EBU}} = 1$, the reaction rate of mixing time scale is:

$$\overline{\dot{\omega}}_{\text{mix}} = \frac{1}{\tau_{\text{mix}}} \min\left(\frac{1}{2}[O_2], fuel \right) \tag{5.132}$$

The effective reaction rate is:

$$\dot{\omega}_{EBU} = \min(\dot{\omega}_{mix}, \dot{\omega}_{kin}) \tag{5.133}$$

where $\dot{\omega}_{kin}$ is the Arrhenius reaction rate.

5.7.2.2 Sub-Grid Scale Diffusion Flamelet Model

According to the assumptions of the sub-grid diffusion flamelet model, the flame is locally stable, maintaining laminar flamelet structures. If one filters the laminar flamelet model equation listed in Section 5.5.1, the filtered component concentrations $\widetilde{Y}_i$ is related to the mixture fraction Z, its fluctuating mean-square value $\widetilde{Z'^2}$, and scalar dissipation rate χ. Therefore:

$$\widetilde{Y}_i = \int_0^1 Y_i(Z, \widetilde{\chi}_0) P(Z) \, dZ \tag{5.134}$$

Assuming that $P(Z)$ is the β distribution probability density function:

$$P(Z) = \frac{Z^{\alpha-1}(1-Z)^{\beta-1}}{B(\alpha,\beta)} \tag{5.135}$$

where:

$$\alpha = \widetilde{Z}\left[\frac{\widetilde{Z}(1-\widetilde{Z})}{\widetilde{Z'^2}} - 1\right]$$

$\beta = \alpha/\widetilde{Z} - \alpha$, $\widetilde{Z'^2} = \widetilde{Z^2} - \widetilde{Z}^2$, and $B(\alpha, \beta)$ denotes the β function:

$$\chi_0 = \chi / \int_0^1 F(Z) P(Z) \, dZ$$

The governing equation of the filtered mixture fraction is:

$$\frac{\partial \bar{\rho}\widetilde{Z}}{\partial t} + \frac{\partial}{\partial x_j}\left(\bar{\rho}u_j\widetilde{Z}\right) = \frac{\partial}{\partial x_j}\left[\bar{\rho}(D+D_T)\frac{\partial \widetilde{Z}}{\partial x_j}\right] \tag{5.136}$$

The governing equation of the fluctuating mean-square value for sub-grid mixture fraction is:

$$\frac{\partial \bar{\rho}\widetilde{Z'^2}}{\partial t} + \frac{\partial}{\partial x_j}\left(\bar{\rho}u_j\widetilde{Z'^2}\right) = \frac{\partial}{\partial x_j}\left[\bar{\rho}(D+D_T)\frac{\partial \widetilde{Z'^2}}{\partial x_j}\right] + 2\bar{\rho}(D+D_T)\frac{\partial \widetilde{Z}}{\partial x_j}\frac{\partial \widetilde{Z}}{\partial x_j} - 2\bar{\rho}\widetilde{\chi} \tag{5.137}$$

where:

$$\widetilde{\chi} = D\overline{\frac{\partial \widetilde{Z}}{\partial x_j}\frac{\partial Z}{\partial x_j}}$$

can be modeled as:

$$\widetilde{\chi} = \frac{\left[\nu + (C_S\Delta)^2 \left|\widetilde{S}\right|\right]}{S_C C_I \Delta^2}\left(\widetilde{Z^2} - \widetilde{Z}^2\right) \tag{5.138}$$

Or:

$$\widetilde{\chi} = \frac{2\varepsilon}{k^{\text{sgs}}}\widetilde{Z''}^2 \tag{5.139}$$

where $D_T = \mu^{\text{sgs}}/S_C$ (S_C denotes the Schmidt number), the dissipation rate $\varepsilon = C_\varepsilon (k^{\text{sgs}})^{3/2}/\overline{\Delta}$; and C_I and C_ε are model coefficients.

5.7.2.3 G-equation Flamelet Model for Premixed Combustion

The filtered G equation that described the premixed flamelet in Section 5.5.2 can be written as:

$$\frac{\partial\left(\overline{\rho}\widetilde{G}\right)}{\partial t} + \frac{\partial\left(\overline{\rho}\widetilde{G}\widetilde{u}_j\right)}{\partial x_j} = -\rho_0 S_L^0 |\nabla G| - \frac{\partial}{\partial x_j}\left[\overline{\rho}\left(\widetilde{u_j G}\right) - \widetilde{u}_j\widetilde{G}\right] \tag{5.140}$$

where S_L^0 denotes the undisturbed laminar flame propagation velocity corresponding to ρ_0. Mass conservation in the flamelet leads to $\rho_0 S_L^0 = \rho_0 S_L$. Using the gradient assumption to close Equation 5.140 results in the LES-G equation:

$$\frac{\partial\left(\overline{\rho}\widetilde{G}\right)}{\partial t} + \nabla \cdot \overline{\rho}\widetilde{G}\widetilde{u}_j = -S^{\text{sgs}} - \nabla \cdot G^{\text{sgs}} \tag{5.141}$$

where the filtered source G^{sgs} can be written as:

$$G^{\text{sgs}} = \overline{\rho}\left(\widetilde{uG}\right) - \widetilde{u}\widetilde{G} \approx \frac{\overline{\rho}\nu_t}{S_c^G}\nabla G \tag{5.142}$$

where ν_t denotes eddy viscosity and S_c^G is the turbulent Schmidt number. The first term on the right-hand side of the Equation 5.140 is an unsolvable source term. Effects of sub-grid turbulence result from the wrinkling of flamelet surfaces generated by sub-grid eddies. It is assumed, based on the flame velocity model, that:

$$S^{\text{sgs}} = \rho_0 S_L^0 |\nabla G| \approx \rho_0 u_t |\nabla \tilde{G}| \tag{5.143}$$

where u_t denotes the turbulent flame propagation velocity.

In fact, as S_L is related to the flame curvature and flow instability, the flamelet stretch rate can be used to consider the effects of these two factors on S_L. Rewriting the last term of Equation 5.141:

$$\nabla \cdot (G^{\text{sgs}}) = \nabla \cdot \bar{\rho} \left(\widetilde{uG} - \tilde{u}\tilde{G} \right) = \bar{\rho} D_t \kappa |\nabla \tilde{G}| \tag{}$$

where $D_t = \nu_t / \text{Pr}_t$; the filtered flame front curvature $\kappa = \nabla \cdot \tilde{n} = \nabla \cdot \left(-\nabla \tilde{G}/|\nabla \tilde{G}| \right)$.

LES-G Equation 5.141 must be combined with LES governing Equations 5.112–5.115 and sub-grid turbulent kinetic energy k^{sgs} Equation 5.123 for solving. Scalar G can be coupled with the thermodynamic parameters by the filtered internal energy e (i.e. $\tilde{e} = C_V \tilde{T} + \Delta h_f \tilde{G}$), where the enthalpy of formation $\Delta h_f = C_p (T_p - T_f)$; T_p and T_f are, respectively, the temperature of burnt gas and fuel; C_v and C_p are, respectively, the isochoric and isobaric specific heat capacity, which are approximately considered as constants; the temperature $\tilde{T} = \left(e - \Delta h_f \tilde{G} \right)/C_V$ changes linearly with G.

The increase of flame speed can be attributed to the increase of flame surface burning area caused by flamelet wrinkling. For strong turbulence, Pocheau [14] suggested that:

$$\frac{u_t}{S_L} = \left(1 + \frac{\beta u_{\text{sgs}}'^{\alpha}}{S_L^{\alpha}} \right)^{1/\alpha} \tag{5.144}$$

The above filtered G Equation 5.141 is valid only at the flame front. For the entire flow field, uniform condition is needed: $|\nabla G| = 1$. This condition applies to the entire flow field away from the flamelet, and $\tilde{G}$ can be taken as the distance function vertical to the flame surface. Thus, the turbulent flame thickness (l_{Ft}, which can be used to express the fluctuations of flame front on vertical direction) can be defined as:

$$l_{Ft} = \left(\overline{G''^2} \right)^{1/2}_{G=G_0} \tag{5.145}$$

where, $\left(\overline{G''^2} \right)^{1/2}$ is the conditional mean square deviation at the flame front $G = G_0$. The Gaussian-type PDF function can be defined as:

$$P(G,x,t) = \frac{1}{\left[2\pi (\overline{G''^2})_0 \right]^{1/2}} \exp \left\{ -\frac{\left[G - \tilde{G}(x,t) \right]^2}{2(\overline{G''^2})_0} \right\} \tag{5.146}$$

The average mass fraction of component i can be expressed as:

$$\widetilde{y}_i(x,t) = \int_{-\infty}^{+\infty} y_i(G,t)P(G,x,t)\,\mathrm{d}G \tag{5.147}$$

The term $\overline{G''^2}$ can be obtained by solving its transport equation, but a simple method is to assume that the SGS flame thickness (l_{Ft}) is a function of laminar flame thickness (l_F), filtered width (Δ), and turbulent fluctuating velocity (u'_{sgs}), i.e., $l_{Ft} = f(l_F, \Delta, u'_{sgs})$. When the flame sheet is very thin, it can be approximately considered that $l_{Ft} \approx Cl$, where l denotes the integral length scale and C is a constant. It can also be considered that $l_{Ft} \approx C_0\Delta + l_F$, with $C_0 \approx 1$.

5.7.2.4 Filtered Mass Density Function Model

Jaberi et al. [15] developed a filtered mass density function model (FMDF) based on the sub-grid probability density function, which is suitable for compressible flows. In this model, the filtered mass density function (F_L) is defined as:

$$F_L(\psi, x, t) = \int_{-\infty}^{+\infty} \rho(x',t)\xi[\psi, \phi(x't)]G(x'-x)\,\mathrm{d}x' \tag{5.148}$$

where:

$$\xi[\psi, \phi(x't)] = \delta[\psi - \phi(x,t)] = \prod_{a=1}^{\sigma} \delta[\psi_a - \phi_a(x,t)] \tag{5.149}$$

where δ is the delta function and ψ denotes the composition domain of the scalar array; $\xi[\psi, \phi(x't)]$ is the "fine-grained" density. The large eddy simulation governing equation of $F_L(\psi, x, t)$ can be derived:

$$\frac{\partial F_L}{\partial t} + \frac{\partial(\langle u_j\rangle_L F_L)}{\partial x_j} = \frac{\partial}{\partial x_j}\left[\langle\rho\rangle_1(\langle D\rangle_1 + D_t)\right] + \frac{\partial}{\partial\psi_a}\left[\Omega_m(\psi_a - \langle\phi_a\rangle_L F_L)\right] - \frac{\partial(\widehat{S_a}F_L)}{\partial\psi_a} \tag{5.150}$$

where $\Omega_m(x, t)$ denotes the frequency of mixing within the subgrid, and can be modeled as:

$$\Omega_m(x,t) = C_\Omega(\langle D\rangle_L + D_t)/(\Delta H)^2 \tag{5.151}$$

where, C_Ω is a model coefficient, ΔH denotes filtering characteristic length, $\langle\ \rangle_1$ denotes the filtered value, $\langle\ \rangle_L$ denotes the Favre-averaged value.

The above FMDF transport equation can be solved according to the Lagrangian Monte-Carlo method. The transport equation for component variable ϕ_a is:

$$\frac{\partial\left(\langle\rho\rangle_1\langle\phi_a\rangle_1\right)}{\partial t}+\frac{\partial\left(\langle\rho\rangle_1\langle u_i\rangle_L\langle\phi_a\rangle_L\right)}{\partial x_i}=\frac{\partial}{\partial x_i}\left[\left(\langle\rho\rangle_1\langle D\rangle_L+D_t\right)\frac{\partial\langle\phi_a\rangle_L}{\partial x_i}\right]+\langle\rho\rangle_1\langle S_a\rangle_L \qquad (5.152)$$

The component $\phi_a=y_a$, $a=1,2,\cdots,N$, $\langle S_a\rangle$ denotes the reaction rate of component a, which can be determined according to the Arrhenius law.

5.7.2.5   Sub-Grid Linear Eddy Model [16]

The linear eddy model (LEM) was first proposed by Kerstein [17] in his research on the mixing process of turbulent flow in 1988, where it was used as a random mixing model to describe the scalar mixing and diffusion in turbulent flows. The LEM is a model with statistical properties, which attempts to describe different physical processes at all scales of the flow in one dimension, such as turbulent mixing, molecular diffusion, chemical reaction, and so on. The reaction and diffusion processes take place in a one-dimensional region, where all the turbulent scales are solvable and no modeling is needed. It can be viewed as one-dimensional direct simulation within the subgrid. Menon first applied the LEM model to the large eddy simulation of combustion. Implementation of the LES-LEM method was divided into two processes. The first process is to carry out the linear eddy calculation independently within each LES cell; the second process is to achieve sub-grid scalar information transport through the LES grid boundary by a spliced process.

The LEM model does not directly solve the filtered scalar equation. Molecular diffusion, turbulent transport, and chemical reactions on small scales and large scales are modeled respectively on each of the corresponding time scales. To illustrate this problem mathematically, the velocity field is divided into $u_i=\widetilde{u}_i+\left(u_i'\right)^R+(u')^S$, where $\widetilde{u}_i$ is the resolvable velocity field of large eddy simulation, $\left(u_i'\right)^R$ is the solvable sub-grid fluctuations of large eddy simulation (which can be obtained from k_{sgs}), and $\left(u_i'\right)^S$ is unsolvable sub-grid fluctuations. Considering the accurate components equation (without any explicit filtering), the equation of component k can be written as the following form:

$$\rho\frac{\partial Y_k}{\partial t}=-\rho\left[\widetilde{u}_i+\left(u_i'\right)^R+\left(u_i'\right)^S\right]\frac{\partial Y_k}{\partial x_i}-\rho D_k\frac{\partial Y_k}{\partial x_i}+\dot{w}_k \qquad (5.153)$$

In LES-LEM, the above equation can be rewritten as:

$$\frac{Y_k^*-Y_k^n}{\Delta t_{LES}}=-\left(\widetilde{u}_i+\left(u_i'\right)^R\right)\frac{\partial Y_k^n}{\partial x_i} \qquad (5.154)$$

$$\frac{Y_k^{n+1}-Y_k^*}{\Delta t_{LES}}=\int_t^{t+\Delta t_{LES}}-\frac{1}{\rho}\left[\rho\left(u_i'\right)^S\frac{\partial Y_k^n}{\partial x_i}+\frac{\partial}{\partial x_i}\left(\rho Y_k^n V_{i,k}\right)^n-\dot{w}_k^n\right]dt' \qquad (5.155)$$

Here, Δt_{LES} is the time step of large eddy simulation. Equation 5.154 describes the Lagrangian transport of resolvable scales in the large-scale three-dimensional scalar field through grid boundaries; Equation 5.155 describes the sub-grid LEM model, where the integral contains three processes that occur within each LES grid: sub-grid mixing, sub-grid molecular diffusion, and chemical reactions. These processes are modeled in a one-dimensional region within each LES grid; the integral of the equation can be written as a one-dimensional form.

Within each LES grid, the following one-dimensional reaction–diffusion equation of components is solved:

$$\rho \frac{\partial Y_k^m}{\partial t^s} = F_{k_s}^m - \frac{\partial}{\partial s}\left(\rho Y_k^m V_{s,k}^m\right) + \dot{w}_k \tag{5.156}$$

Here, t^s denotes the local LEM time scale; the sub-grid region is discretized by N_{LEM} LEM grids in the one-dimensional direction S. The LES solvable value is obtained by the Favre average of all the LEM grids within the sub-grid region. $F_{k_s}^m$ denotes the turbulent stirring of components on the sub-grid scale. A one-dimensional reaction–diffusion equation of temperature within the subgrid can also be expressed in the same way, which is omitted here. Assuming that the gas is thermally perfect, neglecting the calculation of $P^{\text{LEM}}$and assuming that $P^{\text{LEM}} = P^{\text{LES}}$, volume expansion will be caused by heat release, so it is necessary to consider the modeling of the thermal expansion process. Modeling of key processes will be given below.

The sub-grid mixing is modeled by the rearrangement of scalar field within the LEM region. This rearrangement method was called "triplet mapping" by Kerstein (Figure 5.1). Kerstein revealed that this method reproduces the effects of a single eddy in scalar gradient field but does not change the average scalar field. This rearrangement method is only performed in a one-dimensional LEM region, so the implicit assumption is that the small-scale eddies within the sub-grid scale are isotropic. Mathematically, triplet mapping can be defined as a function that maps the initial scalar field on the mixed scalar field:

$$\Psi(x,t) = \begin{cases} \Psi^0(3x-2x_0,t) & x_0 \le x \le x_0 + l/3 \\ \Psi^0(-3x+4x_0+2l,t) & x_0 + l/3 \le x \le x_0 + 2l/3 \\ \Psi^0(3x-2x_0-2l,t) & x_0 + 2l/3 \le x \le x_0 + l \\ \Psi^0(x,t) & \text{otherwise} \end{cases} \tag{5.157}$$

Apparently, the mapping function needs to choose the eddy scale (l) and position (x_0) where stirring occurs. The start position of stirring is randomly generated by the uniform distribution while the rearrangement frequency within per unit length is obtained from the three-dimensional scaling law by Kerstein:

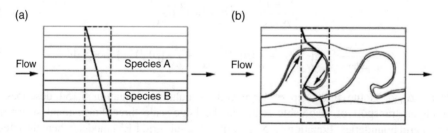

Figure 5.1   Schematic of LEM region triplet mapping method.

$$\lambda = \frac{54\nu \mathrm{Re}_{\bar{\Delta}}\left[(\bar{\Delta}/\eta)^{5/3} - 1\right]}{5C_\lambda \bar{\Delta}^3 \left[1 - (\eta/\bar{\Delta})^{4/3}\right]} \qquad (5.158)$$

where C_λ denotes the scalar turbulence diffusion coefficient, generally taken as 0.067; the time interval between the two stirring is $\Delta t_{\mathrm{stir}} = 1/(\lambda\bar{\Delta})$.

The eddy scale (l) is derived from the following probability density function distribution randomly:

$$f(l) = \frac{(5/3)l^{-8/3}}{\eta^{-5/3} - \bar{\Delta}^{-5/3}}$$

where $\eta = N_\eta \bar{\Delta} \mathrm{Re}_{\bar{\Delta}}^{-4/3}$; N_η is an empirical parameter, taken as $N_\eta \in [1.3, 10.78]$.

Evolution of the scalar field is obtained in a Lagrangian spliced way in the LES region. The spliced way simulates the process of large-scale transport but does not directly solve Equation 5.154. Note that the scalar field of LES is unknown—only the distribution on the LEM region is known. Now, the aim is to transport the scalar from the LEM region of one LES grid to that of another. A sufficiently small LES time step ensures that the scalar is only transported from a LES grid to its adjacent grid, which reduces the complexity of the problem.

The spliced process is shown in Figure 5.2, where one needs to know three physical quantities: (i) the mass to be transferred on the LES grid interface; (ii) the mass flux direction of each grid interface (inflow or outflow); (iii) the order in which the three coordinate axes perform transport operations. In the finite volume method, the mass flux of each LES grid interface is known (from the LES of large-scale field), and the mass flux direction can be given by the velocity direction on grid interfaces. Thus the first two quantities can be automatically known. Numerical operations on scalar transport are equivalent to a three-dimensional convection operator, which is achieved by a sequence of three one-dimensional convection operators. Currently, the operation order of these one-dimensional operators is determined by upward-like methods. In this method, all fluxes leaving the LEM area are through the N_{LEM} grid and those entering are all through the first grid. The largest negative flux (i.e., the mass leaves the LES grid) is the first one leaving the LES grid at the corresponding interface while the largest positive flux (i.e., the mass entering the LES grid) is the last one entering the LES grid. Applications have shown the success of this method.

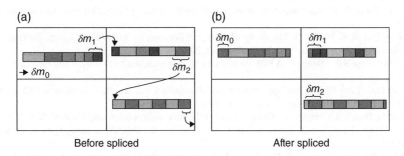

(a) Before spliced

(b) After spliced

Figure 5.2 Schematic of spliced process.

The cross iteration process of LES-LEM can be summarized as follows. First, solve the filtered LES large-scale governing equations and store the mass flux of each surface of grid control volume. Second, use the LEM to calculate within the subgrid, then use the spliced method to calculate the large scale transport of scalar field, and filter the scalar of sub-grid region, such as the component mass fraction:

$$\widetilde{Y}_k = \sum_{i=1}^{N_{LEM}} \rho_i Y_{k_i} / \sum_{i=1}^{N_{LEM}} \rho_i$$

Third, calculate the thermal-chemical properties of fluid using filtered scalars; calculate the filtered internal energy, temperature, and pressure using the filtered component field (obtained from LEM) and density field (obtained from LES); these filtered values are used as initial values of the next LES time step. Thus the iteration within a time step is completed.

The LES-LEM method contains obvious artificial effects. The abnormal diffusion caused by the spliced process and the thereafter redistricting process of LEM region has not been well tackled. In addition, the arrangement of one-dimensional linear eddies on each grid increases the dimensions of scalar field, which may greatly increase the computational cost in three-dimensional problems.

Furthermore, when considering complex chemical reactions, the number of linear eddies arranged in each grid will significantly increase, at the same time one has to consider the transport and reaction processes of various components. This makes the solution process extremely difficult. The LEM model is an empirical model, the realization of turbulent flow field depends on a specific Reynolds number and it cannot reflect the transient characteristics of the flow field and neither can it truly capture typical turbulence characteristics such as energy cascade.

References

[1] Zhou L X. *Multi-Phase Turbulent Reacting Hydrodynamics*. Defense Industry Press (in Chinese), 2002.

[2] Peters N. *Turbulent Combustion*. Cambridge University Press, 2000.

[3] Wirth M, Peters N. Turbulent premixed combustion: A flamelet formulation and spectral analysis in theory and ic-engine experiments. Twenty-fourth (International) Symposium on Combustion, The Combustion Institute, Sydney, Australia, 5–10 July 1992.

[4] Sethian J A. *Level set methods and fast marching methods: evolving interfaces in computational geometry, fluid mechanics, computer vision, and materials science*. Cambridge: Cambridge University Press, 1999.

[5] Dopazo C. Probability density function approach for a turbulent axisymmetric heated jet centerline evolution. *Physics Fluids* 1975, **18**, 397–404.

[6] Pope S B. The probability approach to the modeling of turbulent reacting flows. *Combustion and Flame*, 1976, **27**: 299–312.

[7] Fan Z Q, Sun M B, Liu W D. The Study of Transport Equation for Probability Density Function Model of Turbulent Combustion. *Cruise missiles* (in Chinese), 2010, **30** (3): 90–95.

[8] Anand M S, Pope S B, Mongia H. A PDF method for turbulent recirculating flows, *Lecture Notes Eng.*, 1989, **40**, 672–693.

[9] Haworth D C, Tahry EI S H. PDF approach for multidimensional turbulent flow calculations with application to in-cylinder flows in reciprocating engines. *AIAA Journal*, 1991, **29**:208.

[10] Correa S M, Pope S B. Comparison of a Monte Carlo PDF finite-volume mean model with bluff-body Raman data. Twenty-Fourth (International) Symposium on Combustion, The Combustion Institute, Sydney, Australia, 5–10 July 1992.

[11] Muradoglu M, Jenny P, Pope S B, Caughey D A. A consistent hybrid finite-volume/particle method for the PDF equations of turbulent reactive flows. *Journal of Computational Phys*, 1999, **154** (2), 342–371.

[12] Xu X C, Zhou L X. *Technical Manual of Combustion*. Chemical Industry Press (in Chinese), 2008.

[13] Zhao J X. *Numerical simulation of combustion*. Beijing: Science Press (in Chinese), 2002.

[14] Pocheau A. Scale invariance in turbulent front propagation. *Physical Review E*, 1994, **49**:1109–1122.

[15] Jaberi F A, Colucci P J, Pope S B. Filtered mass density function for large-eddy simulation of turbulent reacting flows. 1999, *Journal of Fluid Mechanics*, **401**:85–121.

[16] Sun M B, Liang J H, Wang Z G. Overview of Subgrid Linear Eddy Models for Turbulent Combustion. *Journal of Combustion Science and Technology (in Chinese)*, 2007, **13** (2):169–176.

[17] Kerstein A R. Linear eddy model of turbulent scalar transport and mixing. *Combustion Science and Technology*, 1988, **60**, 391–421.

6

Heat Transfer Modeling and Simulation

Research into heat transfer in liquid rocket engines consists of examining the thrust chamber heat transfer, gas generator heat transfer, and fuel supply system heat transfer and so on. In fact, heat transfer should be considered for any engine part that is necessary for heat prevention, heat insulation, cooling, or heating because of a heat source or cooling source.

As the temperature of gas in a rocket engine combustion chamber reaches up to 3000–4500 K [1], the whole thrust chamber wall is heated strongly. The most seriously heated part is near the throat of the spout, where the heat flux density can reach up to 10^4–10^5 kW m^{-2}. Therefore, parts which serve under high-temperature environments in engines must be cooled and heat must be insulated to make sure the temperature of a heated part does not exceed the permitted extent of material strength. This depends on cooling technologies. Typical technologies are regenerative cooling, film cooling, transpiration cooling, ablative cooling, and radiation cooling. Different cooling methods can be used, separately or together [2–4].

To make sure rocket engines work normally under the conditions of high heat load, appropriate mathematical modeling should be carried out for the whole internal heat transfer process. Models of heat transfer process in rocket engines in this chapter are based on three forms of heat transfer (i.e., convection, conduction, and radiation), taking internal combustion and fluid characteristics into account.

6.1 Convective Heat Transfer Model of Combustor Wall

The main form of heat transfer in a liquid rocket engine thrust chamber is gas convection heat transfer. In the combustor, the reaction gas convection heat flux is typically more than 80% of the total heat flux, while near the throat it will reach 95% of the total heat flux. Thus,

Internal Combustion Processes of Liquid Rocket Engines: Modeling and Numerical Simulations,
First Edition. Zhen-Guo Wang.
© 2016 National Defense Industry Press. Published 2016 by John Wiley & Sons Singapore Pte Ltd.

determining the convection heat flux value is the most significant work when analyzing the heat transfer of thrust chamber and proper cooling.

The main characteristics of convective heat transfer in a thrust chamber are as follows:

1. The heat flux density is high, the heat flux density of a rocket engine combustor is usually 200–360 times higher than in traditional heat transfer equipment.
2. The variation of heat flux density in flow direction of gas is great—the maximum could be dozens or even hundreds times larger than the minimum value.
3. The distribution of heat flux density in the circumferential direction is uniform; the closer to the jet surface the more uniform it is. The ratio of maximum to minimum could reach five-fold. In the combustor, the heat flux density will even gradually appear anywhere away from the jet surface as combustion continues.
4. The velocity of gas along the surface of the nozzle wall is extremely high, which increases the heat transfer rate required of the heat transfer method. Especially because of the small heated area, the heat flux density at the throat is the highest in the whole thrust chamber.

6.1.1 Model of Gas Convection Heat

The heat flux contribution in the engine thrust chamber poses the difficulty in research into engine thrust chamber heat transfer. It is affected by the injector structure, propellant performance, droplet evaporation and mixing process, and the combustion process. In the thrust chamber, most heat transfer of gas to the combustor wall occurs by convection, while a very small part (5–25%) is by radiation, and heat conduction can be omitted [5]. Therefore, the key issue in the heat conduction of an engine combustor is research into the heat transfer gas convective heat transfer model; the difficulty here is in determining the convective heat transfer coefficient.

Calculation of convective heat flux density is based on relative theories of boundary layer experimental data of a thrust chamber. This method is widely used in heat transfer calculation of engines and power devices.

The American academic Bartz [6] has processed the gas convective heat transfer coefficient in a liquid rocket engine thrust chamber into criterion form of fully developed turbulent heat transfer in the tube:

$$\mathrm{Nu_f} = 0.026\mathrm{Re_f}^{0.8}\,\mathrm{Pr_f}^{0.4} \tag{6.1}$$

where:

Nu is the Nusselt number, $\mathrm{Nu} = hd/\lambda$, $\mathrm{Nu_f} = hd/\lambda_f$, which refers to ratio of surface heat flow to fluid heat conduction;

Re is the Reynolds number, $\mathrm{Re} = \upsilon d/\nu$, $\mathrm{Re_f} = \upsilon d/\nu_f = \rho_f \upsilon d/\eta_f$, which refers to the ratio of inertial forces to viscous forces;

Pr is the Prandtl number, $\mathrm{Pr} = \eta C_p/\lambda$, $\mathrm{Pr_f} = \eta_f C_{pf}/\lambda_p$, which refers to the ratio of media momentum diffusion capacity to thermal diffusivity:

h is the convective heat transfer coefficient,

d is the equivalent diameter,

λ is the thermal conduction rate,

v is gas flow velocity,

ν is kinematic viscosity,

η is dynamic viscosity,

C_p is the specific heat capacity at constant pressure,

T_f is usually called the film temperature (the subscript f refers to the parameters of the film temperature). $T_f = (T_{wg} + T_g)/2$, where T_{wg} is the gas wall temperature (wall temperature in contact with the gas) and T_g is airflow static temperature.

By expanding and processing:

$$h = 0.026 C_{pf} \eta_f^{0.2} (\rho_f v)^{0.8} / Pr_f^{0.6} d^{0.2} \tag{6.2}$$

The convective heat flux density (q_{cv}) is:

$$q_{cv} = h(T_{ad} - T_{wg}) \tag{6.3}$$

where T_{ad} is the adiabatic wall temperature, or the gas recovery temperature.

We now introduce the recovery coefficient (r):

$$r = \frac{T_{ad} - T_g}{T^* - T_g} \tag{6.4}$$

where T^* is the total temperature of gas flow, or the stagnation temperature.

For turbulent flow, we can set:

$$r \approx Pr^{1/3} \tag{6.5}$$

In Equations 6.1 and 6.2, the film temperature (T_f) is the reference temperature, which is very inconvenient. However, we can change the reference temperature into the total temperature, which is much more convenient. Heat taken away by the coolant flow is a very small part of total heat of a gas (typically smaller than 3%), and the heat of regenerative cooling returns to combustor, so we consider total heat of gas to be constant when calculating heat transfer. Thus, thermo-physical properties that consider the total temperature as the reference temperature will not change.

When the reference temperature is changed from film temperature to total temperature, because c_p and Pr change very little with temperature they can be considered to remain unchanged; we then only consider change of ρ and η with temperature, and:

$$\rho \propto 1/T \tag{6.6}$$

$$\eta \propto T^{0.6} \tag{6.7}$$

Meanwhile:

$$T_g = T^* / \left(1 + \frac{\kappa - 1}{2} Ma^2\right) \tag{6.8}$$

$$\rho v = p_c^* A_t / Ac^* \tag{6.9}$$

where:

κ is the specific heat ratio (or isentropic index),
Ma is the Mach number,
p_c^* is the combustor total pressure,
A_t is the throat sectional area,
A is calculated cross-sectional area,
c^* is the characteristic velocity.

Substituting Equation 6.6, we can obtain the convective heat transfer coefficient of gas towards the wall surface as follows:

$$h = \frac{0.026}{d_t^{0.2}} \left(\frac{\eta^{0.2} C_p}{\mathrm{Pr}^{0.6}} \right) \left(\frac{p_c^*}{c^*} \right)^{0.8} \left(\frac{A_t}{A} \right)^{0.9} \sigma \tag{6.10}$$

where d_t is the throat radius, the total temperature T^* is the reference temperature of η, C_p, and Pr, and σ is reference temperature transform coefficient:

$$\sigma = \left[\frac{1}{2} \frac{T_{wg}}{T^*} \left(1 + \frac{\kappa-1}{2} \mathrm{Ma}^2 \right) + \frac{1}{2} \right]^{-0.68} \left(1 + \frac{\kappa-1}{2} \mathrm{Ma}^2 \right)^{-0.12} \tag{6.11}$$

For the throat section, taking the influence of the longitudinal curvature radius into account, the corrective term $(d_t/R_t)^{0.1}$ is added, where R_t is the vertical curvature radius of the throat. Then we can obtain the equation for throat convective heat transfer coefficient (h_t):

$$h_t = \frac{0.026}{d_t^{0.2}} \left(\frac{\eta^{0.2} c_p}{\mathrm{Pr}^{0.6}} \right) \left(\frac{p_c^*}{c^*} \right)^{0.8} \left(\frac{d_t}{R_t} \right)^{0.1} \left(\frac{A_t}{A} \right)^{0.9} \sigma \tag{6.12}$$

where each physical quantity determined by the thermodynamic calculated gas composition approximation can also be worked out:

$$\mathrm{Pr} \approx 4\kappa / (9\kappa - 5) \tag{6.13}$$

$$\eta \approx 1.184 \times 10^{-7} \cdot M_r^{0.5} \cdot T^{0.6} \tag{6.14}$$

where M_r relative molecular mass. There are dimensionless empirical equations in these equations, where η is in Pa.s, and T is in K.

In addition, methods presented in References [7, 8] are commonly used in liquid rocket engine combustion process calculations which are not covered here.

In the engine thrust chamber there are many complex factors that affect convective heat flux density. The major factors are combustor total pressure (p_c^*), throat diameter (d_t), gas total temperature (T^*), average molar mass of the gas (M), specific heat capacity (C_p), dynamic viscosity (η), temperature of the gas wall, geometry of thrust chamber, and so on.

6.1.2 Convection Cooling Model

Convection cooling refers to the coolant that flows along cooling channels to externally cool the thrust chamber. The coolant flow is usually a component of the liquid rocket engine propellants. There are mainly two forms of convection cooling, one is regenerative cooling, which is used by most engines, and the other is emission cooling, which means that propellant components do not return to the combustor after heat absorption; instead, they discharge from a special nozzle to produce thrust.

Cooling reliability requires the components to remain in a permitted thermal state. Under the condition of regenerative cooling, coolant flow $\dot{m}_x$ is limited. Therefore, the first requirement of reliable cooling is that the temperature of coolant (T_{lout}) should not exceed the allowable temperature after all heat that enters into cooling channel is absorbed, i.e., $T_{\text{lout}} \leq T_{\text{lp}}$. For certain components, the maximum allowable temperature is the boiling point, while for others it is the thermal decomposition (e.g., chemical) temperature. For example, hydrocarbon fuels on thermal decomposition (pyrolysis) can produce viscous substances such as tar and solid carbon. Such substances from the coolant flow may deposit on the wall and thereby increase the thermal resistance, which should be avoided.

The second requirement of reliable cooling is that the wall temperature of the combustor should not exceed the allowable limit in all areas of cooling channel. That is to say, following conditions should be met: (i) the temperature of the "hot" well (gas side) should not exceed the allowable temperature under the thermal stability condition of T_{wp}, i.e., $T_{\text{wp}} \leq (T_{\text{wp}})_{\text{thermal stability}}$; (ii) the temperature of the "cold" (coolant side), under coolant boiling condition or cleavage T_{lp}, should meet the condition $T_{\text{wl}} \leq \min(T_{\text{lb}}, T_{\text{lp}})$; (iii) the wall temperature distribution of the material should be in accordance with the conditions permitted strength, $\bar{T}_{\text{wg}} \leq (T_{\text{wp}})_{\text{strength}}$. When calculating regenerative cooling, we should check that these conditions are satisfied, and make calculations based on the cooling method applied, combined with thermo-physical properties of component.

The calculation of coolant heating is based on the stroke of component in the cooling channel. Not only total temperature increase is shown, but also the temperature distribution of coolant in each section. For a section of cooling channel with length d_x, the heat balance equation is (heat dissipation to the surrounding environment can be omitted):

$$\dot{m}_x c_x dT_x = \delta Q \tag{6.15}$$

where, δQ is the heat to be absorbed by coolant, $\dot{m}_x$ is the flow quantity of coolant here, and c_x is the specific heat capacity of the coolant.

For an axial heat flux density of $q(x)$, δQ of an axisymmetric thrust chamber can be worked out according to the following equation:

$$\delta Q = \pi q d\alpha / \cos \alpha \tag{6.16}$$

where d is the inner diameter of the calculated area thrust chamber; α is the inclination between generatrix and the axis of the thrust chamber. Substituting δQ into the heat balance equation gives:

$$dT_x/dx = \pi q d/\dot{m}_x c_x \cos\alpha \qquad (6.17)$$

Numerical integration of the ordinary differential equation, $T_x(x)$ distribution, and outlet temperature of the cooling channel (T_{lout}) or outlet temperature of the cooling channel section can be obtained. When calculating, the coolant inlet temperature of the cooling section (T_{lin}), thrust chamber geometric parameters $d(x)$, $\alpha(x)$, and the relation between $c_x(T)$ and $q(x)$ are needed. In addition, the temperature difference between each section and the inlet temperature is $\Delta T_x = T_{lout} - T_{lin}$.

Suppose the heat flux density is q, the heat transfer coefficient from chamber wall to coolant in each sectional area (h_x) is then:

$$h_x = q/(T_{wl} - T_x) \qquad (6.18)$$

Calculation of heat transfer coefficient is typically determined according to the flux parameter, provided the convective heat flux density is unknown. For turbulent flow of a single-phase fluid in the cooling channel, the following equation applies [9]:

$$Nu = 0.021 Re^{0.8} Pr^{0.4} (Pr_w/Pr)^{0.25} \qquad (6.19)$$

where, Pr_w is the Pr number when the wall temperature is the reference temperature; for other parameters without subscript, the mainstream body temperature is the reference temperature.

Mikheyev gave a single-phase liquid heat transfer coefficient relationship taking into account the heat effect of the initial section:

$$Nu_\Omega = 0.021 Re_\Omega Pr_\Omega^{0.43} \left(\frac{Pr_\Omega}{Pr_{CT}}\right)^{0.25} \varepsilon_1 \qquad (6.20)$$

where ε_1 is the initial heat effect coefficient and the subscript "CT" refers to the physical properties of coolant at the liquid wall temperature.

The Nusselt equation, which is similar to the Mikheyev relationship, can also be applied as:

$$Nu_\Omega = 0.023 Re_\Omega^{0.8} Pr_\Omega^{0.4} \qquad (6.21)$$

In the above empirical relationships, the subscript "Ω" refers to coolant properties at the average temperature of the liquid.

From the above relationships, we can obtain the heat transfer coefficient:

$$h_\Omega = 0.023 \frac{(\rho\omega)_\Omega^{0.8}}{d^{0.2}} k_\Omega \qquad (6.22)$$

where d is equivalent diameter of the cooling channel and $\dot{\omega}$ is the flow rate.

The properties of single-phase gas convection heat and single-phase liquid convective heat are the same. A cryogenic propellant such as liquid hydrogen will change into single-phase supercritical cryogenic hydrogen very quickly when it enters the cooling channel. Since the temperature of hydrogen is too low, and the ratio of wall temperature (T_w) to fluid temperature (T_b) is very high, correction of the coefficients related to wall temperature is important as the above equation is not very accurate. The following shows an appropriate method of correction [10], where each thermal physical parameter takes points that are the mean of those from the mainstream temperature to the wall temperature. Calculation is then made according to the following criterion:

$$Nu_{int} = 0.023 Re_{int}^{0.8} Pr_{int}^{0.4} \tag{6.23}$$

where the subscript "int" refers to mean values. This method is appropriate, but the calculation is very large, so the following correction, which is easier, can be applied:

$$Nu_f = 0.020 Re_f^{0.8} Pr_f^{0.4} (1 + 0.01457 \nu_w / \nu_b) \tag{6.24}$$

where, ν_w is the fluid kinematic viscosity whose reference temperature is wall temperature; ν_b is the kinematic viscosity at the main fluid temperature; the subscript "f" indicates that the film temperature is the reference temperature.

Meanwhile, some scholars [11,12] gave a criterion that includes all kinds of correction factors, because of the large errors involved:

$$Nu_f = 0.062 Re_f^{0.7} Pr_f^{0.4} \phi_T \phi_c \phi_r \tag{6.25}$$

where φ_T is temperature ratio and entrance size correction factor, $\varphi_T = 1 + (x/d)^{-0.7} (T_w/T_b)^{0.1}$; x is the distance from the inlet of the coolant; φ_c is a curvature correction factor (only for concave structures such as the throat, and not for convex ones such as a convergent section):

$$\varphi_c = I^{0.02} \left[1 + \sin \left(\pi \sqrt{\frac{x_c}{L_c + 15d}} \right) \right]$$

where:

$I = Re (d/2R_c)^2$, where R_c is the curvature radius of the corners of coolant channel;
x_c is the axial distance from the starting point of bending;
L_c is the total length of bent section;
φ_r is a correction factor of surface roughness:

$$\varphi_r = \frac{1 + 1.5 Pr^{-1/6} Re^{-1/8} (Pr - 1)}{1 + 1.5 Pr^{-1/6} Re^{-1/8} (Pr f_r / f_s - 1)} \cdot \frac{f_r}{f_s}$$

where f_r and f_s are, respectively, the friction coefficient on a rough surface and a standard surface.

6.2 Heat Conduction Model of Combustor Wall

6.2.1 Fourier Heat Conduction Law

Heat transfer from one part of an object to another only by thermal motion of molecules, atoms, and free electrons without any relative displacement between each part of an objectis called heat conduction. In 1822, the French scientist Fourier established a basic law of heat conduction, based on the quasi-equilibrium assumption, called the Fourier law [13],

For isotropic objects, heat transfer always takes place in the direction perpendicular to the isothermal surface (i.e., normal direction), thus:

$$Q = -\lambda A \frac{\partial T}{\partial n} \tag{6.26}$$

where A is the area, λ is the thermal conduction coefficient (also called thermal conductivity). Generally, thermal conductivity is not constant. It is related to the structure, temperature, pressure, and direction of an object.

By taking one micro-unit in the heat conductor and analyzing it according to the conservation of energy, we can infer a general form of heat-conduction differential equations:

$$\rho c \frac{\partial T}{\partial \tau} = \frac{\partial}{\partial x}\left(\lambda \frac{\partial T}{\partial x}\right) + \frac{\partial}{\partial y}\left(\lambda \frac{\partial T}{\partial y}\right) + \frac{\partial}{\partial z}\left(\lambda \frac{\partial T}{\partial z}\right) + Q' \tag{6.27}$$

where Q' is the heat in a unit volume. In particular, the thermal conduction coefficient λ, specific heat capacity c, and density ρ are constant; the equation can be simplified to:

$$\frac{\partial T}{\partial \tau} = \alpha \left(\frac{\partial^2 T}{\partial x^2} + \frac{\partial^2 T}{\partial y^2} + \frac{\partial^2 T}{\partial z^2}\right) + \frac{Q'}{\rho c} \tag{6.28}$$

That is:

$$\frac{\partial T}{\partial \tau} = \alpha \nabla^2 T + \frac{Q'}{\rho c} \tag{6.29}$$

where, $\alpha = \lambda/\rho c$ is the thermal diffusivity (thermal diffusivity) $(m^2\,s^{-1})$, ∇^2 is the Laplacian, and τ is time.

6.2.2 1D Steady Heat Conduction

6.2.2.1 1D Steady Heat Conduction Analysis

The simplest heat conduction is 1D steady heat conduction whose thermal conductivity is constant. To help in our analysis, we suppose there is no internal heat source for the 1D heat conduction and the heat conduction differential equation is simplified to:

$$\frac{d^2 T}{dx^2} = 0 \tag{6.30}$$

The thermal boundary conditions for a thickness δ is:

$$x = 0, \quad T = T_{w1}$$
$$x = \delta, \quad T = T_{w2}$$

By integration of the simplified equation, we can work out the general solution of this equation, i.e.:

$$T = c_1 x + c_2 \tag{6.31}$$

where c_1 and c_2 are integration constants. Substituting into the boundary conditions:

$$c_2 = T_{w1}$$
$$c_1 = -\frac{T_{w1} - T_{w2}}{\delta}$$

We can now substitute c_1 and c_2:

$$T(x) = T_{w1} - \frac{T_{w1} - T_{w2}}{\delta} x \tag{6.32}$$

The heat flux density can then be found applying the Fourier heat conduction law:

$$q = -\lambda \frac{dT}{dx} = \lambda \frac{T_{w1} - T_{w2}}{\delta} \tag{6.33}$$

6.2.2.2 1D Steady Heat Conduction Analysis with varying Thermal Conductivity

For a general metal, the thermal conductivity can be considered to change linearly with temperature over a very wide temperature range:

$$\lambda_T = \lambda_0 (1 + a\theta) \tag{6.34}$$

where, λ_T and λ_0 are, respectively, the thermal conductivity of materials at the temperature T and T_0; α is the thermal conductivity temperature coefficient; and θ is the temperature difference, $\theta = T - T_0$.

The heat flux density taken by heat conduction is:

$$q = -\lambda dT/dx = -\lambda_0 (1 + a\theta) d\theta/dx$$

$$q = \frac{1}{\delta} \int_0^\delta q \, dx = \frac{1}{\delta} \int_{\theta_2}^{\theta_1} -\lambda_0 (1 + a\theta) \, d\theta$$

$$= \frac{\lambda_0}{\delta} \left(1 + a \frac{\theta_1 + \theta_2}{2} \right) (\theta_2 - \theta_1) \tag{6.35}$$

$$= \lambda (T_{w2} - T_{w1})/\delta$$

where λ is the thermal conductivity of the material at the average wall temperature, $\bar{T}_w = (T_{w2} + T_{w1})/2$; δ is the wall thickness; and T_{w2} and T_{w1} are inner and outer surface temperatures, respectively. When the thermal conductivity changes linearly with temperature, we can assume that the thermal conductivity is constant, but λ_w in the equation should be the thermal conductivity of the material at the average wall temperature.

6.2.3 2D Steady Heat Conduction

Considering a steady condition in which there is heat exchange between the wall surface and gas and coolant (no internal heat source included), the equation of 2D heat conduction is:

$$\frac{\partial}{\partial x}\left(\lambda \frac{\partial T}{\partial x}\right) + \frac{\partial}{\partial y}\left(\lambda \frac{\partial T}{\partial y}\right) = 0 \tag{6.36}$$

The boundary conditions are:
gas wall:

$$\lambda \frac{\partial T}{\partial n} = -h_g\left(T_g - T_{wg}\right)$$

liquid wall:

$$\lambda \frac{\partial T}{\partial n} = -h_l(T_{wl} - T_l)$$

where $\partial T/\partial n$ is the normal temperature gradient; h_g and h_l are heat transfer coefficient of gas wall and liquid wall, respectively.

Taking into account the gas radiant heat, the gas wall heat flux (h_g) should include gas convection heat transfer (q_{cv}) and radiation heat transfer, i.e.:

$$h_g = (q_{cv} + q_r)/\left(T_g - T_{wg}\right) \tag{6.37}$$

Even more simply, if λ is a constant, this equation can be changed into the Laplace equation:

$$\frac{\partial^2 T}{\partial x^2} + \frac{\partial^2 T}{\partial y^2} = 0 \tag{6.38}$$

6.2.4 Unsteady Heat Conduction

The alterative cooling method for a thrust chamber is unsteady heat conduction. When this method applies, the thrust chamber has not reached the thermal balance, the temperature rises continuously with working time, and the longest working time of the thrust chamber depends upon the endothermic capacity of its units.

A thrust chamber that absorbs heat transferred to the wall through its heat capacity is called a thermal capacity thrust chamber or heat sink style thrust chamber. This is a kind of passive thermal protection. As the working time increases, the wall temperature rises gradually, which is unsteady heat conduction; the heat conduction equation is:

$$\frac{\partial T}{\partial \tau} = \alpha \left(\frac{\partial^2 T}{\partial x^2} + \frac{\partial^2 T}{\partial y^2} + \frac{\partial^2 T}{\partial z^2} \right) \qquad (6.39)$$

The boundary condition in the equation above is the same, and when $\tau = 0$, $T = T_{w0}$ (T_{w0} is the initial temperature of the wall).

Standard numbers that can be used to determine the unsteady conduction of the wall temperature are given by the Fourier criterion $\left(F_o = \alpha\tau/\delta^2 \right)$. The larger F_o is the higher the wall temperature. An effective method to decrease wall temperature and lengthen service life is to increase the wall thickness (δ).

6.3   Radiation Heat Transfer Model

6.3.1   Basic Law of Radiation

6.3.1.1   Basic Concept [14]

An object can emit electromagnetic waves of different wavelengths to space. Electromagnetic waves of different wavelengths have different effects: those with a wavelength within 0.38–1000 μm have thermal radiation effects; most of energy is within 0.76–20 μm of the IR section. The wavelength of visual light is 0.38–0.76 μm, so most of the thermal radiation waves cannot be seen by the naked-eye.

When radiation energy G strikes the surface of an object, part of it is absorbed when it enters the sub-surface (we call this G_α); another part is reflected, which we call it G_p (reflection includes mirror reflection and diffuse reflection), and the remaining part goes through the object, which we call it G_τ (Figure 6.1). We define the absorption rate as:

$$\alpha = \frac{G_\alpha}{G}$$

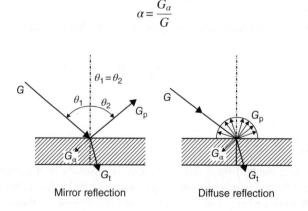

Mirror reflection   Diffuse reflection

Figure 6.1   Absorption, reflection, and transmission of radiation.

the reflection rate is:

$$\rho = \frac{G_\rho}{G}$$

and the penetration rate is:

$$\tau = \frac{G_\tau}{G}.$$

Then:

$$\alpha + \beta + \tau = 1$$

The reflection rate is zero, because gases have no reflection ability for radiation energy.

An absorption rate of $\alpha = 1$ means that all radiation energies of different wavelengths can be absorbed by the object, which is then called a blackbody. A reflection rate of $\rho = 1$ means that an object can reflect all radiation energies, which is then called a mirror body or white body. A penetration rate of $\tau = 1$ means that all radiation energies can penetrate through the object, which is then called a transparent body.

The emissive power is the sum of all radiation energies of all the wavelengths emitted in all directions to the hemisphere on a unit surface area in the unit time, i.e.:

$$E = \frac{F}{A} \tag{6.40}$$

The emissive power is in units of w m^{-2}. The term F is the sum of all radiation energies of all the wavelengths emitted in all directions to the hemisphere. If the surface of the radiation object is a blackbody, it is expressed as E_b.

The monochromatic emissive power (E_λ) is a radiation energy of a specified wavelength emitted in all directions to the hemisphere on a unit surface area in the unit time, the units of which are W m^{-2} μm, i.e.:

$$E_\lambda = \frac{dF}{d\lambda} \tag{6.41}$$

In planimetry, the size of space in a certain direction is expressed with a plane angle (in Radians). Similarly, the size of space in a certain direction can be expressed with a solid angle or infinitesimal solid angle of 3D space; their definitions are:

$$\Omega = \frac{A_c}{r^2}, \quad d\Omega = \frac{dA_c}{r^2} \tag{6.42}$$

In the spherical coordinates of Figure 6.2, r is the radius of a space ball, φ is the longitude angle, and θ is the latitude angle. We can then obtain the following equation:

$$dA_c = rd\theta \cdot r\sin\theta d\varphi \tag{6.43}$$

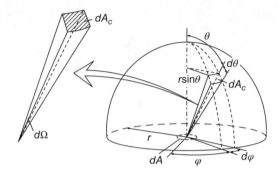

Figure 6.2 Relation between infinitesimal solid angle and hemisphere space.

Substituting into Equation (6.42), the infinitesimal solid angle is:

$$d\Omega = \sin\theta d\theta d\varphi \tag{6.44}$$

Another important concept is the directional radiation intensity, the blackbody of which can be foreseen. Because of symmetry, energies from an infinitesimal blackbody area to unit solid angles of different longitude angle directions in the space at the same latitude angle are the same. Therefore, we can study the distribution of blackbody radiation in different directions of space as long as the distribution pattern of radiation energies in different latitude angles is known. It is supposed that radiation energy from an infinitesimal blackbody area (d_A) to an infinitesimal solid angle surrounding the space latitude angle (θ) is $d\Phi(\theta)$; experimental measurement shows that:

$$\frac{d\Phi(\theta)}{dA\,d\Omega} = I\cos\theta \tag{6.45}$$

Here, I is a constant, having no relation with θ. Another form of this equation is:

$$\frac{d\Phi(\theta)}{dA\,d\Omega\cos\theta} = I \tag{6.46}$$

Here, $dA\cos\theta$ can be considered as the area in the direction of θ. This area is called the visual area (Figure 6.3). The physical quantity on the left-hand side of the equation above is the energy that has fallen into a unit solid angle in any direction of space, energy that was emitted from a unit visual area of a blackbody; this quantity is the directional radiation intensity.

The radiation heat transfer and relative position between two surfaces has a very significant relation. Usually, the percentage of radiation that falls into surface 2 as radiation energy emitted from surface 1 is the angular coefficient of surface 1 to 2, $X_{1,2}$.

First, let us look at the angular coefficient of one infinitesimal surface (dA_1) to another infinitesimal surface (dA_2) (Figure 6.4), which is expressed as $X_{d1,d2}$, where the subscripts "$d1, d2$" represent dA_1 and dA_2, respectively. According to the definition:

$$X_{d1,d2} = \frac{\text{Radiation energy fallen into } DA_2 \text{ from } DA_1}{\text{Total radiation emitted from } DA_1}$$

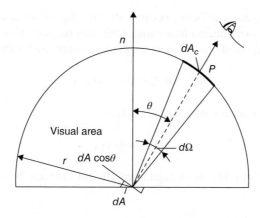

Figure 6.3   Schematic diagram of the visual area.

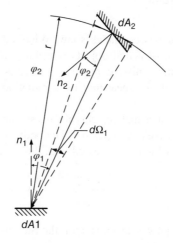

Figure 6.4   Radiation between two infinitesimal surfaces.

$$= \frac{L_{b1} \cos\varphi_1 dA_1 d\Omega_1}{E_{b1} dA_1} = \frac{dA_2 \cos\varphi_1 \cos\varphi_2}{\pi r^2} \tag{6.47}$$

Similarly:

$$X_{d2,d1} = \frac{dA_1 \cos\varphi_1 \cos\varphi_2}{\pi r^2} \tag{6.48}$$

Thus:

$$dA_1 X_{d1,d2} = dA_2 X_{d2,d1} \tag{6.49}$$

This is the equation for relativity of the angular coefficient between the two infinitesimal surfaces, showing that $X_{d1,d2}$ and $X_{d2,d1}$ are not independent—they are subject to the equation above.

The relativity of angular coefficient between the two limited surfaces A1 and A2 can be obtained by analyzing the radiation heat transfer between surfaces of two blackbodies shown in Figure 6.4. The heat transfer between two surfaces is expressed with $\Phi_{1,2}$. Then:

$$\Phi_{1,2} = A_1 E_{b1} X_{1,2} - A_2 E_{b2} X_{2,1} \tag{6.50}$$

When $T_1 = T_2$ and net radiation heat transfer is zero:

$$A_1 X_{1,2} = A_2 X_{2,1} \tag{6.51}$$

This is the equation of relativity of angular coefficient between the two limited surfaces.

6.3.1.2 Actual Object Radiation

Emissive Power
Research on a blackbody encompasses Planck's law, Wien's displacement law, and Stefan–Boltzmann's law. The radiation of actual objects is usually different from that of absolute blackbodies, the monochromatic emissive power (E_λ) varies irregularly with wavelength and temperature, and it can only be measured by a radiation spectrum test at a specified temperature.

In fact, the emissive power of an actual object is always smaller than that of a blackbody, and the ratio of an actual object emissive power (E) to a blackbody emissive power (E_b) is called the blackness of the actual object, which is expressed with the symbol ε. Therefore:

$$\varepsilon = \frac{E}{E_b} \tag{6.52}$$

where the blackbody emissive power depends upon the Stefan–Boltzmann law:

$$E_b = \sigma_0 T^4$$

where σ_0 is the Boltzmann constant (5.67×10^{-8} W m^{-2} K^{-4}).

According to the definition above, emissive power of an actual object is:

$$E = \varepsilon \sigma_0 T^4 \tag{6.53}$$

Notably, ε is not only related to the texture and surface condition of an object but also to temperature of the object, especially for metals. Consequently, simply taking it as a constant will introduce an error. In contrast, for most non-metal materials ε does not have a significant relationship with surface condition provided it is within 0.85–0.95.

Importantly, in hemisphere space the radiation intensity of an actual object is not the same as that of absolute blackbody. Instead, the radiation intensity of an actual object is significantly different in various directions, that is to say that blackness is also related with radiation direction.

Absorption Ability

Generally, the actual object absorption rate (α) depends upon two factors: (i) the characteristics of the absorbing object itself and (ii) the wavelength of radiation emitted to the object. The percentage absorption of radiation of a specified wavelength is called the monochromatic absorption rate, α_λ. Typically, α_λ varies with wavelength. For example, glass can hardly absorb visible light and infrared rays whose wavelengths are less than 2.5 μm; it can be considered to be transparent. However, the absorption rate of infrared rays whose wavelengths are more than 4 μm is nearly 1, showing that glass is not transparent. Consequently, glass allows visible light and infrared rays whose wavelengths are short to pass through it and enter, for example, a greenhouse, but prevents objects in the greenhouse emitting long-wavelength infrared rays through it to the external surroundings.

Kirchhoff's Law

Kirchhoff's Law shows the relation between actual object emissive power (E) and absorption ratio (α). This law can be used to study the export of radiation heat transfer between two surfaces. If two parallel plates are very close to each other, radiation energy from one plate can fully fall onto the other plate (Figure 6.5). If plate 1 is the surface of a blackbody, the emissive power, absorption ratio, and surface temperature are, respectively, E_b, α_b ($=1$), and T_1. If plate 2 is the surface of any object, the emissive power, absorption ratio, and surface temperature are E, α and T_2, respectively. Now, let us observe the energy balance. In a unit area of plate 2, the energy emitted during a unit time is E; this energy is absolutely absorbed when it emits to the blackbody surface. Meanwhile, the energy from blackbody surface 1 is E_b. When this energy falls on plate 2, only αE_b of it can be absorbed, while the remaining part, $(1-\alpha)E_b$, is reflected back to plate 1, and is completely absorbed by blackbody surface 1. The difference between expenditure and income of plate 2 is the heat flux density of radiation heat transfer:

$$q = E - \alpha E_b \tag{6.54}$$

When the system is under the condition $T_1 = T_2$ (i.e., the condition of thermal equilibrium), $q = 0$ and Equation 6.54 is changed into:

$$\frac{E}{\alpha} = E_b \tag{6.55}$$

Extending this relationship to any object, we can acquire the following relation:

$$\frac{E_1}{\alpha_1} = \frac{E_2}{\alpha_2} = \cdots = E_b \tag{6.56}$$

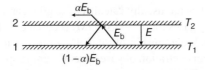

Figure 6.5   Schematic diagram of Kirchhoff's law.

We can also express this with:

$$\alpha = \frac{E}{E_b} = \varepsilon \tag{6.57}$$

These are two forms of mathematical equation of Kirchhoff's Law. Under the condition of thermal equilibrium, the ratio of radiation of any object itself to the absorption rate of emission from a blackbody to it constantly equals the emissive power of the blackbody. We can also express this simply as: under the condition of thermal equilibrium, the absorption rate of radiation from a blackbody to any object equals the emission rate of this object at the same temperature.

6.3.1.3 Radiation Characteristics of Combustor Media [15]

During numerical calculation of the combustion radiation heat transfer, the radiation characteristics of media in the combustor directly affect the final calculation of radiation heat transfer; justifiability, then, the radiation characteristic parameters of the media are one of issues that must be solved.

Because of complexity of media in the combustor (i.e., combustion products) solution of the radiation characteristics of the product is very difficult. What is more, the radiation characteristics are related to composition, volume fraction, and radiation wavelength of the combustion products, making this work more complex and difficult. An actual fuel burning issue should be solved according to the specified case. This section only simply covers the radiation characteristics of gaseous combustion products during the calculation of liquid rocket engine combustion values.

Gaseous combustion products such as steam, CO, CO_2, and so on have a very tiny scattering effect on radiation energy (gas scattering coefficient ($K_{s,g}$) approximately equals zero) and can always be omitted during the numerical calculations. However, gaseous products have a very strong selective absorption and emission for radiation, so that the variation of gas radiative properties (absorption coefficient, $K_{a,g}$) with radiation wavelength must be considered.

To calculate the absorption and emission of gaseous combustion products within a thermal radiation wavelength, the entire spectrum must be divided into many small frequency bands; we suppose that the absorption and emission of each gas component are uniform within these frequency bands or vary smoothly according to a certain functional relationship. During the engineering calculation, the spectrum absorption rate and emission rate are usually calculated by narrow-band analog or wide-band analog. During the numerical calculation of combustion radiation, a media is usually considered as a gray body, and the gas radiation characteristics have tiny effect on entire radiation heat transfer. Thus, for radiation characteristics of gas products, accurate spectral properties cannot improve the accuracy of radiation heat transfer of numerical solution, and the gas radiation characteristics are usually calculated by gray gas weighting and analogue.

The weighted sum of gray gases model (WSGGM) is based on a Hottel chart. In this chart, a curve showing the functional relationship of total emission rate, total absorption rate and gas temperature, pressure, and concentration of gases over the entire spectrum is fitted using appropriate polynomials to different temperatures and pressures; the equation is specified as the weighed sum of total emission rate, total absorption rate of a few kinds of gray gases.

For the radiative properties of the gaseous products of combustion, when WSGGM stimulation is used in numerical calculation, combusting gases are normally divided into I kinds of gray gases ($I = 2$: steam and CO_2 gas), and the emission rate (ε) is:

$$\varepsilon = \sum_{i=0}^{I} a_{\varepsilon,i}(T)[1 - \exp(-K_i p s)] \tag{6.58}$$

where $a_{\varepsilon,i}(T)$ and K_i are, respectively, the weighing coefficient and absorption coefficient of emission coefficient of i kinds of gray gas; p and T are local gas pressure and temperature, respectively; s is the length of the light path (another possibility is to take the characteristic size of calculation grids).

In calculating the combustion heat transfer, according to Hottel chart, from the composition and temperature of gas, we can infer the value of $a_{\varepsilon,i}$ and K_i together with their equations, where $a_{\varepsilon,i}$ can be extended approximately with a polynomial of temperature:

$$a_{\varepsilon,i}(T) = \sum_{j=1}^{I} b_{\varepsilon,i,j} T^{j-1} \tag{6.59}$$

where $b_{\varepsilon,i}$ is a coefficient of temperature polynomial for emissive gases; $b_{\varepsilon,i}$ and K_i are determined according experimentally.

For calculation of the combustion radiation, if all i gases have a relation of $K_i p s \ll 1$, the total emissivity of combusting gas is:

$$\varepsilon = \sum_{i=0}^{I} a_{\varepsilon,i} K_i p s$$

The absorption coefficient of gas is:

$$K_{a,g} = -\frac{\ln(1-\varepsilon)}{s} \tag{6.60}$$

6.3.2   Empirical Model of Radiation Heat Flux Density Calculation

Because the service condition of a liquid rocket engine has a high temperature and high pressure, radiation heat transfer of the engine is very strong. In a liquid oxygen/kerosene rocket engine, the strongly radiating combustion products are mainly CO_2 and H_2O; the radiation of other gases can be omitted.

When only steam and carbon dioxide are considered, the emissivity of gas can be calculated based on the following equation:

$$\varepsilon_g = C_{H_2O} \varepsilon_{H_2O}^* + C_{CO_2} \varepsilon_{CO_2}^* - \Delta\varepsilon \tag{6.61}$$

where $\varepsilon^*_{CO_2}$ and $\varepsilon^*_{H_2O}$ are determined under the ideal condition that partial pressures of carbon dioxide and steam are extrapolated to zero under the total pressure of gas $p = 10^5$ Pa. The term $\Delta\varepsilon$ is a correction brought in because of partial overlap of the radiation band of water and carbon dioxide. When the temperature is higher than 1000 K, the correction is calculated based on the following equation:

$$\Delta\varepsilon = \varepsilon_{H_2O}\varepsilon_{CO_2} \tag{6.62}$$

Gases cannot be considered as gray bodies because gas radiation is selective, thus the absorption ratio (α_g) of gases does not equal the emissivity (ε_g). For a mixed gas containing steam and carbon dioxide, the absorption ratio of shell radiation can be expressed with:

$$\alpha_g = C_{H_2O}\alpha^*_{H_2O} + C_{CO_2}\alpha^*_{CO_2} - \Delta\alpha \tag{6.63}$$

where $\alpha^*_{H_2O}$, $\alpha^*_{CO_2}$, and $\Delta\alpha$ can be determined via the following empirical equations:

$$\alpha^*_{H_2O} = \left[\varepsilon^*_{H_2O}\right]_{T_w, p_{H_2O}s(T_w/T_g)} \left(\frac{T_g}{T_w}\right)^{0.45} \tag{6.64}$$

$$\alpha^*_{CO_2} = \left[\varepsilon^*_{CO_2}\right]_{T_w, p_{CO_2}s(T_w/T_g)} \left(\frac{T_g}{T_w}\right)^{0.65} \tag{6.65}$$

$$\Delta\alpha = [\Delta\varepsilon]_{T_w} \tag{6.66}$$

where T_w is the wall temperature of gas shell; the subscript in brackets is a parameter used to determine the quantity in the brackets. Thus, the heat flux density (q) of heat transfer between gas and shell is:

$$q = \varepsilon_g E_{b,g} - \alpha_g E_{b,w} = 5.67 \left[\varepsilon_g\left(\frac{T_g}{100}\right)^4 - \alpha_g\left(\frac{T_w}{100}\right)^4\right] \tag{6.67}$$

6.3.3 Numerical Simulation of Combustion Heat Radiation

6.3.3.1 Basic Equation of flame Radiation Heat Transfer

The basic equation that is usually used in numerical simulation of combustion heat transfer is the continuum energy equation, i.e.:

$$\frac{\partial}{\partial t}(\rho\varphi) + \frac{\partial}{\partial x_i}(\rho u_i \varphi) = \frac{\partial}{\partial x_i}\left(\Gamma_\varphi \frac{\partial\varphi}{\partial x_i}\right) + S_\varphi + Q_r \tag{6.68}$$

The dependent variable φ refers to the enthalpy $C_g T$ of continuum, Γ_φ is the turbulent transport coefficient of enthalpy, and the item of heat transfer source in the equation contains the gas phase reaction heat source (S_φ), and Q_r in this equation is an item of radiation heat transfer source. In each step of discrete calculation of the continuum energy equation, the items of radiation heat transfer sources (Q_r) for each space infinitesimal are items of multiple integrals. In each calculation step in solving the discrete continuum energy equation, multiple integrals item of radiation heat transfer source (Q_r) for every infinitesimal are necessary, in order to complete closure of the combustion temperature field numerical equations.

This section covers the numerical simulation of items of radiation heat transfer source, Q_r. Because of strong flame radiation heat transfer during combustion, Q_r is usually higher than flow items and diffusion items. In the combustion heat transfer with flame, radiation heat transfer is about 90%. Meanwhile, because there is strong heat transfer from flame to surrounding walls and media, the temperature of media in the flame is very uneven. Therefore, flame radiation heat transfer significantly affects the flow and chemical reaction of combustion. For numerical calculation of the energy equation for combustion process, numerical calculation of the combustion and flame radiation heat transfer must be coupled to the energy equation of all flowing through the combustion. This is to close the energy equation of combustion, in order to carry out the numerical simulation.

Radiation heat transfer of a combusting flame is related to the temperature of media in each part of space, radiation absorption capacity, and scattering ability of media. In addition, radiation heat transfer of a combusting flame is also related to the optical properties, size, density, and radiation wavelength compositions of media. According to the fundamentals for radiation heat transfer of absorbing media and scattering media, there is monochromatic thermal radiation whose wavelength is λ and radiation intensity is I_λ. When transmitting in the s direction, the rate of variation of radiation intensity within an interval ds in this direction is:

$$\frac{dI_r(r,s)}{ds} = -(K_{a,\lambda} + K_{E,\lambda})I_\lambda(r,s) + K_{a,\lambda}I_{b,\lambda} + \frac{K_{s,\lambda}}{4\pi}\int_0^{4\pi} I_r(r,\Omega)\Phi(\Omega,\Omega')d\Omega' \qquad (6.69)$$

where:

r is the position vector of radiation intensity;
s is the direction vector of radiation intensity;
s is the transmission distance of radiation intensity;
$K_{a,\lambda}$ and $K_{s,\lambda}$ are monochromatic absorption coefficient and scattering coefficient, respectively, of media when the wavelength of radiation is λ;
$I_{b,\lambda}$ is the blackbody monochromatic radiation intensity when the wavelength is λ;
Ω is the direction of space infinitesimal solid angle in the s direction;
Ω' is the direction of space infinitesimal incident solid angle on this point (infinitesimal unit);
$\Phi(\Omega, \Omega')$ is the scattering phase function, showing the directional distribution of radiation energy in the space after scattering of incident radiation energy of infinitesimal unit;
$d\Omega'$ is the solid angle between infinitesimal unit and space infinitesimal of surrounding radiation heat transfer.

The first item on the right-hand side of Equation 6.69 is the decrease of radiation intensity I_λ due to absorption and scattering of infinitesimal unit media, the second item is the volume radiation of infinitesimal unit media, and the third item is the increase of radiation intensity in the *s* direction due to the scattering intensity in all directions after the radiation intensity from surrounding space to infinitesimal unit scattering to all directions.

For numerical simulation of the combustion radiation heat transfer, it is extremely difficult to calculate the radiation intensity on all infinitesimal units accurately, because the compositions of combusting gas media are complex. In numerical calculation of engineering situations, media in the combustor are usually considered as gray bodies, and the absorption coefficients and scattering coefficients of all triatomic gases in infinitesimal unit media are weighted accumulated to infer a total absorption coefficient (K_a) and scattering coefficient (K_s) of flame media; by integrating the radiation wavelength λ from zero to infinity, the above equation can be simplified to:

$$\frac{\mathrm{d}I}{\mathrm{d}s} = -(K_a + K_s)I + K_a I_b + \frac{K_s}{4\pi}\int_0^{4\pi} I\phi(\Omega, \Omega'')\mathrm{d}\Omega' \tag{6.70}$$

where I is the total radiation intensity of an infinitesimal unit in the *s* direction, which is the function of position *r* and radiation direction *s*. The total radiation intensity of a blackbody is:

$$I_b = \frac{n^2\sigma T_g^4}{\pi} = \frac{E_b}{\pi} \tag{6.71}$$

where E_b is the emissive power of the blackbody, and the above two equations are basic equations of transmission of combustion radiation energy.

To obtain a value of radiation heat transfer rate (Q_r) of each infinitesimal interested in the entire numerical calculation in the combustor, it is necessary to calculate the integration of radiative energy transfer equation in 0–4π of Ω' in all directions of space to infer the radiation intensity of each infinitesimal. Thus, the relationship between Q_r and infinitesimal radiation intensity I can be worked out. On the other hand, the radiation heat flux density on each infinitesimal interface can be worked out directly from radiative the energy transfer equation by a certain simplification, in order to infer the relationship between Q_r and radiation heat flux density. In addition, the Q_r of each infinitesimal in flow field can be statistically calculated directly. These different means form all kinds of simulating methods of combustion radiation heat transfer covered in the next section.

6.3.3.2 Simulating Methods for Combustion Radiation Heat Transfer

Typical simulating methods for simulating the combustion radiation heat transfer are the heat flow method, zoning method, Monte Carlo method, discrete transfer (DT) method, and discrete ordinate (DO) method [16].

Heat Flow Method

The heat flow method simplifies the complex hemisphere thermal radiation into uniform radiation heat q^+ and q^- perpendicular to the interface, and infers the simplified radiation heat differential equations on this basis. Then discretization is applied to work out the radiation heat transfer rate (Q_r) of each infinitesimal.

The basis of this method is that via radiation heat of an infinitesimal unit in all directions on each interface this interface can be hemisphere integrated; then the hemispherical emissive power (q) in infinitesimal area of this interface can be inferred. The term q is simplified into uniform radiation heat perpendicular to the interface. For example, radiation heat in the x direction on interface of infinitesimal unit is defined as:

$$q_x^+ = \int_0^{2\pi} I \cos\theta d\Omega \qquad (6.72)$$

To avoid integration, we define q_x^+ as a variable along the x direction, and the equation can be written into a transfer equation of radiation heat transfer energy along the x direction:

$$\frac{dq_x^+}{dx} = -(K_a + K_s)q_x^+ + K_a E_b - \frac{K_s}{2}\left(|q_x^+| + |q_x^-|\right) \qquad (6.73)$$

Similarly, differential equation of radiation heat in the opposite direction of x is:

$$\frac{dq_x^-}{dx} = (K_a + K_s)q_x^- - K_a E_b - \frac{K_s}{2}\left(|q_x^+| + |q_x^-|\right) \qquad (6.74)$$

According to the simplifying above we take the mean of q_x^+ and q_x^- as dependent variable:

$$q_x = \frac{1}{2}\left(|q_x^+| + |q_x^-|\right)$$

of radiation heat differential equations. Subtracting Equations 6.73 and 6.74, the differential equation of radiation heat will be inferred:

$$\frac{d}{dx}\left(\frac{1}{K_a + K_s} \times \frac{dq_x}{dx}\right) = K_a(q_x - E_b) \qquad (6.75)$$

The radiation heat transfer rate ($Q_{r,x}$) of a unit volume in the interval dx is:

$$Q_{r,x} = -\left(\frac{dq_x^!}{dx} - \frac{dq_x}{dx}\right) = -2K_a(q_x - E_b) \qquad (6.76)$$

By similar simplification in three directions of an infinitesimal unit in 3D space the equation of radiation heat differential equation and radiation heat exchange rate (Q_r) of a unit volume of infinitesimal unit can be inferred; consequently, it will not be introduced in detail.

The heat flow method is very easy to understand. One of the most obvious characteristics is that the grids are the same as those of finite difference or finite element methods, and the radiation heat differential equations inferred have the same form as the general differential equations. Thus, it is very convenient to solve simultaneous equations with the vapor momentum equation and energy equation, and so on, which requires less calculation time.

However, the disadvantage of this method is that very large errors are introduced by simplifying the complex hemisphere thermal radiation on an interface of infinitesimal unit into uniform radiation heat perpendicular to an infinitesimal unit interface and solving with general transmission differential equations. In fact, on each infinitesimal unit interface there is not only radiation heat transfer perpendicular to this interface but also radiation heat transfer in other directions. Then, radiation integration from space is only simplified into radiation heat along three-dimensional coordinate axes without considering the space distribution and relationship of radiation intensities. This results in no relationship among radiation heats in each direction, so the equation is unreal physically. Meanwhile, non-isotropic scattering materials cannot be calculated with this method. In brief, the accuracy of this calculation is poor.

Zoning Method

The calculation of radiation heat transfer rate with the zoning method is based on the theory that radiation heat transfers from each area directly with the surrounding space. To apply this method, the space of the combustor is divided into solid areas, and the walls of combustor are divided into surface areas (Figure 6.6). We suppose that the temperature and physical property parameters are uniformly distributed in each solid area and surface areas, and write energy balance equation in each sub-field to get a set of equations with unknowns of heat fluids or temperatures and solve the numerical solution.

Now, we will briefly explain a calculation method for any solid area infinitesimal (V_j) and radiation heat transfers of all surrounding areas in the combustor. Suppose that the radiation heat transfer of a unit volume in a small area is $Q_{r,j}$ and the radiation heat of all solid areas gained by the V_j is expressed as $Q_{V \rightarrow V_j}$, and $V = \sum V_i$, $A = \sum A_i$. The radiation heat sent out from V_j is Q_{V_j}, so the radiation heat exchange rate sent out from V_j of this solid area is:

$$V_j Q_{r,j} = Q_{V \rightarrow V_j} + Q_{A \rightarrow V_j} - Q_{V_j} = \sum_i Q_{V_i \rightarrow V_j} + \sum_i Q_{A_i \rightarrow V_j} - Q_{V_j} \tag{6.77}$$

Here, we can suppose that the temperature and physical property parameters in each solid area and surface area are uniform, then the radiation heat transfer (Q_{Vi}) of a unit volume in a small area to surrounding space V_i is:

$$Q_{V_i} \approx \oint_{V_i} 4K_a \sigma T_{gi}^4 dV_i = 4K_{ai} \sigma T_{gi}^4 V_i \tag{6.78}$$

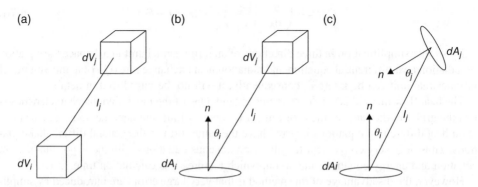

Figure 6.6   Schematic diagram of calculation of zoning method: (a) Solid area to solid area; (b) solid area to surface area; (c) surface area to surface area.

For convenience, we also suppose that media scattering in the combustor is so small that they can be omitted. Thus, during integration along the path from one solid area (V_i) to another solid (V_j), the radiation energy share to the distance r and absorbed by intermediate media can be simply written as $1 - e^{-K_a r}$ in lieu of actual:

$$1 - e^{-\int_0^r K_a \, dr}$$

The remaining share can be written as $e^{-K_a r}$; K_a is the absorption coefficient of media along the path. Then, we can get the direct radiation heat $Q_{V_i \to V_j}$ from any solid area V_i to another solid area V_j:

$$Q_{V_i \to V_j} = \sigma T_{gi}^4 \int_{V_i} \int_{V_j} \frac{K_{ai} K_{aj}}{\pi r^2} e^{-K_a r} \, dV_i dV_j \tag{6.79}$$

The multiple integral in this equation is the "direct radiation exchange area" from one solid area (V_i) to another solid area (V_j).

Similarly, we can obtain the direct radiation heat ($Q_{A_i \to V_j}$) from any surface area (A_i) to another solid area (V_j) in the combustor:

$$Q_{A_i \to V_j} = \varepsilon_{si} \sigma T_{si}^4 \int_{A_i} \int_{V_j} \frac{K_{aj}}{\pi r^2} e^{-K_a r} \cos\eta \, dV_j dA_i \tag{6.80}$$

where, ε_{si} is wall emissivity of surface area (A_i); η is the angle between the ray from one point in the area to infinitesimal volume (dV_j) and normal of infinitesimal wall. Thus, multiple integration in this equation is the "area of direction radiation exchange" from area (A_i) to solid area (V_j) or "area of directional radiation exchange". Based on this equation $Q_{V_i \to V_j}$ is the energy of V_i directional radiation that is absorbed by V_j. In fact, we can also calculate V_j to obtain the energy reflected to other areas from V_i. Then the part absorbed by V_j after radiation emission from this small area to another area is reflected to V_j is added, and "area of total directional radiation exchange" is formed.

Similarly, the direct radiation heat ($Q_{A_i \to A_j}$) from surface area A_i to surface area A_j can also be inferred. According to the derivation of angular coefficient in Section 6.3.1, the following equation can be inferred:

$$Q_{A_i \to A_j} = \varepsilon_{si} \varepsilon_{sj} \sigma T_{si}^4 \int_{A_i} \int_{A_j} \frac{\cos\theta_i \cos\theta_j}{\pi r^2} e^{-K_a r} \, dA_i dA_j \tag{6.81}$$

It is defined that "area of total directional radiation exchange" from solid area V_i to solid area V_j in the combustor is expressed as the symbol $G_i G_j$, surface area A_i to solid area V_j is $S_i G_j$, solid area V_i to surface area A_j is $G_i S_j$, surface area A_i to surface area A_j is $S_i S_j$. The $G_i G_j$, $S_i G_j$, $G_i S_j$, and $S_i S_j$ can all be inferred by each multiple integration.

Consequently, we can write the radiation heat transfer rate (Q_r) of this solid area V_j as $Q_r = V_i Q_{r,j}$, and:

$$V_i Q_{r,j} = \sum_i G_i G_j \sigma T_{gi}^4 + \sum_i S_i G_j \sigma T_{gi}^4 + 4K_{ai}\sigma T_{gi}^4 V \tag{6.82}$$

The calculation method of radiation heat transfer between any surface area A_j and all surrounding areas in the combustor is: Suppose the radiation heat transfer of a unit area in an area is $Q_{r,j}$, the radiation heat of all surface areas obtained by them is expressed as $Q_{A \to Aj}$. Radiation heat from this surface area is Q_{Aj}, thus the radiation heat transfer rate from surface area A_j is:

$$A_j Q_{r,j} = Q_{A \to A_j} + Q_{V \to A_j} - Q_{A_j} = \sum_i Q_{A_i \to A_j} + \sum_i Q_{V_i \to A_j} - Q_{A_j} \tag{6.83}$$

Then we can derive by the same method as above:

$$A_i Q_{r,j} = \sum_i S_i S_j \sigma T_{gi}^4 + \sum_i S_i G_j \sigma T_{gi}^4 + 4K_{ai}\sigma T_{gi}^4 A \tag{6.84}$$

We can then obtain an expression of radiation heat transfer rate in each area of combustion flow field in the combustor, including wall boundary, and then solve the solution simultaneously with the energy equation.

In conclusion, it is preferable in principle to the calculate radiation heat transfer with the zoning method. When the temperature in infinitesimal area of calculation domain is uniform and there are not too many zones, this method has very high accuracy; the typical calculation result is usually used as a reference by which to check the accuracy of other simulation methods of radiation heat transfer.

In the calculation, radiation transmission is changed into solving a set of nonlinear algebraic equations. If the simulation domain in the whole combustor is divided into $M_v + M_s$ solid area and surface area, there are $M_v + M_s$ equations in this set of equations. Meanwhile, "directional radiation exchange area" between each two areas should be calculated and mutual reflection is taken into account prior to solving this set of equations. If we solve a set of linear equations about the "area of total directional radiation exchange", many multiple integration calculations are needed, which is quite time-consuming; to finally solve nonlinear equations of radiation transmission will need an extremely large calculation memory and long calculation time. Meanwhile, to reduce amount of calculation, only very limited calculation zones can be divided in the combustor, but each zone is in fact quite large and temperatures in each zone is quite uneven, which produces errors; but the absorption coefficient (K_a) will show very large variation, making it is more difficult to calculate multiple integration if the difference between each solid area temperature and physical property parameter is too large. Consequently, too large an error will appear when applying method above. On the other hand, the grids needed for the zoning method do not fit grids that are used to solve other equations with finite difference or finite element methods, so the calculation becomes more difficult. In addition, this method is also not suitable for solving radiation with a complex boundary surface. Because of above difficulties, this method has not been used widely in engineering calculations of combustor radiation heat transfer so far.

Other Methods

To ensure that the essence of radiation heat transfer from infinitesimal unit to surrounding directions can be interpreted, and no complex multiple integration is needed, the Monte Carlo (MC) method divides radiation energy from each infinitesimal unit to surrounding space into aliquots by spatial angle; this addresses the respective disadvantages of the zoning and heat flow methods. Therefore, the spatial angle of each aliquot emits an energy beam. By tracking energy beams from all infinitesimal units in the combustor, the radiation heat transfer rate of each infinitesimal unit can finally be determined using the probability of absorption of radiation energy. With this method, direct multiple integration with multiple summation is avoided, and the distribution and relations of thermal radiations in the hemisphere is taken into account.

The Monte Carlo method overcomes major disadvantages of the zoning and heat flow methods, and can adapt the radiation of objects with complex configuration. However, the convergence is poor for numerical calculation, and so calculation requires too much time.

The main idea behind the discrete transfer (DT) method is to consider the boundary grid surface as the absorption resource and emission resource of radiation, disperse radiation energy from boundary grid surface to hemisphere into finite energy beams, track the characteristic ray of each discrete energy beam, and solve the radiation transfer equation along characteristic rays. After these energy beams have gone through the internal grids and been absorbed and scattered by media, they are finally absorbed when they reach another boundary surface; in this way, the radiation energies entering and going out on each boundary surface are balanced. After solving radiation heat source, the radiation heat transfer rate (Q_r) of infinitesimal unit on each grid can be acquired, so that the energy equation of flow field numerical calculation in the overall combustion can be closed, and we can solve the numerical solution of all sets of the equations.

Generally speaking, discrete transfer method is in fact an improvement and simplification of the Monte Carlo method, which changes the random sampling decision of energy beams emission direction on infinitesimal into a three-dimensional distribution. On the other hand, this method tracks and calculates the absorption of energy beam infinitesimal by infinitesimal between the two boundary surfaces; this is not the same as Monte Carlo method, which needs to track until the calculation beam energy on a certain position of probable absorption is suddenly absorbed and to consider reflection on the walls after energy beams are gradually absorbed. Therefore, the discrete transfer method overcomes disadvantages of the Monte Carlo method, such as statistical error and the costly calculation time. Increasing the number of discrete directions can accordingly improve the calculation accuracy. Although calculation time is increased, the necessary computer memory will not be increased. The discrete transfer method needs a smaller computer memory than the zoning method, and the DT method can calculate radiation whose boundary configuration is complex.

However, the discrete transfer method simplifies radiation energies in each solid angle of tiny space into rays, which results in an incomplete balance between wall radiation heat and space radiation heat source; consequently, an abnormal result will probably be acquired, which is called the radiation effects of a radiation heat transfer calculation. Meanwhile, this method cannot be used for multi-dimensional radiation with scatterings well, because this method cannot deal with incident scattering item. Besides, for numerical calculation of radiation heat transfer using this method, just like the pseudo-proliferation in calculation fluid dynamics, different degrees of "false scatterings" usually exist.

The main idea behind the discrete ordinate (DO) method is to disperse the radiation transmission equation into radiation transmission equations in several directions for solution by numerical calculation, to acquire the radiation intensity (I) of each infinitesimal unit. Then, calculate Q_r of each infinitesimal unit that is considered as a radiation source item of the energy equation and close overall set of flow field equations.

Research results of the discrete ordinate method show that this method can calculate radiation with scatterings as well as other existing methods and is very easy to associate with flow equation and solve the solution because the incident scattering item can be conveniently dealt with. Therefore, in system simulation of scattering media, the discrete ordinate method may be a very promising calculation simulation of radiation heat transfer.

Compared with the discrete transfer method, this method deals with radiation heat transfer of multi-dimensional scattering media system better. However, because of the discreteness of the radiation transmission equation on spatial solid angle, for radiation intensity or heat flux density, there is some degree of inconsistency with the true distribution physically. In other words, there are still radiation effects in the calculation. There is some degree of "false scatterings" in numerical calculation of radiation heat transfer using this method.

References

[1] Liu G Q, *The theory of the liquid rocket engine*, Beijing Chinese Aerospace Press, 2005 (in Chinese).
[2] A. Kumakawa, F. Ono and N. Yatsuyanagi (1995) Combustion and Heat Transfer of LO2/HC/Hydrogen Tripropellant. AIAA paper 95–2501.
[3] D. Linne, M.L. Meyer, and D.C. Braun (2000) Investigation of Instabilities and Heat Transfer Phenomena in Supercritical Fuels at High Heat Flux and Temperatures. National Aeronautics and Space Administration Glenn Research Center. AIAA paper 2000–3128.
[4] M. Popp, G. Schmidt (1994) Heat Transfer Investigation for High Pressure Rocket Combustion Chambers. American Institute of Aeronautics and Astronautics. AIAA paper 94–3102.
[5] A.Y. Chen and L. Dang (2002) Characterization of Supercritical JP-7's Heat Transfer and Coking Properties. American Institute of Aeronautics and Astronautics. AIAA paper 2002–0005.
[6] Bartz D.R. (1957) A simple equation for rapid estimation of rocket nozzle convective heat transfer coefficients, *Jet Propulsion*, 27(1): 49–51.
[7] Levlev, B.M. (1953), *Liquid Propellant Engines*, Oborongiz, Moscow.
[8] Frolov, P.F. (1955) *Liquid Propellant Engines*, The Defense Industry.
[9] P. Couto (2002) Analysis of Supercritical Start-Up Limitations for Cryogenic Heat Pipes with Parasitic Heat Loads. American Institute of Aeronautics and Astronautics. AIAA paper 2002–3095.
[10] Tao W Q. (2009) *The multi scale numerical simulation of the flow and heat transfer problems: Method and Application*, Beijing Science Press (in Chinese).
[11] Niu L, Cheng H E, Li M H. (2002) The influence of the ratio of height to width and the roughness to the flow of the regenerative cooling channel. *Journal of Shanghai Traffic University*. 36(11), 1612–1615 (in Chinese).
[12] B. Hitch (1998) Enhancement of Heat Transfer and Elimination of Flow Oscillations in Supercritical Fuels. American Institute of Aeronautics and Astronautics. AIAA paper 98–3759.
[13] Zhang J Z (2009) *Advanced Heat Transfer*, Beijing: Science Press (in Chinese).
[14] Dan H P, *The Numerical Calculation of the Infrared Radiation Characteristics and transmission*, Press of Harbin Institute of Technology, 2006 (in Chinese).
[15] Xu X C, *Combustion Technology Handbook*, Beijing: Press of Chemical Industry, 2008 (in Chinese).
[16] Yang T, *The Combustion Theory of the Rocket Engine*, Changsha: Press of National University of Defense Technology, 2008 (in Chinese).

7

The Model of Combustion Instability

In the combustion chamber of liquid rocket engines there exists intense exothermic reactions between fuel and oxidizer, producing high temperature and high pressure gas. The Laval nozzle converts the thermal energy of the gas into kinetic energy, generating counterforce to push the rocket forward. However, in many cases the combustion chamber does not work as expected—high-frequency oscillations of the combustion chamber pressure occur. The amplitude of such oscillations may reach 10–1000% of the steady-state combustion chamber pressure, and the oscillation frequency varies from hundreds of Hz to 15,000 Hz or higher. This phenomenon is called combustion instability.

Almost every rocket engine has experienced combustion instability problems during the development process. These problems seriously undermine the work of the engine and rocket systems, prompting us to explore the phenomenon of combustion instability. Combustion instability is the oscillatory combustion phenomenon caused by the coupling of combustion processes and system fluid-dynamic processes or acoustic oscillations, accompanied by periodic oscillation of the gas pressure, temperature, and velocity. Only when the damping process occurring in the system is strong enough, so that the dissipation of the oscillation energy is faster than the energy provided by the combustion, will the oscillations decay.

7.1 Overview

In the early 1940s, the Jet Propulsion Laboratory in California discovered rocket engine combustion instability. The development process of almost all hydrogen-oxygen rocket engines such as the RL-10, J-2, SSME, in the United States had encountered the problem of combustion instability [1]. In the late 1950s, the United States began to develop the Saturn-V launch vehicle. Until the early 1970s, more than 2000 full-size simulation tests were carried out in

Internal Combustion Processes of Liquid Rocket Engines: Modeling and Numerical Simulations,
First Edition. Zhen-Guo Wang.
© 2016 National Defense Industry Press. Published 2016 by John Wiley & Sons Singapore Pte Ltd.

the trial process of its main engine F-1 (LOX/kerosene) to solve the problems of combustion instability that troubled designers [2, 3]. In the 1980s, the Arianna launch vehicle of the European Space Agency (ESA) also encountered launch failure due to high-frequency combustion instability appearing in the first-stage Viking engine. To strengthen the understanding of engine combustion instability, the ESA completed a comprehensive research project to study the influences of propellant atomization, combustion, and gas dynamics on the combustion instability. The former Soviet Union also experienced combustion instability problems in the development process of liquid rocket engines [4]. China has also encountered combustion instability problems in the development process of the Long March series of rockets [5].

Combustion instability is a special kind of unstable combustion phenomenon. Due to the complexity of the combustion process, the early combustion instability theoretical model is relatively simple. It can only provide some conclusions on the mechanism and cannot be used for engineering analysis; examples are the Crocco–Cheng sensitive time lag model [6] and Priem–Habiballah numerical analysis model [7]. More researchers have relied on experiments to solve the combustion instability problems through a lot of tests to repeatedly modify the design. For example, it may take numerous combustion tests to determine the shape and location of the partitions and tune and other issues. The test work is heavy and expensive, consuming a large amount of manpower and material resources, and sometimes still cannot identify the potential factors affecting the combustion instability. Along with the development of the combustion instability research project of ESA in the 1980s, there has been new progresses in theoretical and experimental fields. The great improvement seen in computer technology makes it possible to analyze combustion instability by numerical methods. In recent years, many published works also show that numerical analysis is the main direction of today's combustion instability study. In China, Zhuang Fengchen academician, Liu Weidong, Zhao Wentao, Nie Wansheng, Zhang Baojiong, Hong Xin, and Huang Yuhui and so on have carried out research work on combustion instability [8–31].

7.1.1 Behavior of Combustion Instability

Combustion instability has different manifestations. Since the pressure measurement is very convenient, the periodic change of pressure is frequently used to characterize combustion instability. The position of pressure measurement can be located in the combustion chamber or in the propellant feed system. It is notable that the pressure oscillation caused by combustion instability is different from the normal pressure fluctuation caused by steady-state combustion. There are two differences between them: (i) When combustion instability occurs, the chamber pressure oscillations are obviously periodic, and the oscillation energy is concentrated on a few inherent frequencies, and there is a connection between the oscillations at different positions in the combustion chamber. In the normal steady-state combustion, although there are often certain degrees of pressure fluctuations, the fluctuations are usually random and the oscillations at different positions are unrelated, and the oscillation energy is dispersed; the total effect of these fluctuations tends to zero during a certain time interval. (ii) When combustion instability occurs, the amplitude of the chamber pressure oscillation is large, usually greater than 5% of the average chamber pressure, and sometimes it may be up to several tens of percent or more. The amplitude of random perturbations in steady-state combustion is often small.

Monitoring of the temperature and heat transfer can also successfully demonstrate the occurrence of combustion instability. A thermocouple buried in the wall can record the rapid rise of wall temperature.

Combustion instability also leads to oscillations of the axial position of the Mach diamond region in the exhaust plume, which can be measured by high-speed photography. Oscillations of the Mach diamond region are usually consistent with the oscillations of chamber pressure in frequency. Sometimes optical technology is used to monitor the luminosity changes of exhaust plumes. However, it may be very weak. It is estimated that the response amplitude of the optical oscillations is only 0.1% of that of chamber pressure oscillations. Measurement results show that the periodic change of flow or thrust is also a sign of combustion instability.

7.1.2   Classification of Combustion Instability

Typically, the excitation mechanisms of various combustion instabilities are different. Accordingly, different methods are required to control or eliminate the combustion instability. Historically, combustion instabilities are classified according to their frequency range, but between the so-called low frequency, intermediate frequency, and high frequency there is no clear borderline. Combustion instabilities classified only according to the frequency will cause much confusion. A better classification method is to connect the combustion instability with its effects, the most important coupling mechanism, and the eliminating device.

7.1.2.1   Low-Frequency Instability

The frequency of low-frequency combustion instability is usually below 200 Hz; it is mainly caused by coupling of the combustion process in the combustion chamber and the flow process of the propellant feed system, and it is usually related to the ignition quality and injection speed of the propellant entering the combustion chamber. Ignition quality includes ignition delay time, flame propagation speed, and flame stability characteristics. The combustion chamber, scale of the propellant pipeline, and the flow rate and mixing ratio of the propellant have a key role in low-frequency oscillations. Coupling of the combustion process and the injector structure can also cause low-frequency instability; the injector may work like a diaphragm and produce an "Oiler"-type oscillation, causing inhomogeneous propellant injection and atomization, resulting in low-frequency instability. Some other situations can also result in coupling between the combustion (or chamber pressure) and structure system and cause low-frequency instability. For example, the perturbation of chamber pressure makes the cooling jacket bend, causing pressure oscillations of the propellant contained in the cooling jacket. This coupling can lead to low-frequency instability.

When low-frequency combustion instability occurs, the wavelength of the gas oscillation is usually much larger than the characteristic length of the chamber or the supply system. Therefore, it can be considered that the pressure oscillation of combustion chamber is uniformly distributed at any instant, and it can be seen as the oscillations of the whole gas field in the combustion chamber. Meanwhile, the pipeline of the propellant supply system or liquid collection chamber also exhibits oscillations. This instability is often a sine wave with low amplitude at the beginning, which then develops linearly into a higher amplitude.

Of the different types of combustion instabilities, low-frequency instability is probably the easiest one to deal with from a viewpoint of theoretical and experimental analysis or development. From the standpoint of theoretical analysis, the combustion chamber can be simulated by using a concentrated volume element, and the combustion is represented by a simple constant time delay—the resistance of propellant supply system is neglected, although the inertia and capacity of the supply system may be important in the analysis. The combustion time delay is defined as the time required for the liquid propellant to be completely vaporized and consumed. An experiential average value can often be obtained for each propellant. The time delay usually referred to is the flight time of the component with the worst volatility from the injector surface to the impinging point. Because it is a major part of the total time delay, methods used to eliminate low-frequency instabilities include increasing the injector pressure drop, increasing fluid inertia, as well as reducing the volume of the combustion chamber, and so on. Among the approaches used to change the time delay, some are successful, but some are problematic since they may degrade system performance or cause high-frequency instability though they can successfully eliminate low-frequency instability.

7.1.2.2 High-Frequency Instability

High-frequency instability is a result of combustion processes coupled with the combustor acoustic oscillations—it is also known as a resonant combustion or acoustic instability, and the oscillation frequency is usually above 1000 HZ. When high-frequency combustion instability occurs, for the measured dynamic pressure in the combustion chamber at different locations, the relationship between the oscillation frequency and the phase of each point is often consistent with the natural modes of acoustic modes of the combustion chamber. Thus, according to the acoustic characteristics of the combustion chamber, high-frequency instability can be divided into axial (longitudinal) or horizontal (radial and tangential) mode. The above various modes of high-frequency combustion instability can be divided according to their order of resonance into the first-order vibration mode, second-order vibration mode, and so on, such as first-order radial vibration mode, second-order longitudinal vibration mode, and third-order tangential vibration mode.

Concerning the mechanism of high-frequency instability, current points of view include ignition time lag, sensitive chemical time lag, physical time lag, detonation process, the changes of chemical reaction rate caused by the fluctuations of pressure or temperature, the "explosion" when the droplets are heated to beyond the critical temperature and critical pressure, and the jet flow, liquid fan or the crushing and mixing of liquid droplets [1].

To maintain high-frequency instability there must first be an oscillating energy, the energy to maintain the high-frequency instability of liquid rocket engine comes from the combustion of the propellant; secondly, the oscillation energy must be added at an appropriate time phase related to oscillating pressure. Therefore, methods to eliminate high-frequency instability usually fall into two categories: (i) change the propellant spray combustion field or pressure wave characteristics, so that the energy released by the combustion fluctuations is less than the oscillation energy required to maintain oscillation, such as baffle devices; (ii) change the dynamic energy loss or damping, making it greater than the energy obtained from combustion response, such as with various different types of damping devices.

7.1.2.3 Intermediate-Frequency Instability

Intermediate-frequency combustion instability is the oscillation caused by coupling between the combustion process in the combustion chamber and a portion of flow processes of the propellant supply system. The frequency range is usually 200–1000 Hz, lying between high and low frequency oscillations.

When intermediate-frequency combustion instability occurs, it is often accompanied by a gradually increased combustion noise with a specific frequency, and its amplitude increases slowly. Besides gas oscillations, fluctuations also usually appear in the propellant supply system; the frequency and phase of gas oscillation is often not consistent with the inherent acoustic modes of the combustion chamber, which is different from the high-frequency combustion instability. On the other hand, it is also different from the low-frequency combustion instability. Because its frequency is slightly higher, the wavelength of gas oscillation is close to or slightly larger than the characteristic length of the combustion chamber, so fluctuations in the combustion chamber and the supply system pipeline cannot be ignored; The pressure oscillation in the combustion chamber will change spatially, and cannot be seen as a whole gas field like that in low-frequency combustion instability. Intermediate-frequency combustion instability may also lead to oscillations of the propellant mixture ratio and decrease of engine performance.

7.1.3 Characteristics of Combustion Instability

In experiments and actual cases of liquid rocket engine, combustion instability phenomena show the following characteristics [32]:

1. *Periodic:* When combustion instability occurs, the pressure oscillations of the combustion chamber have obvious periodicity, oscillation energy is concentrated on the oscillation of several inherent frequencies, and there is a certain relation between the gas oscillations of the combustion chamber at different locations.
2. *Destructive:* A large amplitude oscillation of combustion chamber pressure will produce an explosion when the pressure exceeds the pressure limits of the combustion chamber wall. Combustion instability leads to high-frequency oscillations in the combustion chamber, damaging the boundary-layer structure of chamber walls, strengthening the heat transfer between gas and chamber walls, usually resulting in the combustion chamber being burnt; combustion instability changes the performance parameters such as specific impulse and thrust of rocket engines, and thus deteriorates the rocket accuracy.
3. *Complicated:* At present, there are many types of rocket engines. Different propellant combinations, injector forms, engine sizes, and operating parameters have very complicated influences on the combustion instability. Up to now, determinate universal laws from experiments are still very few.
4. *Random:* Many experimental results show that the characteristics of combustion instability is not repeatable, the distribution of experimental results far exceed the prediction accuracy at fixed conditions. For engines manufactured in the same batch, some may be stable, while others may be unstable. Typically, the stability and performance of the engine are better in the development stage, but become very instable just in identification or flight. These phenomena indicate that combustion instability is very sensitive to some structural parameters,

state parameters, and initial states; these parameter sensitivities are even beyond the current accuracy of manufacturing and testing.

5. *Nonlinear:* Self-excited oscillations in the combustion chamber increase exponentially from small amplitudes to reach a certain magnitude, and then the growth slows down, eventually forming a limit cycle. In the absence of initial disturbance, the combustion chamber is stable, but the combustion becomes unstable after introducing a certain magnitude of explosive bombs. Sometimes the oscillation amplitude in the combustion chamber is very large, the time history of pressure oscillation is then no longer sinusoidal, but with shock waveform, which is also the result of nonlinear interactions.

6. *Difficult to control:* Through decades of practice, it has been discovered that the tune and the baffle are effective means to suppress combustion instability, and have been widely used in rocket engines. However, there are still no design criteria for determining the location of tune and baffle, structure, size, and area. Effective design of tune and baffle still requires a lot of repeated experiments and revised verification. In the case of combustion instability, the development of liquid rocket engines is always a costly job.

7.2 Acoustic Basis of Combustion Instability

In almost all cases, there is a remarkable similarity between the observed frequency of combustion instability and the inherent acoustic modes of the combustion chamber, which people to conclude that combustion instability is closely related to acoustics. To illustrate the interaction of combustion and acoustic field, we will first discuss the Rayleigh criterion for acoustic oscillations arising from heat or mass supply, and then discuss the acoustic vibration modes of the combustion chamber and the self-oscillation effects within rocket engines.

7.2.1 Rayleigh Criterion for Acoustic Oscillations Arising from Heat or Mass Supply

In all heat engines, the combustion process is used to heat the working fluid. Thus, the combustion process can be replaced by a hypothetical heating process. Then, the interactions of combustion and acoustic field can be simplified as the interaction of heating and acoustic field.

As early as 1777, B. Higgins had noted that sound was generated when putting a flame into a tube open at both ends, which was called "the singing flame" [33]. Thereafter, many people found similar phenomena. Rayleigh was the first to explain this phenomenon and gave the so-called Rayleigh criterion. The Rayleigh criterion states that if the combustion rate related to some mechanism oscillates with large enough amplitudes, and the phase difference between oscillations of combustion rate and pressure is small enough, this mechanism can stimulate combustion instability. The Rayleigh criterion enjoys a fundamental role in the study of the rocket engine combustion instability, and has been widely used in the experimental analysis and active control of combustion instability. However, the Rayleigh criterion does not indicate any specific physical mechanism of instability excitation.

In a combustion system, unstable heat release transferring energy to the sound field does not necessarily lead to combustion instability. According to the Rayleigh criterion, only when the phase angle between the motion of the work fluid participating in the thermal process and the heat exchange is suitable, thermoacoustic oscillations be sustained. In fact, only when the phase difference between oscillations of sound pressure and heat release rate lies in the range 0–90°, the fluctuation of heat release rate will strengthen the pressure oscillation, and when the supply rate of energy from periodic heat release process to sound field is greater than the diffusion and damping rate of energy by the boundary conditions of combustion chamber will the combustion process become unstable. Determination of the conditions of combustion instability given by the Rayleigh criterion can be described by the following equation:

$$\int_V \int p'(x,t)q'(x,t)\,\mathrm{d}t\,\mathrm{d}V \geq \int_V \int \sum_{i=1} L_i(x,t)\,\mathrm{d}t\,\mathrm{d}V$$

where, $p'(x, t)$ is the pressure fluctuation, $q'(x, t)$ is the fluctuation of heat release rate, and $L_i(x, t)$ is energy loss of the i-th acoustic.

7.2.2 Acoustic and Acoustic Oscillations

Before introducing the acoustic vibration modes of the combustion chamber, it is necessary to give a brief review and description of the related concept of acoustic wave and acoustic oscillations.

Acoustic Wave

The term acoustic wave generally refers to the mechanical wave propagating in various media at any frequency. The acoustic frequencies that can be felt by the human ear are in the range 20–2000 Hz.

The propagation of a mechanical wave must have two conditions: wave sources and media. When the wave propagation direction is parallel to the vibration direction of the medium particle, it is called a longitudinal wave or density wave; when the wave propagation direction is perpendicular to the vibration direction of medium particle, it is called transverse wave. Liquids and gases can only transmit longitudinal waves, so that only longitudinal waves exist in liquid rocket engines.

Acoustic Oscillations

In the process of wave propagation, the vibration of the medium results in a periodic change in parameters, such as local density, temperature, and pressure, and their instantaneous values fluctuate around their stable values. This phenomenon is called acoustic oscillation.

When the perturbation is small, the local parameters in the medium can be approximately expressed as:

$$p = \bar{p} + p', \quad \rho = \bar{\rho} + \rho', \quad u = \bar{u} + u'$$

where:

$\bar{p}$, $\bar{\rho}$, and $\bar{u}$, respectively, denote mean pressure, mean density, and mean velocity;
p' denotes the sound pressure;
ρ' denotes the acoustic vibration density;
u' denotes the acoustic vibration rate; these small perturbations can be described by linear wave equations.

Planar Harmonic Wave

The harmonic wave is the simplest, most basic, and important fluctuation pattern. For small perturbation waves (linear acoustic waves), any complex acoustic wave is the linear superposition of some harmonic waves. The wave equation for a harmonic wave is:

$$\frac{\partial^2 p'}{\partial t^2} = \bar{a}^2 \frac{\partial^2 p'}{\partial x^2} \tag{7.1}$$

where $\bar{a}$ denotes the mean sound velocity.

The first order harmonic wave can be expressed by trigonometric function as:

$$p'(x,t) = \hat{p}\cos(kx - \omega t - \theta) \tag{7.2}$$

where:

ω denotes the angular frequency;
$k = \omega/\bar{a}$ denotes the harmonic number, which is referred to as the wave number;
θ denotes initial phase angle;
$\hat{p}$ denotes the amplitude of sound pressure.

Harmonic waves are expressed in plural as:

$$p' = \hat{p}e^{-i(kx - \omega t - \theta)} = \tilde{p}e^{-i(kx - \omega t)} \tag{7.3}$$

where $\tilde{p} = \hat{p}e^{i\theta}$ is called the complex amplitude of sound pressure and $\hat{p}$ denotes real amplitude of acoustic pressure. For a planar harmonic wave:

$$u' = \frac{i}{\bar{\rho}\omega}\frac{\partial p'}{\partial x}$$

the sound pressure is proportional to sound vibration rate ($u' \approx p'$).

Sound Energy Density (ε)

Acoustic oscillations involve two kinds of energy, one is the kinetic energy of a gas micelle and the other is the potential energy of gas volume deformation.

Taking a small enough volume element V in the sound field with an original volume of $\bar{V}$, for a planar sound wave, the kinetic energy obtained by this volume element due to acoustic disturbances is:

$$\varepsilon_u = \frac{1}{2}(\bar{\rho}\bar{V})u'^2 \tag{7.4}$$

where u' denotes the sound vibration rate of the gas micelle. Due to sound disturbance, the volume element has the potential energy:

$$\varepsilon_p = -\int p'\,dV = \frac{1}{2}\frac{\bar{V}p'^2}{\bar{\rho}\bar{a}^2} \tag{7.5}$$

The overall sound energy within volume element is the sum of kinetic energy and potential energy, i.e.:

$$\Delta\varepsilon = \Delta\varepsilon_u + \Delta\varepsilon_p = \frac{\bar{V}}{2}\bar{\rho}\left(u'^2 + \frac{p'^2}{\bar{\rho}^2\bar{a}^2}\right) \tag{7.6}$$

The acoustic energy per unit volume is called the acoustic energy density (e), i.e.:

$$e = \frac{\Delta\varepsilon}{\bar{V}} = \frac{1}{2}\bar{\rho}\left(u'^2 + \frac{p'^2}{\bar{\rho}^2\bar{a}^2}\right) \tag{7.7}$$

This equation is applicable to any shape of sound field and any form of sound waves.

7.2.3 Acoustic Modes in the Combustion Chamber

The combustion chamber of a liquid rocket engine can be approximately regarded as a cylindrical tune, so the classical wave equation can be used to describe the pressure oscillations with time and space in the combustion chamber.

7.2.3.1 One-Dimensional Longitudinal Mode (Basic Wave Equation)

From the one-dimensional Euler equations, the wave equation can be derived as follows:

$$\begin{cases} \dfrac{\partial\rho}{\partial t} + \dfrac{\partial}{\partial x}(\rho u) = 0 \\[2mm] \rho\dfrac{\partial u}{\partial t} + \rho u\dfrac{\partial u}{\partial x} + \dfrac{\partial p}{\partial x} = 0 \\[2mm] \rho C_p\dfrac{\partial T}{\partial t} + \rho u C_p\dfrac{\partial T}{\partial x} - \dfrac{\partial p}{\partial t} - u\dfrac{\partial p}{\partial x} = 0 \\[2mm] p = \rho RT, a^2 = \gamma RT \end{cases} \tag{7.8}$$

Multiplying the continuity equation by C_pT and adding the energy equation and the state equation to eliminate temperature (T), one obtains the equation of pressure:

$$\frac{\partial p}{\partial t} + \gamma p \frac{\partial u}{\partial x} + u \frac{\partial p}{\partial x} = 0 \tag{7.9}$$

For small amplitude oscillations, $p = \bar{p} + p'$, $\rho = \bar{\rho} + \rho'$, $u = \bar{u} + u'$; $\bar{p}$ and $\bar{\rho}$ are known quantities; $\bar{u}$ is a function of x; p', ρ', and u' are functions of x and t. Linearization of the above formula leads to:

$$\frac{\partial p'}{\partial t} + \gamma \bar{p} \frac{\partial u'}{\partial x} = -\left(\bar{u} \frac{\partial p'}{\partial x} + \gamma p' \frac{\partial \bar{u}}{\partial x} \right) \quad \text{(Acoustic pressure equation)} \tag{7.10}$$

Similarly, the momentum equation can be linearized:

$$\frac{\partial u'}{\partial t} + \frac{1}{\bar{\rho}} \frac{\partial p'}{\partial x} = -\left(\bar{u} \frac{\partial u'}{\partial x} + u' \frac{\partial \bar{u}}{\partial x} \right) \quad \text{(Acoustic vibration rate equation)} \tag{7.11}$$

In the two formula above, the left-hand side denotes acoustic terms; the right-hand side denotes the effects of mean flow in the combustion chamber, which is a small perturbation, equivalent to a forcing function. Taking the differential of the sound pressure equation (Equation 7.10) with time, and substituting the rate equation of acoustic vibration (Equation 7.11), one obtains:

$$\frac{\partial^2 p'}{\partial x^2} - \frac{1}{\bar{a}^2} \frac{\partial^2 p'}{\partial t^2} = \frac{\bar{u}}{\bar{a}^2} \frac{\partial^2 p'}{\partial x \partial t} - \bar{\rho} \frac{\partial^2}{\partial x^2}(\bar{u}u') + \frac{\gamma}{\bar{a}^2} \frac{\partial \bar{u}}{\partial x} \cdot \frac{\partial p'}{\partial t} \tag{7.12}$$

The above formula is the one-dimensional small perturbation wave equation with average flow. To facilitate the calculation, p', u' are usually expressed in the plural form: $p' = p'(x)e^{i\omega t}$, $u' = u'(x)e^{i\omega t}$. Substituting them in the wave equation, one obtains an inhomogeneous ordinary differential equation (Helmholtz equation):

$$\frac{d^2 p'(x)}{dx^2} + k^2 p'(x) = h \tag{7.13}$$

$$h = ik \frac{\bar{u}}{\bar{a}} \frac{dp'(x)}{dx} + i\frac{\gamma k}{\bar{a}} p'(x) \frac{d\bar{u}}{dx} - \bar{\rho} \frac{d^2}{dx^2}(\bar{u}u'(x)) \tag{7.14}$$

where $k = \omega / \bar{a}$ denotes the complex wave number; h is called the forcing function.

To solve the equation one must first determine the boundary conditions and know the average flow velocity ($\bar{u}$). For liquid rocket engines, the boundary conditions are often not easy to determine, and the $\bar{u}$ is not constant as it is a mass-adding flow. Therefore, it is difficult to obtain an analytical solution of Equation 7.13. Instead, one may let $h = 0$ to simplify it to one-dimensional harmonic equation.

7.2.3.2 Transverse Mode (Three-Dimensional Wave Equation)

From the development history of the rocket engine, one can see that combustion instability to first appear is usually the transverse mode, which is the most destructive kind of vibration mode. To analyze the transverse vibration mode one must use the three-dimensional wave equation, the derivation of which is the same as that of the one-dimensional longitudinal wave equation.

For a cylindrical combustion chamber, mimicking the process of establishing the one-dimensional wave equation, the three-dimensional wave equation can be written as [34]:

$$\nabla^2 p' - \frac{1}{a^2}\frac{\partial^2 p'}{\partial t^2} = h \tag{7.15}$$

In cylindrical coordinates, the above formula becomes:

$$\frac{\partial^2 p'}{\partial t^2} = \bar{a}^2 \left[\frac{1}{r}\frac{\partial}{\partial r}\left(r\frac{\partial p'}{\partial r}\right) + \frac{1}{r^2}\frac{\partial^2 p'}{\partial \theta^2} + \frac{\partial^2 p'}{\partial x^2}\right] + h\bar{a}^2 \tag{7.16}$$

Only when $h = 0$ and the boundary conditions satisfy (Classic Tune):

$$x = 0, \quad u'_x = 0, \quad \frac{\partial p'}{\partial x} = 0$$

$$X = L, \quad r = R, \quad u'_r = 0, \quad \frac{\partial p'}{\partial r = 0}$$

can a particular solution of the wave equation be obtained via integration:

$$p'_{l,m,n} = J_m\left(\frac{\pi\alpha_{m,n}r}{R}\right)\cos\left(\frac{l\pi x}{L}\right)\left[B_{l,m,n}\cos\left(\omega_{l,m,n}t + m\theta + \theta^+_{l,m,n}\right)\right.$$
$$\left. + C_{l,m,n}\cos\left(\omega_{l,m,n}t - m\theta + \theta^-_{l,m,n}\right)\right] \tag{7.17}$$

where:

l, m, n denotes the order of vibration mode in the x, θ, r direction;
the term:

$$J_k\left(\frac{\rho\alpha_{m,n}}{R}r\right)$$

is the kth-order Bessel function of the independent variable r, denoting the distribution of sound pressure along radial direction;
$\omega_{l,m,n}$ denotes the angular frequency of acoustic vibration;
$B_{l,m,n}, C_{l,m,n}, \theta^+_{l,m,n}, \theta^-_{l,m,n}$ denote constants determined by initial conditions;
$\cos((l\pi/L)x)$ denotes the distribution of sound pressure along longitudinal direction;

the two items within [] represent two oppositely traveling waves in a tangential direction, which will form tangential stable standing waves when certain conditions are satisfied.

The tune's natural frequency formula of various acoustic vibration modes is:

$$f_{l,m,n} = \frac{\omega_{l,m,n}}{2\pi} = \frac{\bar{a}}{2}\sqrt{\left(\frac{l}{L}\right)^2 + \left(\frac{\alpha_{m,n}}{R}\right)^2} \tag{7.18}$$

Note: l, m, n are integer numbers, $\alpha_{m,n}$ is the root of the equation of:

$$\frac{d}{dr}[J_m(\pi\alpha)] = 0$$

determined by m, n, representing the order of vibration mode in tangential and radial directions.

It can be seen that only the harmonic wave with the above frequencies may appear in the sound chamber. Any existing sound wave must be a linear superposition of several natural vibration modes.

According to the relationship between sound pressure and the speed of acoustic vibration, one can obtain:

$$u'_{l,m,n} = \frac{1}{\bar{\rho}\omega_{l,m,n}} \cdot \nabla \hat{p}_{l,m,n} \sin\omega_{l,m,n}t \tag{7.19}$$

According to the expression of natural frequencies, when one or two of l, m, n are equal to zero, one can determine the natural frequencies of different vibration modes (Figure 7.1).

1. Longitudinal vibration mode:
   When $m = n = 0$, $l \neq 0$, then $\alpha_{m,n} = 0$:

$$f_l = \frac{l\bar{a}}{2L} \quad l = 1, 2, 3 \cdots$$

That is, for the longitudinal vibration mode, frequencies of higher modes are an integer multiple of the base frequency.

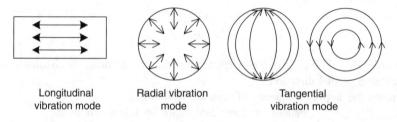

| Longitudinal
vibration mode | Radial vibration
mode | Tangential
vibration mode |

Figure 7.1 Combustor acoustic vibration modes.

2. Radial vibration mode:

When acoustic waves spread along the radial direction, they reflect on the cylinder walls to form standing waves. Letting $l = m = 0$, and $n \neq 0$, one can get the frequency of radial vibration mode:

$$f_n = \alpha_{o,n}\bar{a} \Big/ 2R \qquad n = 1, 2, 3 \cdots$$

where the coefficient $\alpha_{o,n}$ is determined by vibration mode order n, when $n = 1$, $\alpha_{0,1} = 1.22$; when $n = 2$, $\alpha_{0,2} = 2.233$.

3. Tangential vibration mode:

When the initial disturbance is not symmetric about the centerline, it will trigger the tangential vibration mode, and $l = n = 0$, $m \neq 0$:

$$f_m = \alpha_{m,o}\bar{a} \Big/ 2R$$

where coefficients $\alpha_{1,0} = 0.586$ and $\alpha_{2,0} = 0.972$. The tangential vibration mode may be a standing wave mode or traveling wave mode.

7.2.4 Self-Excited Oscillations in Rocket Engines

Self-Excited System

Acoustic vibration modes of classical tunes have been described earlier, where the wave equations are derived from the inviscid Euler equations. If the wall is completely rigid, the energy will not decay after initial disturbances. Actually, the sound waves disappear quickly due to the viscous damping and wall loss in tunes.

When combustion instability occurs in rocket engines, the pressure oscillations are sustainable and stable, even growing, and the energy source is the heat released from combustion. Therefore, combustion instability can be understood as the mutual coupling between the process of combustion energy release and acoustic vibration process. As long as the combustion energy is available, even if it is a small part, it is sufficient to compensate for various acoustic energy losses, and amplify the weak disturbance to form unstable combustion. Consequently, the engine combustion flow field can be regarded as a self-excited system. It can be determined by the Rayleigh criterion whether the unstable heat release process within the system will amplify the vibrations of sound pressure.

Loss of Acoustic Energy

Acoustic energy loss can be divided into three forms: internal loss, boundary loss, and structural damping. Internal loss is mainly caused by the effects of hysteresis and viscous and the particulate relaxation. Boundary loss is mainly caused by nozzle damping and wall losses.

1. Hysteretic or viscous losses:

For isentropic flows, the Laplace definition of sound speed gives:

$$\bar{a} = \sqrt{dp/d\rho} \qquad\qquad (7.20)$$

That is: $dp = \bar{a}^2 d\rho$, since $p = \bar{p} + p'$, then $p' = \bar{a}^2\rho'$, where $\bar{a}^2$ is a real number, therefore $p' \sim \rho'$ are in the same phase.

For a real gas, the compression and expansion of the gas are always related to the viscosity. We can rewrite $p' = \bar{a}^2\rho'$ as:

$$p' = \bar{a}^2\rho' + R\frac{\partial\rho'}{\partial t} \tag{7.21}$$

where R denotes the effective viscosity of various damping; $p' = \tilde{p}e^{i\omega t}$, $\rho' = \tilde{\rho}e^{i\omega t}$. Then, $\tilde{\rho} = \tilde{p}/(\bar{a}^2 + i\omega R)$. Obviously, $\tilde{\rho}$ and $\tilde{p}$ are not in-phase; the phase lag angle between them is:

$$\theta = tg^{-1}\frac{\omega R}{\bar{a}^2} \tag{7.22}$$

That is, the density perturbation lags behind the pressure perturbation.

2. Damping of nozzle:

One of the differences between the combustion chamber tune and the cylindrical cavity with closed, rigid wall is that the nozzle end of the combustion chamber is not closed and rigid; the nozzle will discharge energy in the form of radiation and convection, thereby resulting in damping. For different vibration modes, the damping of the nozzle is different.

Nozzle damping is important in rocket engines. The velocity gradients in the nozzle can effectively reflect acoustic waves. For pure transverse waves, a supersonic nozzle is equivalent to a rigid wall. For longitudinal waves, a wave cannot be totally reflected by the nozzle, part of the acoustic energy flows out from the nozzle and this loss results from an interaction between the sound field and the average flow, which can be represented by nozzle admittance or response function.

3. Wall losses:

An oscillating boundary layer will emerge on the combustion chamber wall, but the thickness of the oscillating velocity and temperature boundary layer is much smaller than that of the average velocity and temperature boundary layer. Therefore, the velocity and temperature gradients in this region are very large. The velocity and temperature oscillations will decay through friction and heat transfer.

4. Structural damping:

The combustion chamber shell is not an absolutely rigid body, so transmission and reflection of acoustic energy coexist in the solid/gas interface. When the traveling wave of plane sound vibration projects vertically onto the solid phase from the gas phase, part of the acoustic energy will be reflected back to the gas phase. When oscillations occur in the combustion chamber, standing-wave acoustic energy will be transmitted to the chamber shell at a certain rate. This exchange of energy will change the oscillation modes in the chamber cavity, and result in acoustic radiation to the surrounding environment and acoustic energy dissipation in a solid internal. The greater the viscoelastic coefficient of shell, the larger the damping effect.

5. Droplet damping:

Droplets accelerate or decelerate with the gas oscillations (compression or expansion). Since the inertia of a droplet is relatively large, the oscillation rate of droplets and gas are at a different phase. As droplet density is usually much larger than the gas density, it is difficult to keep up with the oscillation motion of the gas (or temperature), so the droplet oscillation can provide very strong damping effects. The damping is far greater than the gas viscous losses, wall damping, and so on; thus, this is one of the main factors that keep the engine work stably.

7.3   Response Characteristics of Combustion Process in Liquid Rocket Engines

It has been shown in the previous section that the rocket engine combustion chamber can be regarded as a self-excited oscillation system. The energy source of the system is the energy released from the combustion process. The oscillators are the combustion chamber and the propellant supply system. Only when there is a certain response and feedback process between the energy source and the oscillator can the oscillation process be maintained. The key question of combustion instability research is to determine the relationship between the energy release and the response and feedback of oscillation system, i.e., the mechanism of combustion instability. This is a problem not yet solved. In this section, a qualitative analysis will be given of the characteristics of each sub-process of combustion process in the liquid rocket engine to determine its response characteristic.

In the work process of liquid rocket engines, systems and sub-processes related to combustion are (i) the propellant supply system; (ii) spray atomization process; (iii) the evaporation process; (iv) the mixing process; and (v) the chemical reaction process. These processes are all related to the engine combustion instability, but their effects are not equal. The following will give a qualitative analysis of the response characteristics of several major systems or processes.

7.3.1   Response Characteristics of the Propellant Supply System

7.3.1.1   Response Mechanism

The propellant supply system is generally in response to low-frequency pressure oscillations in the combustion chamber. Due to the lower frequency oscillation of its own and the quick propagation of disturbances within the combustion chamber, the relaxation time in the combustion chamber is much smaller than for the oscillating cycle. Therefore, the combustion chamber can be regarded as a system with a lumped parameter, and the spatial distributions of parameters such as pressure and temperature are neglected. This is an overall combustion instability.

In a liquid rocket engine, because of the fluid inertia in the propellant supply system and atomization and combustion of propellants, there exists a significant relaxation time difference of combustion chamber pressure in response to mass flow rate changes of propellant. If this time difference satisfies certain conditions, it will stimulate low-frequency combustion instability. The reason for this is as follows. If the injection pressure remains constant, the increase of combustion chamber pressure by excitation will cause a decrease in injection pressure drop, and result in a decrease of propellant flow rate into the combustion chamber, which leads to a decrease of combustion chamber pressure and increase of injection pressure drop, and the flow entering the combustion chamber increases and, subsequently, combustion chamber pressure increase again, If there is no damping, this oscillation will continue, thereby forming a low-frequency oscillation cycle.

7.3.1.2 Analysis Model

There are three main methods to describe the effects of supply system on combustion stability: the first is to restrict the boundary conditions, such as assuming the injection pressure or flow rate is kept constant, which is mainly used in the situation where the combustion chamber work state does not change intensely with the propellant supply system characteristics; the second is to use the transfer function of the supply system, which is obtained by linearizing the dynamic equations of each system component—this method can solve the problem in either the time or frequency region; the third method is to use nonlinear differential equations. The second approach is the most effective and most widely used.

The transfer function of the propellant supply system can be derived from a set of equations of each component. Each variable can be expressed as the sum of its steady-state value and disturbance value and then substituted into the equations. Subtracting the steady-state equation, one obtains the equations of the disturbances, Laplace transformation of the perturbation equations gives the transfer function of the system:

$$G(s) = \frac{\dot{m}_i'(s)}{\Delta p'(s)}$$

Let $s = i\omega$, then one can determine the relationship between pressure oscillation and flow rate oscillation $G(\omega)$. The frequency response function can be used to analyze combustion stability.

Now we will take the injection admittance as an example to show the calculation method of admittance. Injection admittance is defined as the ratio of injection rate and the change rate of combustion chamber pressure [5]:

$$Y_i = -\frac{q_{mi}'/\bar{q}_m}{p'/\bar{p}} \tag{7.23}$$

Accordingly, the injection admittance of oxidizer and fuel are defined as:

$$Y_{io} = -\frac{\bar{p}}{p'} \cdot \frac{q_{mio}'}{\bar{q}_m} \tag{7.24}$$

$$Y_{if} = -\frac{\bar{p}}{p'} \cdot \frac{q_{mif}'}{\bar{q}_m} \tag{7.25}$$

where:

Y_{io} denotes injection admittance of oxidant;
Y_{if} denotes injection admittance of fuel;
q_{mio}' denotes the disturbance value of oxidizer injection rate;
q_{mif}' denotes the disturbance value of fuel injection rate.

A propellant supply system usually consists of pipes, valves, liquid collection chamber, turbine pumps, propellant tank, and so on. If the admittance of the sub-systems is known, we can use the admittance ratio of the components to derive the injection admittance of the system:

$$Y_i = Y_0 \left(\frac{Y_1}{Y_0}\right) \left(\frac{Y_2}{Y_1}\right) \cdots\cdots \left(\frac{Y_m}{Y_{m-1}}\right) \left(\frac{Y_i}{Y_m}\right) \qquad (7.26)$$

7.3.2   Response Characteristics of Spray Atomization Process

The propellant spray atomization process is closely related to the stability characteristics of the combustion chamber. If the travel time of propellant from injection surface to the point of impingement approaches the half cycle of a longitudinal vibration mode oscillation (or an odd number of half cycles), it may stimulate longitudinal instability. If the injection pressure drop of the liquid injector is too large, it may lead to the transverse high-frequency combustion instability. Propellant injection velocity, flow area, and mass distribution along the injector face may affect stability.

The main factors affecting injection process response are the influence of upstream conditions on spray propellants form, small perturbation effects under combustion chamber state, burst of drops, and jet-flow caused by shocks. Among them, the influence of upstream disturbance on injection process may have the following form [1]:

1. Flow oscillation.

    Bunching of propellant downstream of the injector may cause flow oscillation, which produces a "klystron" effect, promoting the sinusoidal flow rate of the downstream position to become a steep peak waveform. This could be the reason for low and intermediate frequency instability. The steep pulse-type flow rate change caused by the klystron amplification can explain the saw-tooth wave of chamber pressure at low and intermediate frequency instability. After this amplification, the low-frequency and low-amplitude oscillations may result in periodic chamber acoustic resonance.

2. Hydraulic jump.

    Hydraulic jump describes the phenomenon whereby the propellant jet changes from a cylindrical jet into a shrubby jet, which is related to injection pressure, length-diameter ratio of nozzle, structure shape of nozzle inlet, and propellant properties. The hydraulic jump is one of the possible factors that stimulate the "sonic boom" burning phenomenon. When the injector works within the transition zone between cylindrical and shrubby liquid jet, localized pressure disturbances appear. These disturbances spread by unburned propellant bag and can be magnified into steep peak waves. Hydraulic jump only occurs within a certain range of injector pressure drop; therefore, the injector design can be changed to eliminate the hydraulic jump.

3. Injector vibrations.

    Coupling between the actual mechanical vibration of injector jet plane and intermediate flow component structural vibration may also cause a significant change in the spray, and cause instability. The atomization process of a liquid propellant is discussed in Chapter 2 of the present book. Liquid jets must first be expanded into a liquid membrane or cylindrical jet before atomization. Under the effects of surrounding aerodynamic forces, the disturbance waves of the liquid membrane and jet surface will increase and eventually lead to liquid membrane breakup to form droplets. Since the breakup of the liquid membrane is caused by surface wave increase, the propellant droplet shedding occurs in a certain periodicity and this cycle is related to the shedding frequency of liquid filament.

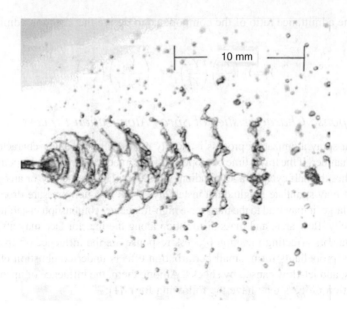

Figure 7.2   Liquid filament shedding in the atomization process.

Figure 7.2 shows the liquid filament shedding in the atomization process. The periodic liquid silk shedding and instability combustion are similar, but the surface disturbance wave frequency generally of the atomization process is around 10^5 Hz, which is much higher than the frequency of the combustion chamber. Therefore, the gas in the combustion chamber can be considered to be steady for surface waves.

7.3.3   Response Characteristics of Droplet Evaporation Process

Sirignano considered that the rate controlling process of combustion is the main factor in determining the stability of combustion [35]. From the characteristic time of each sub-process of combustion process—the characteristic time of the mixing process is 10 μs to 1 ms; the characteristic time of the evaporation process is about 1 ms—the characteristic time of chemical reactions under high pressure and temperature is at least one order of magnitude smaller than for the evaporation.

The evaporation characteristic time is the longest, so the response characteristics of droplet evaporation process to pressure changes are likely to be an important factor leading to instability combustion. The droplet evaporation response characteristics under the pressure oscillation environment have been discussed in Section 7.3.3 and, thus, are omitted here.

7.4   Sensitive Time Delay Model $n-\tau$

7.4.1   Combustion Time Delay

In a real liquid rocket engine, the process that converts propellants from the initial state into combustion products at the end is a continuous process. It is extremely difficult to make a quantitative description of this continuous transformation process. Thus, in many analytical models, this continuous process is simplified into step processes. It is assumed that each of propellant micro-element does not experience any significant energy release and volume changes within a period of time, then it is converted immediately into products in the combustion chamber. This time interval is called the time delay. The time delay mentioned in the following analysis model uses this concept.

In early analysis of combustion instability, the time delay was usually seen as a constant. However, the rate of a propellant energy release process is affected by various factors in the combustion chamber; consequently, the experienced time of this process—the time delay— apparently will also be affected. The higher the process rate, the smaller the corresponding time delay. Thus, the time delay is not constant. Therefore, it can be assumed that the time delay is divided into constant time delay and variable time delay:

$$\tau_t = \tau_i + \tau \tag{7.27}$$

where τ_t is the total time delay; τ_i is the constant time delay, representing the experienced time of the process less affected by various factors in combustion chamber such as atomization and mixing process; τ is the variable time delay, also known as sensitive delay, representing the experienced time of each process more greatly influenced by these factors.

7.4.2 Sensitive Time Delay Model

The sensitive time delay model was first proposed by Crocco and Cheng [6]. The basic idea of this model is based on the N-S equation of spray and gas–liquid two-phase flow, using the small perturbation linear analysis method and establishing the relationship between combustion chamber pressure oscillations (p') and the droplet evaporation rate ($\dot{m}'$). The change in regulation of pressure disturbance over time is then obtained and the stability of the combustion process is determined.

We propose the hypothesis:

1. The gasification process of propellant is a mutation process, namely, the droplets remain unchanged in a certain period of time, but suddenly turn into vapor after the lag τ.
2. The energy release of propellant is determined by the motion equations of droplets and time delay.
3. In the sensitive time delay with state change of propellant, pressure plays a major role.
4. The gas is an ideal gas.

7.4.2.1 Relationship between Evaporation Rate $\dot{m}$, Time Delay τ, and Pressure p

Assume that the rate (f) of the combustion preparing process for propellant a micro-element is a function of pressure (p), temperature (T), and other factors:

$$f(p, T \cdots) = f(\bar{p}, \bar{T}) + p' \frac{\partial f}{\partial p} + T' \frac{\partial f}{\partial T} + \cdots \tag{7.28}$$

Further, assume that various factors such as temperature are only related to pressure, i.e.:

$$T = T(p), \cdots$$

Hence:

$$f(p, T \cdots) = f(\bar{p}, \bar{T}, \cdots) \left[1 + p' \frac{1}{f} \left(\frac{\partial \bar{f}}{\partial p} + \frac{\partial \bar{f}}{\partial T} \frac{dT}{dp} + \cdots \right) \right]$$

$$= f(\bar{p}, \bar{T}, \cdots) \left(1 + n \frac{p'}{\bar{p}} \right) \tag{7.29}$$

where:

$$n = \frac{\bar{p}}{f} \left(\frac{\partial \bar{f}}{\partial p} + \frac{\partial \bar{f}}{\partial T} \frac{dT}{dp} + \cdots \right)$$

denotes pressure interaction index.

If consider the evaporation rate ($\dot{m}$) of the liquid phase as a process rate, then:

$$\dot{m} = \bar{\dot{m}} \left(1 + n \frac{p'}{\bar{p}} \right) \tag{7.30}$$

That is:

$$\frac{\dot{m} - \bar{\dot{m}}}{\bar{\dot{m}}} = n \frac{p'}{\bar{p}}$$

namely:

$$\frac{\dot{m}'}{\bar{\dot{m}}} = n \frac{p'}{\bar{p}}$$

is actually a response function.

In fact, the pressure interaction index (n) is the amount related to process, and is not only related to the time delay but also related to local parameters $p, T, \cdots$ of the gas phase. This makes it difficult to determine n.

We will derive the change relationship between the sensitive time delay and combustion chamber pressure perturbation below.

During the delay period that the liquid propellant micro-element experienced between entering the combustion chamber to becoming combustion products, a certain amount of energy will be absorbed by the liquid propellant. Only when the energy accumulated during the preparation of the combustion reaches a certain level can the propellant be transformed into the combustion products, i.e.:

$$\int_{t-\tau}^{t} f(t') dt' = E_a \tag{7.31}$$

where f denotes the rate of combustion preparing process for liquid propellant micro-element and E_a denotes the energy to be obtained during the sensitive time delay.

Similarly, under steady combustion conditions:

$$\int_{t-\bar{\tau}}^{t} \bar{f}(t')\,\mathrm{d}t' = E_a \tag{7.32}$$

Typically, in the combustion chamber of liquid rocket engines, the velocity of droplets or gas is much smaller than the speed of sound, therefore it is considered that the steady-state value of the gas pressure and temperature and other parameters are uniform and constant, i.e., they does not change in space and time. Thus, the steady-state value of process rate of propellant micro-element is also constant, which can be obtained as:

$$\frac{\mathrm{d}\tau}{\mathrm{d}t} = -n\frac{p'(t)-p'(t-\tau)}{p} \tag{7.33}$$

where $\mathrm{d}\tau/\mathrm{d}t$ represents the time change rate of time delay and is associated with the perturbation of combustion chamber pressure. It plays an important role in the theoretical analysis of combustion instability.

7.4.2.2 Model Equations

Since the propellant combustion process within a liquid rocket engine combustion chamber is very complex, accurate quantitative analysis of the combustion instability is very difficult; semi-empirical simplified analysis methods are usually used.

Assume that the combustion chamber is completely full of combustion gas and unreacted propellant droplets, and the latter act as gas sources. In addition, assume that the combustion gas is a uniform, inviscid (except for the existence of droplet drag), hot ideal gas, and assume that the liquid phase is fully dispersed throughout the combustion chamber. The changes of energy (the sum of the internal energy and the kinetic energy) then are negligible.

One can use non-dimensional two-phase flow conservation equations for analysis. By selecting the gas state of the injector surface in steady-state operation as the reference state, one can carry out non-dimensionalization for gas pressure, temperature, density, specific enthalpy, and other parameters. Other parameters can use the following reference values: the speed of sound as the reference velocity, chamber length or radius as the reference length, the ratio of reference length and speed of sound as the reference time, the ratio of the product of reference density and sound velocity and reference length as the reference gas generation rate, and so on. In addition, notably, in the energy conservation equation used here, the internal energy term includes chemical energy. Thus, one obtains the non-dimensional conservation equations of two-phase flow:

$$\begin{cases} \dfrac{\partial \rho}{\partial t} + \nabla\cdot(\rho V) = -\dfrac{\partial \rho_1^0}{\partial t} - \nabla\cdot(\rho_1^0 V_1) = \dot{m} \\[2mm] \rho\dfrac{\partial V}{\partial t} + \rho V\cdot\nabla V + \dfrac{1}{\gamma}\nabla p = (\dot{m} + k\rho_1^0)(V_1 - V) \\[2mm] \rho_l\dfrac{\partial V_1}{\partial t} + \rho_1 V_1\cdot\nabla V_1 = k\rho_1^0\left(\vec{V} - V_1\right) \\[2mm] \dfrac{\partial}{\partial t}\left(\rho T_s - \dfrac{\gamma-1}{\gamma}p\right) + \nabla\cdot(\rho T_s V) = \dot{m}e_{1,s} \end{cases} \tag{7.34}$$

where:

ρ denotes the non-dimensional gas density;

ρ_1 is not liquid density, but non-dimensional liquid mass concentration;

$\overrightarrow{V}, \overrightarrow{V_l}$ denote the gas velocity vector and liquid velocity vector;

$\dot{m}$ is the generation rate of gas phase per unit volume;

T_s is the gas stagnation temperature;

γ is the gas specific heat ratio;

k is the momentum exchange coefficient;

$e_{l,s}$ is the non-dimensional liquid energy; $e_{l,s} = 1$ based on the assumptions.

Boundary conditions: for liquid phase, if one assumes that the propellant injection process is not influenced by combustion oscillation, only the non-dimensional jet velocity (u_{li}) and jet density (ρ_{li}^0) need to be given. For the gas phase, solid wall surface boundary conditions make the normal component of gas velocity equal to zero. Since the gas velocity reaches the speed of sound at the nozzle throat, the disturbances downstream of the sound speed interface cannot affect upstream regions, so the upstream regions of the nozzle throat are divided into two parts in the analysis: combustion chamber and nozzle convergent section. Combustion occurs in the combustion chamber, and the Mach number is low, but combustion does not continue in the nozzle convergent section and the Mach number gradually increases to 1. Through a study of the oscillation characteristics in the nozzle convergent section [1], one can obtain the relationship between flow disturbances at the nozzle entrance, which can be used as boundary conditions of the combustion chamber flow at the entrance of the nozzle.

7.4.2.3 Linear Instability Analysis

The combustion instability analysis model can be divided into two types: linear and nonlinear. Linear instability refers to those instabilities for which the oscillation amplitude will continue to grow with time under any small disturbance. In contrast, a nonlinear instability grows in amplitude only when the disturbance is higher than a finite value; the oscillation will decay when the disturbance is below this critical value. We briefly describe the analysis method of linear instability below.

In linear theories, it is assumed that the disturbance value is very small. This means that only the linear term of disturbances needs to be considered in the analysis. Then, the relationship between perturbation values is linear so that the mathematical derivation is greatly simplified. Linear analysis only deals with the conditions under which linear instabilities may occur; it does not involve the final state after disturbance growth.

The superposition principle can be used in linear analysis, namely, the sum of several solutions is also the solution of the equation. Thus, any oscillation can be decomposed into Fourier components with different frequencies. These components satisfy their own equations, and the analysis of some components will enable us to determine the stability of the system.

To study the decay or growth of disturbance with time, the equation can be linearized, and the disturbance value of each parameter is expressed as an exponential function of time. For example, the pressure can be expressed as:

$$p = \bar{p} + p'e^{st} = \bar{p} + p'e^{(\lambda + i\omega)t} \tag{7.35}$$

where $\bar{p}$ denotes steady-state pressure and p' denotes the amplitude of pressure perturbation. Other parameters are also expressed in a similar form.

Expressing all parameters in the form of Equation 7.35 and substituting them into Equation 7.34, and then subtracting the corresponding steady-state equations and omitting higher-order terms, one can derive the conservation equations expressed by perturbation amplitude of the parameters:

$$s\rho' + \bar{\rho}\nabla\cdot V' + V'\cdot\nabla\bar{\rho} = \dot{m}' - \nabla(\rho'\bar{V}) \tag{7.36}$$

$$s\rho_1' + \nabla\left(\bar{V}_1\rho_1^{0'}\right) = -\dot{m}' - \nabla\left(\bar{\rho}_1^0 V_1'\right) \tag{7.37}$$

$$s(\bar{\rho}V' + \bar{V}\rho') + \frac{\nabla p'}{\gamma} = -s\left(\rho_1^0 V_1' + V_1\rho_1^{0'}\right)$$

$$-\nabla\cdot\left(2\bar{\rho}\bar{V}V' + 2\bar{\rho}_1^0\bar{V}_1 V_1' + \bar{V}\bar{V}\rho' + \bar{V}_1\bar{V}_1\rho_1^0\right) \tag{7.38}$$

$$sV_1' + (\bar{V}_1\cdot\nabla)V' + (V_1'\cdot\nabla\bar{V}_1) = k(V' - V_1') \tag{7.39}$$

$$s\left(\bar{\rho}T_s' - \frac{\gamma-1}{\gamma}p'\right) + \nabla\cdot(\bar{\rho}\bar{V}T_s') = 0 \tag{7.40}$$

where:

$$T_s' = T' + (\gamma-1)\bar{V}\cdot V'$$

Using the concept of time delay, and assuming that the process rate is not affected by state change in the combustion chamber, one can get:

$$\dot{m}' = n[1 - \exp(-s\bar{\tau})]\bar{\dot{m}}\frac{p'}{\bar{p}} \tag{7.41}$$

From the analysis in the previous section, if the sensitive time delay is only a function of pressure, or other parameters affecting the time delay are all associated with pressure, the process rate relation can be expressed as:

$$\frac{f}{\bar{f}} = 1 + n\frac{p - \bar{p}}{\bar{p}} \tag{7.42}$$

For horizontal high-frequency combustion instability, the sensitive time delay changes not only with pressure but also with the uneven distribution of mixing ratio caused by gas displacement. This non-uniform distribution of mixing ratio is most significant near the injector surface. Considering the sensitivity of the rate of combustion process to gas displacement, one can obtain:

$$\frac{f}{\bar{f}} = 1 + n\frac{p - \bar{p}}{\bar{p}} + m_r\xi_r + m_\theta\xi_\theta \tag{7.43}$$

where m_r, m_θ denote radial and tangential displacement sensitivity index; ξ_r, ξ_θ denote radial and tangential gas displacement.

Therefore, considering the displacement effects, Equation 7.41 becomes:

$$\dot{m}' = \bar{m}[1 - \exp(-s\bar{\tau})]\left(n\frac{p'}{\bar{p}} + m_r\xi'_r + m_0\xi'_\theta\right) \tag{7.44}$$

So far, the non-dimensional perturbation equations of combustion chamber gas oscillation have been established. They can be used to analyze the linear stability of the system. Below we discuss the method for solving these equations.

Substituting Equation 7.39 into Equations 7.36 and 7.38, one can obtain the following equations:

$$\frac{sp'}{\gamma} + \nabla \cdot V' = -sX + \nabla \cdot Y + \dot{m}'$$

$$sV' + \frac{\nabla p'}{\gamma} = -sZ - \nabla \cdot W \tag{7.45}$$

where:

$$X = (\gamma - 1)\bar{\rho}\bar{V} \cdot V' + (1 - \bar{T})\rho'$$

$$Y = -\bar{V}p' + (1 - \bar{\rho})V' - (\gamma - 1)\bar{\rho}\bar{V}(\bar{V} \cdot V) - (1 - \bar{T})V\rho'$$

$$Z = \bar{V}\rho' + \bar{\rho}_1^0 V'_1 + \bar{V}_1\rho'_1 - (1 - \bar{\rho})V' \tag{7.46}$$

$$W = 2\bar{\rho}\bar{V}V' + 2\bar{\rho}_1^0\bar{V}_1 V'_1 + \bar{V}\bar{V}\rho' + \bar{V}_1\bar{V}_1\rho'_1$$

When the velocity of gas or liquid droplets is far less than the speed of sound, the terms in the right-hand side of Equations 7.46 are the same magnitude as $\bar{u}_e$, or higher-order terms when compared with perturbation values of various parameters. Here, A denotes the non-dimensional nozzle inlet velocity and is first-order. Therefore, the disturbance value of each parameter can be expanded into the following series form:

$$p' = p_0 + p_1 + p_2 + \ldots$$

$$V' = V_0 + V_1 + V_2 + \ldots$$

where subscripts "0", "1", and "2", respectively, represent zeroth-order, first-order, and second-order solutions. Here, p_1/p_0 and V_1/V_0 are the same magnitude as $\bar{u}_e$, while p_2/p_0 and V_2/V_0 are the same magnitude as $\bar{u}_e^2$, and so on.

Substituting the above series into Equations 7.46, and using the first two terms of these series according to its magnitude, respectively, one obtains the following two equations:

$$\begin{cases} \dfrac{sp_0}{\gamma} + \nabla \cdot V_0 = 0 \\[4mm] sV_0 + \dfrac{\nabla p_0}{\gamma} = 0 \end{cases} \tag{7.47}$$

and:

$$\begin{cases} \dfrac{sp_1}{\gamma} + \nabla \cdot \boldsymbol{V}_1 = -s\boldsymbol{X}_1 + \nabla \cdot \boldsymbol{Y}_1 + \dot{m}_1 \\[2mm] s\boldsymbol{V}_1 + \dfrac{\nabla p_1}{\gamma} = -s\boldsymbol{Z}_1 - \nabla \cdot \boldsymbol{W}_1 \end{cases} \tag{7.48}$$

Obviously, the zeroth-order Equations 7.47 are the acoustic equations; after eliminating $\boldsymbol{V}_0$ one can get the following wave equation:

$$\nabla^2 p_0 - s^2 p_0 = 0 \tag{7.49}$$

The complex frequency (s) in the above formula is also expanded into series form:

$$s = s_0 + s_1 + \cdots\cdots$$

Thus, the zeroth-order and first-order equations can be accordingly turned into:

$$\begin{cases} \dfrac{s_0 p_0}{\gamma} + \nabla \cdot \boldsymbol{V}_0 = 0 \\[2mm] s_0 \boldsymbol{V}_0 + \dfrac{\nabla p_0}{\gamma} = 0 \end{cases} \tag{7.50}$$

and:

$$\begin{cases} \dfrac{s_0 p_1}{\gamma} + \nabla \cdot \boldsymbol{V}_1 = -s_0 \boldsymbol{X}_1 + \nabla \cdot \boldsymbol{Y}_1 + \dot{m}_1 - \dfrac{s_1 p_0}{\gamma} \\[2mm] s_0 \boldsymbol{V}_1 + \dfrac{\nabla p_1}{\gamma} = -s_0 \boldsymbol{Z}_1 - \nabla \cdot \boldsymbol{W}_1 - s_1 \boldsymbol{V}_0 \end{cases} \tag{7.51}$$

Below we will discuss the solutions of longitudinal and transverse modes [1].

Solutions of Longitudinal Vibration Mode
From Equation 7.50 one can obtain the zeroth-order equations of the longitudinal vibration mode:

$$\begin{cases} \dfrac{s_0 p_0}{\gamma} + \dfrac{du_0}{dx} = 0 \\[2mm] s_0 u_0 + \dfrac{d}{dx}\left(\dfrac{p_0}{\gamma}\right) = 0 \end{cases} \tag{7.52}$$

Its solution is:

$$\begin{cases} p_0 = p_{0A} \cos \omega_0 x \\[2mm] u_0 = -i\dfrac{p_{0A}}{\gamma} \sin \omega_0 x \\[2mm] s_0 = i\omega_0 = i\dfrac{q\pi}{L_c} \end{cases} \tag{7.53}$$

where s_0 denotes the intrinsic value of the complex frequency s, and can be determined accord-ing to the requirements $u_0 = 0$ in the injection surface $x = 0$ and nozzle inlet $x = L$; p_{0A} denotes antinodes amplitude; q denotes the order of longitudinal vibration mode.

From the first-order Equation 7.51 one can obtain the first-order correction (u_1) as follows:

$$\frac{\gamma u_1(x)}{p_{0A}} = \gamma\left(K_p - 1\right)\bar{u}(x)\cos\omega_0 x - \gamma K_p\omega_0\int_0^x \bar{u}(x')\sin\omega_0(x - 2x')\,dx'$$

$$+ \omega_0\int_0^x \bar{u}(x')\left[\sin\omega_0 x - (2 - \gamma)\sin\omega_0(x - 2x')\right]dx' \qquad (7.54)$$

$$- \frac{k}{2}\int_0^x \bar{\rho}_l(x)\left[\cos\omega_0 x - \cos\omega_0(x - 2x')\right]dx' - s_1 x\cos\omega_0 x$$

where $K_p = n[1 - \exp(-s\bar{\tau})]$.

We introduce the following first order nozzle admittance conditions [1]:

$$u_1(L_c) = -Ap_0(L_c) \qquad (7.55)$$

where A denotes the nozzle admittance coefficient and L_c denotes the nozzle inlet coordinates.

The above formula is a complex equation and can be used to determine the real part λ_1 and the imaginary part $i\omega_1$ of s_1. Since $\lambda_0 = 0$, λ_1 is the amplification factor. One can distinguish the combustion stability according to the sign of λ_1: If λ_1 is negative, the system is stable; when λ_1 is positive, the combustion instability occurs.

Substituting the above equation into Equation 7.54, and letting $\lambda_1 = 0$, one can get the stability boundary conditions:

$$n(1 - \cos\omega_0\bar{\tau}) = G_R \qquad (7.56)$$

and:

$$G_R = \frac{\gamma(\bar{u}_e - A_R) - (2 - \gamma)\omega_0\int_0^{L_c}\bar{u}\sin 2\omega_0 x\,dx + \frac{k}{2}\int_0^{L_c}\bar{\rho}_l(1 - \cos 2\omega_0 x)\,dx}{\gamma\left(\bar{u}_e + \omega_0\int_0^{L_c}\bar{u}\sin 2\omega_0 x\,dx\right)} \qquad (7.57)$$

where A_R denotes the real part of the nozzle admittance coefficient (A).

Figure 7.3 illustrates a typical stability limit determined by Equation 7.56. In this figure, the horizontal coordinate denotes the ratio of sensitive time delay and oscillation cycle, while the longitudinal coordinate denotes the ratio of pressure interaction index and G_R. As can be seen from the figure, when the sensitive time delay τ is equal to an odd multiple of the half oscillation cycle the possibility of instability is the largest. For the possible combustion instability, the ratio of n/G_R has a minimum value of 0.5. Only when this ratio is greater than 0.5, can combustion instability possibly occur. Therefore, it can be considered that terms that increase the value of G_R have stabilizing effects, such as droplet drag, nozzle damping, and so on. In addition,

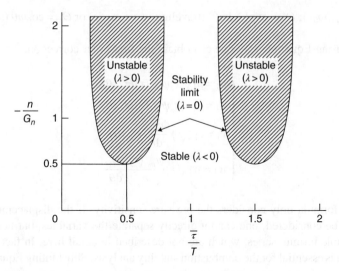

Figure 7.3   Typical stability limit of the longitudinal vibration mode.

according to Equation 7.57 one can further analyze the influence of the combustion zone axial distribution, length of the combustion chamber, nozzle inlet Mach number, and other parameters on the combustion stability.

Solutions of Transverse Vibration Mode
For the transverse vibration mode, one can obtain the corresponding zeroth-order equations by using cylindrical coordinates:

$$
\begin{cases}
\dfrac{s_0 p_0}{\gamma} + \dfrac{\partial u_0}{\partial x} + \dfrac{1}{r}\dfrac{\partial(r v_0)}{\partial r} + \dfrac{1}{r}\dfrac{\partial w}{\partial \theta} = 0 \\[2mm]
s_0 u_0 + \dfrac{1}{\gamma}\dfrac{\partial p_0}{\partial x} = s_0 v_0 + \dfrac{1}{\gamma}\dfrac{\partial p_0}{\partial r} = s_0 w_0 + \dfrac{1}{\gamma r}\dfrac{\partial p_0}{\partial \theta} = 0
\end{cases}
\tag{7.58}
$$

where u, v, and w denote the velocity components in the x, r and θ directions. Obviously, the zero-order equation of transverse vibration mode is an acoustic wave equation, and its solutions can be obtained as follows:

$$
\begin{cases}
p_0 = p_{0A}\psi_{mn}(r)\Theta_m(\theta) \\[2mm]
u_0 = 0 \\[2mm]
v_0 = -\dfrac{p_{0A}}{\gamma s_0}\dfrac{\mathrm{d}\psi_{mn}}{\mathrm{d}r}\Theta_m(\theta) \\[2mm]
w_0 = -\dfrac{p_{0A}}{\gamma s_0}\dfrac{\psi_{mn}}{r}\dfrac{\mathrm{d}\Theta_m}{\mathrm{d}\theta}
\end{cases}
\tag{7.59}
$$

where $\psi_{mn} = J_m(\omega_0 r)$; $\Theta_m = \exp(\pm im\theta)$ (traveling-wave mode) or $\Theta_m = \cos m\theta$ (standing-wave mode).

Similarly, using Equation 7.49 one can obtain the first-order correction:

$$\begin{cases} p_1 = p_{1A}(x)\psi_{mn}(r)\Theta_m(\theta) \\ u_1 = u_{1A}(x)\psi_{mn}(r)\Theta_m(\theta) \\ v_1 = v_{1A}(x)\dfrac{d\psi_{mn}}{dr}\Theta_m(\theta) \\ w_1 = w_{1A}(x)\dfrac{\psi_{mn}(r)}{r}\dfrac{d\Theta_m}{d\theta} \end{cases} \tag{7.60}$$

The above formula only considers the pressure sensitivity. If the displacement sensitivity also needs to be considered, one cannot directly separate the variables, but needs to expand them into double infinite series, which are not described in detail here. In fact, $u_{1A}(x)$ in the above formula is essential for the combustion stability analysis. Substituting Equation 7.59 into 7.49 leads to:

$$u_{1A}(x) = \int_0^x \frac{\dot{m}_1}{\psi_{mn}\Theta_m}\,dx' - \frac{p_{0A}}{\gamma}\left[(\gamma+1)\bar{u}(x) + 2s_1 x + k\int_0^x \bar{p}_1\,dx'\right] \tag{7.61}$$

Similarly, when calculating the parameters at $x = L_c$, one can introduce the nozzle admittance condition:

$$u_{1A}(L_c) = -\frac{p_{0A}}{\gamma}\varepsilon \tag{7.62}$$

where ε denotes nozzle combination admittance parameter.

Substituting Equation 7.62 into Equation 7.61, and letting $\lambda_1 = 0$, one can obtain the relation of stability boundary, which is identical to the longitudinal vibration mode of Equation 7.56, but G_R is determined by the following formula:

$$G_R = 1 + \frac{1}{\gamma} - \frac{\varepsilon_R}{\gamma\bar{u}_e} + \frac{k}{\gamma\bar{u}_e}\int_0^{L_c} \bar{p}_1\,dx \tag{7.63}$$

where ε_R denotes the real part of ε.

It can be seen from the above equation that the droplet drag can serve to improve the combustion stability. This is similar to the case of the longitudinal vibration mode. However, since ε_R is usually a small positive value, the effect of nozzle slightly reduces the combustion stability.

According to the Mach number in the combustion chamber and the axial distribution of the combustion zone, as well as nozzle admittance calculation, one can use Equation 7.63 to calculate G_R. Thus, for various transverse modes, one can determine the corresponding stability limits. Figure 7.4 shows a typical stability limit of the transverse vibration mode.

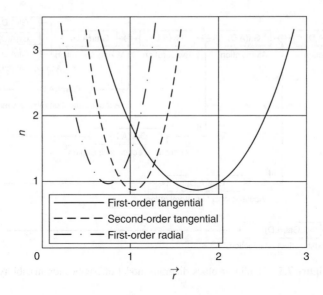

Figure 7.4 Typical stability limit of the transverse vibration mode.

7.5 Nonlinear Theory for Combustion Stability in Liquid Rocket Engines

Combustion instability is actually a kind of nonlinear vibration related to combustion and acoustic processes. The most important characteristic of nonlinear vibration is that its transfer functions of processes are not only related to frequencies but also related to amplitudes. Injection, atomization, vaporization, mixing, and chemical reaction of liquid rocket engine propellant can be seen as series-wound processes. Consequently, the global diagram of combustion instability is formed by these segments (Figure 7.5) [32]. In each segment, A denotes the pulse amplitude and D denotes the derivative to time, equivalent to the transfer function S of a linear segment. The acoustic process of the combustion chamber can be seen as a group of band-pass filters. G_1, G_r, and G_t in Figure 7.5 indicate axial, radial, and tangential filters, respectively. The band-pass of the filter is located on inherent vibration frequencies of the combustor. The fluctuating pressure field, velocity field, and temperature field will affect injection, atomization, vaporization, mixture, and combustion to form four closed loops (loops I, II, III, and IV). Only when a closed loop is formed, can large amplitude oscillations be stimulated. These four closed loops represent four different coupling mechanisms.

For each segment of the block diagram model in Figure 7.5, several theoretical, experimental, and numerical studies have been carried out during recent decades. The combustion instability driven mechanism integrated from several segments is undoubtedly very complex, but there is only one key concept, i.e., "positive feedback". Positive feedback is the reason for the evolution and development of an instability. Only sufficient positive feedback can lead to combustion instability. The key to finding and determining the incentive mechanism of combustion instability is to find the positive feedback mechanism in Figure 7.5.

In the field of reacting fluid dynamics, there are two known general positive feedback processes. One is the autocatalytic process and the other is the temperature sensitive process. The autocatalytic process refers to the case in which one of the reactants can promote its own

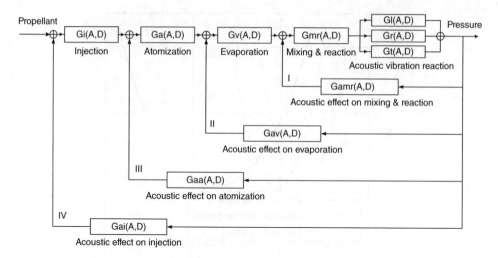

Figure 7.5   Nonlinear block diagram model of combustion instability.

synthesis or promote the synthesis of their matrix, so that any small perturbations of the react-ants can be significantly enlarged. Examples of autocatalytic processes are widespread in bio-chemical and reaction–diffusion systems. The temperature sensitive effect refers to the case in which the temperature can act as a vibration generator in the case of heat release or local heat release, and the main expression is the Arrhenius law:

$$k(T) \propto \exp\left(-\frac{E}{RT}\right)$$

The highly nonlinear dependence of the reaction rate constant on temperature together with heat release effect forms a positive feedback mechanism. An accidental increase of T will accelerate the reaction, and then much faster heat release further increases the temperature of the mixture. In the theory of adiabatic stirred tank reactors, sustained oscillations in the tempera-ture-related system have been encountered. It can be assumed that the interactions of the two positive feedbacks and coherent spatial-temporal mechanisms in a combustion chamber may lead to unstable combustion. Unfortunately, in previous studies of liquid rocket engine combus-tion instability, chemical kinetics has long been neglected and not been systematically studied.

However, some facts remind us that the chemical kinetics may be a very important candidate driving mechanism for combustion instability in a liquid rocket engine. First, the combustion process is an amplifier. Significant oscillations cannot be produced without combustion. It is hardly imaged that injecting water into hot air can lead to large oscillations. Second, as shown in the experiment of Meyer *et al.* (Figure 7.6), under some supercritical conditions it is doubtful that atomization will exist in the combustor, or if it does it is relatively weak. Probably, then, there is no vaporization process under supercritical conditions. In addition, no intrinsic differ-ence between the combustion instability under supercritical condition and under subcritical conditions has been found. Third, although there are some differences, gas propellant combus-tion has already shown some similar behavior to liquid propellant. Fourth, since the propellants are injected cryogenically or at room temperature, in the chamber there must exist regions

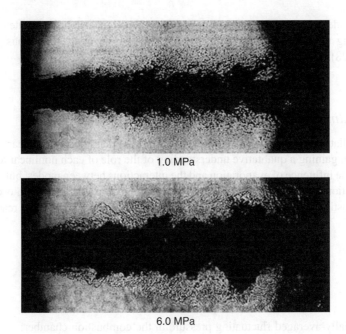

1.0 MPa

6.0 MPa

Figure 7.6 Shadow images of liquid nitrogen and helium coaxial injection under subcritical (1.0 MPa) and supercritical (6.0 MPa) chamber pressure conditions [36].

where the reactants are well mixed, and the temperature is low, unlike other areas of the combustion chamber. Chemical kinetics may become the control processes of turbulent spray combustion in those regions. Fifth, experiments on a hydrogen/oxygen/hydrocarbon tripropellant engine show that the addition of hydrogen in the hydrocarbon/oxygen flame is helpful in improving the combustion stability of hydrocarbon/oxygen flame. The most likely reason for this is that hydrogen affects the hydrocarbon/oxygen chemical reaction process.

Combustion instability is a complex volume distributed nonlinear vibration phenomenon. Numerous experiments show that its typical nonlinear characteristics are as follows: (i) When the oscillation is small, the amplitude of the combustion chamber pressure grows linearly until it reaches a certain amplitude and then the growth slows down and forms a limit cycle; (ii) When subjected to small perturbations, the combustion process may be stable, but when subjected to large perturbations, the combustion is likely to become unstable; (iii) In most cases the pressure oscillations are sinusoidal in shape, but sometimes appear as a "shock wave" of pressure peak; (iv) There exist nonlinear interactions between acoustic modes of combustion chamber pressure, sometimes there exist more than one acoustic oscillation mode simultaneously, and sometimes only one main vibration mode exists, but when this main mode is suppressed, the other oscillations may become unstable; (v) After the combustion becomes unstable, ordered temporal and spatial oscillations are formed in the combustion chamber, which is a dissipative structure, and this phenomenon cannot be explained by equilibrium thermodynamics.

This section combines nonlinear dynamics, linear non-equilibrium thermodynamics, and other linear science to establish a field oscillator model, uniform reactor model, and modes interaction model to carry out research into the combustion instability excited by chemical kinetics, trying to give a uniformly reasonable explanation of the numerous experimental phenomena [28].

7.5.1 Nonlinear Field Oscillator Model

The field oscillator model is the most commonly used simple model in nonlinear vibration. It is very helpful in gaining a qualitative understanding of the role of each nonlinear term. If we do not consider the influence of evaporation and the interactions between modes, but only consider chemical reactions and single mode oscillation, we can get the following field oscillator equation of combustion stability from the basic equations of fluid with chemical reactions:

$$\frac{\partial^2 p_n}{\partial t^2} + \frac{\omega_n}{Q_n}\frac{\partial p_n}{\partial t} + \omega_n{}^2 p_n = \frac{\partial r^n}{\partial t} \tag{7.64}$$

where:

p_n is the spatially-averaged fluctuating pressure in the combustion chamber;
t is the time;
ω is the circular frequency;
Q is the acoustic quality factor of the combustion chamber, representing the ability of the combustion chamber to store sound energy;
r is the spatially-averaged chemical reaction rate;
n as subscript and superscript represents the nth acoustic mode.

If r^n is a differentiable analytic function of p_n, and is not directly related to the time derivative of p_n and time t, r^n can be expanded into Taylor series of p_n:

$$r^n = r_0^n + r_p^n p_n + \frac{1}{2}r_{pp}^n p_n^2 + \frac{1}{6}r_{ppp}^n p_n^3 + \cdots \tag{7.65}$$

Substituting Equation 7.65 into Equation 7.64, one gets:

$$\frac{\partial^2 p_n}{\partial t^2} = \left[\omega_n/Q_n - \left(r_p^n + r_{pp}^n p_n + \frac{1}{2}r_{ppp}^n p_n^2\right)\right]\frac{\partial p_n}{\partial t} + \omega_n^2 p_n = 0 \tag{7.66}$$

The above formula is actually the famous Van der Pol equation, which was first used to describe the tube self-excited oscillations, and later became the classic example of nonlinear vibration together with the Duffing equation. Simple analysis of the above equation can uncover the different roles played by the first three coefficients about pressure sensitivity of combustion in Equation 7.65:

1. Whether the small perturbations can increase mainly depends on the first coefficients of the pressure perturbation sensitivity of a two-phase turbulent flame in the combustion chamber;

only when the sensitivity is greater than the losses, namely, $r_p^n > \omega_n/Q_n$, can combustion instability be formed, which is known as the threshold condition of combustion instability.

2. As the quadratic term has even symmetry properties, its contribution to the nonlinear vibration amplitude is neutral, but it has an effect on the average of burning rate in combustion chamber. When the quadratic term is positive, the high-frequency pressure oscillations in the combustion chamber can increase the local average burning rate, thus increasing the combustion efficiency and the average pressure of the combustion chamber.

3. Whether the combustion instability can develop into finite-amplitude oscillations, namely, whether the limit cycle is stable, mainly depends on the third coefficient r_{ppp}^n of the Taylor series. When it is negative, the amplitude of combustion instability reaches a certain value and no longer grows, forming a stable limit cycle, which is called the gain saturation of unstable combustion.

Suppose the local heat release rate (r) in the combustion chamber is controlled by the Arrhenius law, then:

$$r^n \propto \exp\left(-\frac{E}{RT}\right) \propto \exp\left(-\frac{E}{RT_0(1+\delta T/T)}\right) \propto \exp\left\{-\frac{E}{RT_0[1+(\gamma-1)p_n/\gamma P_0]}\right\} \quad (7.67)$$

One gets:

$$r_p{}^n \propto \frac{E}{RT_0(1+x)^2} r^n \simeq \frac{E}{RT_0} r^n \quad (7.68)$$

In above formula, γ is the specific heat ratio, E is the activation energy, and R is the universal gas constant:

$$x = \frac{\gamma-1}{\gamma}\frac{p_n}{P_0}$$

From Equation 7.68 it is seen that, when chemical equilibrium is not reached, neither a large activation energy nor a large reaction rate distribution (r^n) is beneficial to combustion stability.

7.5.2 Continuous Stirred Tank Reactor Acoustic Model

Currently, there is no combustion instability model for studying the interactions between multistep chemical reactions and acoustic processes. This section will combine the continuous stirred tank reactor (CSTR) model and the pulse combustion model in combustion theory to build a new combustion instability model that can be used to study combustion instability with multi-step chemical reaction kinetics and acoustic effects (Figure 7.7).

The fuel and oxidant enter the combustor from the inlet. The unreacted propellant and the combustion products are discharged from the tail pipe. Since the tail pipe is very long, it is easy to arouse acoustic oscillations. The combustion chamber pressure oscillations near the end of the tail pipe will affect the gas flow rate exhausted from the combustor. The volume flow rate

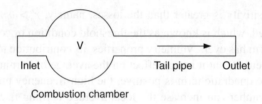

Figure 7.7   Schematic of a continuous stirred tank reactor acoustic model (CSTRA).

exhausted from combustor again does work on acoustic oscillations in the tail pipe. Thus, the acoustic process of the tail pipe and the chemical reaction process in the combustion chamber will interact. Assuming that the parameters of propellant in the combustion chamber are homogeneous, the dynamic process in the combustion chamber is controlled by multi-step chemical reaction kinetic equations, and the dynamic process in the tail pipe is described by the pressure field oscillator equation, one can derive the following equations:

$$V\frac{dC_j}{dt} = Q_0 C_{0,j} - QC_j - V\sum_i \varpi_{ji} ri \tag{7.69}$$

$$V\sum_j C_j c_{vj}\frac{dT}{dt} = Q_0\sum_j \left[(h_{0,j}-h_j)C_{0,j}\right] - V\sum_i (\Delta H_i r_i) - K_{HEAT}(T-T_a) + RTV\frac{dC_{TOT}}{dt} \tag{7.70}$$

$$P = C_{TOT}RT \tag{7.71}$$

$$Q = K_{EFF}\sqrt{P-P_{OUT}} \tag{7.72}$$

$$P_{OUT} = P_{TAIL} = p \tag{7.73}$$

$$\frac{\partial^2 p}{\partial t^2} + K_{DISS}\frac{\partial p}{\partial t} + \Omega^2 p = K_{PLUS}\frac{\partial}{\partial t}(P_{OUT}Q) \tag{7.74}$$

where:

V is the volume of combustion chamber;
C is the molar concentration;
ϖ is the coefficient of reaction equation;
r is the chemical reaction rate;
c_v is the constant volume specific heat;
h is enthalpy;
ΔH is the chemical enthalpy;
K_{HEAT} is the heat transfer coefficient between the gas and the wall;
R is the universal gas constant;
T is temperature;
C_{TOT} is the total molar concentration;
P is the combustion chamber pressure;
Q is the volume flow rate entering into the nozzle from the combustion chamber;
K_{EFF} is the relational coefficient between volume flow and pressure drop;

P_{OUT} is the transient pressure at the head of the nozzle;

P_{TAIL} is the steady pressure at the head of the nozzle;

p is the acoustic pressure at the head of the nozzle;

K_{DISS} is the dissipation factor of sound pressure in the nozzle;

Ω is the acoustic natural frequency of the nozzle;

K_{PLUS} is the working coefficient of combustion chamber on the head of the nozzle;

subscripts: "0" denotes steady state value, "j" denotes species and "i" denotes chemical reactions.

Equation 7.69 is the mass conservation equation of propellants, the component increment in the combustor is the sum of convection and chemistry reaction. Equation 7.70 is the energy conservation equation, the temperature increment is the sum of convection, reaction heat, the thermal diffusion to the wall (temperature is T_a), and mass increment. Equation 7.72 shows that the volumetric flow (Q) from the combustor to the nozzle is a function of the difference between pressure in the combustor and the pressure in nozzle. Equation 7.74 is the field oscillator equation of the nozzle, the incentive of pulsating pressure comes from the work pulsation of the product of combustion chamber pressure and volume flow rate.

Take the acetaldehyde oxidation multi-step reaction system as an example. Acetaldehyde is selected because it is a pure substance, the chemical reaction system of acetaldehyde is simpler than for other hydrocarbons, and acetaldehyde is an intermediate product of hydrocarbon combustion. Meanwhile, chemical kinetics experiments have confirmed that the oxidation reaction of acetaldehyde at low pressure contains abundant oscillations. Based on the experimental results, Cavanagh et al. [37] published numerical results of acetaldehyde oxidation at low pressure. The acetaldehyde oxidation reaction dynamics model of Lignola was used, which includes 30 components and 71 basic reactions. The specific heat and enthalpy value of each component is from a NASA report. The engineering software MATLAB was used to simulate the ordinary differential Equations 7.69–7.74. Numerical results show that the heat transfer coefficient (K_{HEAT}) from combustion chamber to wall is an important bifurcation parameter. According to the value of K_{HEAT}, the reactor shows four states (Figure 7.8):

1. In region I, when the heat transfer coefficient on the wall surface of the combustion process is small, the combustion has two steady modes, combustion state and extinguishing state,

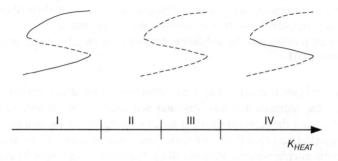

Figure 7.8 With increasing heat transfer coefficient of the combustion zone to the wall surface, combustion appears as four different states.

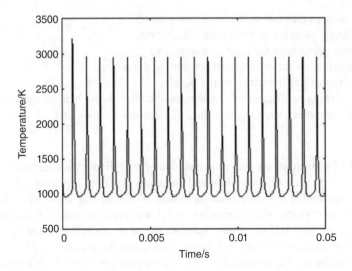

Figure 7.9   In region II, a combustion chamber temperature without a nozzle exhibits large oscillations.

   like a typical homogeneous reactor systems, and will not experience combustion oscillations.

2. In region II, the two steady modes become unstable and Hopf bifurcation occurs in a medium temperature range. High-frequency, large-amplitude self-excited oscillation appears in the reactor and the oscillating pressure wave over time is in a shock wave form (Figure 7.9). With increasing K_{HEAT}, the amplitude of this oscillation may be reduced and the frequency may gradually increase. Within a certain range of K_{HEAT}, oscillations appear as a double-period phenomenon. That is, a large fluctuation appears after several smaller fluctuations. The acoustics process in the nozzle has no obvious effects on the large amplitude combustion oscillations.

3. In region III, with the continuing increase of K_{HEAT}, the oscillations decay into a sine waveform (Figure 7.10). The acoustic processes in the combusting tail pipe and the combustion oscillation in the combustion chamber will produce a complex interaction, similar to the interaction of two spring oscillators.

4. In region IV, when K_{HEAT} is greater than a certain value, the self-excited oscillations cannot be sustained, combustion becomes stable. However, if the acoustic dissipation coefficient in the nozzle is rather small, combustion oscillations can be maintained with the help of the nozzle acoustic process and the oscillation frequency is equal to the acoustic natural frequency of the nozzle.

   These results indicate that, when the combustion zone is maintained within a certain temperature range, the temperature-sensitivity and self-catalytic mechanisms in hydrocarbon and oxygen multi-step chemical systems can generate complex combustion oscillations. These results give a good explanation of the "pressure peak" and "sinusoidal" and other combustion oscillation experiment phenomena. Heat transfer to the environment from the combustion zone has a decisive role in maintaining the combustion zone within this proper temperature range. In a real rocket engine combustion chamber, the gas heat transfer to wall is small when compared with the total heat release of combustion, but the evaporation process of droplet groups absorbs

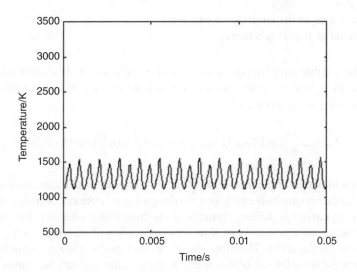

Figure 7.10 In region III, the amplitude of combustion chamber temperature without a nozzle is substantially weakened.

large amounts of heat, so that suitable temperature and mixing conditions may be formed in the some regions.

7.5.3   Spatio-Temporal Interaction Dynamic Model

Theoretically, there are unlimited acoustic modes in the engine combustion chamber, and any mode may become unstable when its plus is larger than damping. However, these modes are not mutually independent, and their interactions have an impact on whether the mode will oscillate and on the oscillation amplitude. Identifying the mode that is most likely to produce an instability can help predesign the tunes and baffles of a combustion chamber, and prevent the phenomenon that when one mode is suppressed another is strengthened. The field oscillator model and continuous stirred tank reactor acoustic model are all lumped parameter models, which cannot reflect the spatial oscillation characteristics of combustion instability. Therefore, this section establishes the spatio-temporal interaction model to study the interactions between different acoustic oscillations.

For the cylindrical combustion chamber, the pulsating pressure field can be represented by the following series of natural acoustic modes:

$$p = \sum_{m=0}^{\infty}\sum_{n=0}^{\infty}\sum_{q=0}^{\infty} p_{mnq}(t) J_n\left(\pi\alpha_{mn}\frac{r}{R}\right)\cos\frac{q\pi x}{L}[K_1\cos(n\theta-\omega t-\varphi_1) + K_2\cos(n\theta-\omega t-\varphi_2)] \quad (7.75)$$

where:

m, n, and q, respectively, denote the order of radial, tangential, and longitudinal vibrations; L denotes the effective length of the combustion chamber;

R denotes the radius of the combustion chamber;

J is the first kind of Bessel functions.

Because the one that often happens in liquid rocket engines is the first order tangential combustion instability, here we only consider the interaction of two countermove first-order tangential traveling-wave modes. Let:

$$p = p_1^+(t) J_0\left(\pi\alpha_{01}\frac{r}{R}\right)\cos\left(\theta - \omega_1^+ t - \varphi_1^+\right) + p_1^-(t) J_0\left(\pi\alpha_{01}\frac{r}{R}\right)\cos\left(\theta - \omega_1^- t - \varphi_1^-\right) \tag{7.76}$$

There is one important difference between the traveling-wave and standing-wave vibration modes. That is, the pressure wave node of a traveling wave is rotational, therefore the consumption of vibration energy is uniform throughout the combustion chamber; the pressure wave node of a standing wave is stationary, thus the consumption of vibration energy at antinodes is larger than that at the nodes. Therefore, the energy and species transport equation of tangential traveling-wave vibration can be considered as tangentially uniform. Assuming that the heat release rate has a form of Equation 7.65, and using the temporal and spatial average technology similar to Galerkin method, one can obtain the ordinary differential equations of first-order tangential traveling-wave modes over the time:

$$\frac{dp_1^+}{dt} = \frac{1}{2}\left(a^+ p_1^+ + b^+ p_1^{+^3} + c^+ p_1^{-^2} p_1^+\right) \tag{7.77}$$

$$\frac{dp_1^-}{dt} = \frac{1}{2}\left(a^- p_1^- + b^- p_1^{-^3} + c^- p_1^{+^2} p_1^-\right) \tag{7.78}$$

where b^+ and b^- are called self-saturation coefficients, while c^+ and c^- are called co-saturation coefficients. Although the above equation is derived from a special case of tangential mode, it is equally applicable to other modes. The interaction of the amplitude of two tangential vibration modes contains third-order terms, but is not influenced by second-order terms.

Assuming $X = p_1^{+^2}$ and $Y = p_1^{-^2}$ and multiplying p_1^+ and p_1^- on both sides in the above formula, one gets:

$$\frac{dX}{dt} = a^+ X + b^+ X^2 + cXY \tag{7.79}$$

$$\frac{dY}{dt} = a^- Y + b^- Y^2 + c^- XY \tag{7.80}$$

For steady state:

$$\frac{dX}{dt} = \frac{dY}{dt} = 0 \tag{7.81}$$

By carrying out bifurcation analysis of Equations 7.79 and 7.80, we can see that the characteristics of the equations are mainly decided by the relative size of self-saturation and co-saturation coefficients. When the self-saturation coefficient is stronger than the

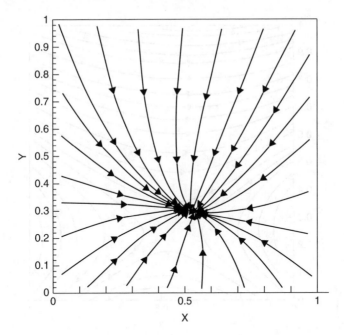

Figure 7.11 Case of weak coupling.

co-saturation coefficient, the coupling effect between different modes is weak, different modes can oscillate at the same time, but their amplitudes are smaller than those of single-mode oscillations. This is called the weak coupling case (Figure 7.11). If the self-saturation coefficient is weaker than the co-saturation coefficient, the co-saturation effect between modes is dominant, the coupling effect between modes is strong, and there is only one oscillation mode at this time. This is called the strong coupling case (Figure 7.12). These two situations are known as cooperation and competition between modes. Because the sound energy that can be provided by the propellant premixed region is limited, there exist competition and cooperation between different acoustic modes, which is consistent with experiments.

7.5.4   General Thermodynamic Analysis of Combustion Instability

The Boltzmann principle of equilibrium thermodynamics considers that in a system containing more than 10^{10} molecules the probability of coherence is zero. A rocket engine combustion chamber is an open system, there are a large number of propellants injected into the combustion chamber per unit time. Obviously, equilibrium thermodynamics cannot be used to analyze the orderly dissipative structure of combustion instability. In contrast, non-equilibrium thermodynamics consider that "non-equilibrium and nonlinear is the source of order", and there are several necessary conditions to produce a self-organized: to be far from equilibrium and exceeding the unstable branch point of thermodynamics; coherent spatiotemporal behavior and the dynamic process forming a feedback; enough heat and mass flow.

When the perturbation and deviation of a combustor working state from steady state are very small, a linear non-equilibrium thermodynamics method can be used for analysis. An important

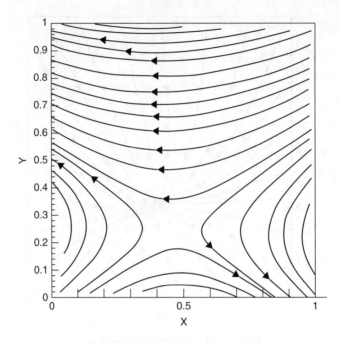

Figure 7.12   Case of strong coupling.

consequence of linear non-equilibrium thermodynamics is the theorem of minimum entropy production. The theorem of minimum entropy production can be expressed as follows: for the steady states near equilibrium, the relationship between the force and flow is approximately linearly symmetric, equilibrium state is stable and the entropy production is minimal at this state. From this theorem, we know that if the combustion in the chamber reaches or approaches equilibrium everywhere it is stable to small perturbations. Some existing combustion instability analysis models have used the result of thermodynamic calculation, the premise of which is that equilibrium is achieved everywhere in the combustion chamber. Therefore, it is impossible to predict combustion instability with this model.

The theorem of minimum entropy production stands with the condition that the relationship between force and flow is approximately linearly symmetric. Statistical mechanics shows that all transport phenomena can be satisfactorily described by the following relationship as long as the change scale of macro-gradient (l_h) is much larger than the mean free path (l_r):

$$l_h \gg l_r \tag{7.82}$$

For transport phenomena, the characteristic that the relationship between force and flow is approximately linearly symmetric is generally established. The evaporation process is a phase change process, but now the quasi-steady-state evaporation theory used in analysis of combustion instability considers that the evaporation process is controlled by diffusion processes. Thus the evaporation process controlled by Fick's diffusion law is also a linear process, which cannot lead to combustion instability. In other words, the quasi-steady evaporation theory widely used in the past does not contain the excitation mechanism of high-frequency combustion instability.

However, the situation is completely different for chemical reactions. The chemical reaction rate determined by the Arrhenius law is a strongly nonlinear function, and a symmetrical relationship between force and flow is applicable only in regions near equilibrium or in the case that the activation energy is very low. For most chemical reactions, the theorem of minimum entropy production is not true.

If there is another state near the combustion chamber design point, at which entropy production is smaller, the combustion will experience instability. According to the ideal thermodynamic cycle of a combustion chamber, the higher the temperature of the combustion process, the smaller the entropy production. Therefore, the fluctuation of burning rate and temperature with the same phase can make entropy production smaller. This shows that the Rayleigh criterion is consistent with the theorem of minimum entropy production.

When the work state of the combustion chamber is far from equilibrium, linear non-equilibrium thermodynamics are no longer applicable—only nonlinear non-equilibrium thermodynamics can be used.

7.6   Control of Unstable Combustion [32]

The ultimate goal of the combustion instability research is to control and utilize combustion oscillation. Using tunes and baffles to change the damping characteristics of the combustion chamber is a commonly employed control technology in rocket engine combustion instability. However, there are no design guidelines of tune and baffle to date. The determination of the form, number, location, and size of tune and baffle still needs numerous tests. So it may be very difficult to really enforce passive control of combustion instability (as in the case of F-1 rocket engine development).

Active control of combustion instability is another promising possibility, but there are still many practical problems that need to be solved. One problem is that most current active control strategies are based on linear control theory, but combustion instability is a very strong nonlinear phenomenon, and there are many differences between the control of nonlinear systems and linear systems.

In addition, both passive and active controls are from the perspective of acoustic suppression but do not consider the control of combustion instability from the incentive mechanism of combustion instability. In this section, we will use the earlier established concepts and methods to discuss the basic passive and active control methods of combustion instability, and to explore whether there is a third combustion instability control method.

7.6.1   Passive Control

Using tunes and baffles to change the damping characteristics of the combustion chamber is the commonly employed passive control technique in rocket engine combustion instability. It is very effective in inhibiting transverse high-frequency instabilities.

As illustrated in Figure 7.13, baffles are installed on an injection plane. One can separate the mixing, atomization, and initial burning process of propellant conducted from the injection plane into several regions, so that the sound vibration characteristics of the gas in the combustion chamber are changed and the damping effect on the sound vibration is increased. This can effectively inhibit the tangential and radial high-frequency combustion instabilities.

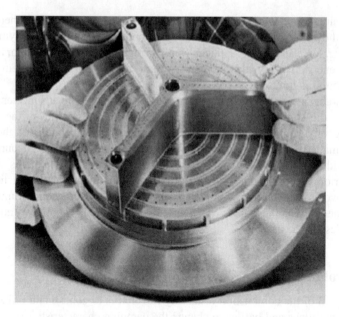

Figure 7.13 An injector equipped with baffles [40].

Figure 7.14 Schematic of tune.

Figure 7.14 shows a tune configuration. Tune is a quarter-wavelength resonator, arranged around the injector in a manner of an annular groove, which can change the acoustic vibration characteristics of the combustion chamber and dissipate the propagation energy of sound waves, inhibiting high-frequency combustion instabilities. The structure of a tune is simple and easy to produce, and cooling is needed. It is an effective combustion stabilization device for a thrust chamber. It can be used alone, or used together with baffles. Tunes and baffles are often used simultaneously in large thrust engines, where the baffle is used to inhibit low-order vibration modes and the tune is used to inhibit high-order vibration modes.

Traditionally, the main function of baffles is to change the acoustic characteristics of the combustion chamber and the function of tunes is to absorb sound energy. Thus, many studies on tune mainly discuss the ability to absorb sound energy, and also use the sound absorption ability as an indicator during the optimization process of tune. In general, the theoretical calculations of acoustic absorption ability of tunes are consistent with the data of cold tests, but are less successful when applied to combustion tests.

According to the field oscillator equation of combustion instability:

$$\frac{\mathrm{d}^2 p_n}{\mathrm{d}t^2} + \frac{\omega_n}{Q_n}\frac{\partial p_n}{\partial t} + \omega_n^2 p_n = \rho c^2 \int \sum_i \sigma_i \frac{\partial r_i}{\partial t} \Psi_n \mathrm{d}V \tag{7.83}$$

One can see that there are two ways to inhibit the oscillations: (i) reduce the acoustic quality factor (Q_n) of the combustion chamber and (ii) change the distribution of the acoustic natural vibration mode (Ψ_n) of the combustion chamber, thereby reducing its correlation with the distribution of combustion sensitive regions. The combination of these two parameters in Equation 7.83 can be used to optimize the structure, position, and number of tunes and baffles. For tune, it is more important to change Ψ_n. Therefore, it is inappropriate to only use the sound energy absorption ability as the optimal criteria, the two aspects must be considered simultaneously.

7.6.2   Active Control

Active control of vibration originally was a cross discipline of solid mechanics, automatic control, computer, material and testing technology, and other subjects, and a high technology in the field of "vibration engineering". Now it has become one of the hot research fields of vibration engineering. The active control of combustion stability is the vibration active control in the field of two-phase flow combustion. Active control systems have been applied in many different combustion systems, such as rocket engines, ramjet engines and afterburners, and so on.

In contrast with passive control, active control keeps monitoring and evaluating the state of the combustion chamber. An active control system includes sensors, observers, controllers, and actuators. Sensors are usually pressure or optical sensors, and the focus of future research is to improve the viability of sensors in the harsh operating environments of actual systems. Sensor signals are passed to the observer to determine the state of the system, the controller then sends out control signals that are received and implemented by the actuator according to the observer information. Actuators used in the laboratory include vibrators, speakers, and high-frequency modulated fuel nozzles. However, the vibrator usually has great mass, low-frequency response, and the motion range of speaker is relatively small, thus only a high-frequency modulated fuel nozzle has the potential to be applied to actual systems.

Current strategies of active control in combustion instability almost all use anti-resonant methods. That is, when the combustion chamber becomes unstable, it imposes force on the combustion process by adjusting the flow rate of propellant. The frequency of the force is equal to the natural frequency of the combustion oscillation, and the phase is opposite, so that the two oscillations offset each other. When the chamber pressure is stable, the external force is cancelled. However, since the nature of the combustion process is unstable, the combustion chamber pressure may oscillate again after a period of time, and the active control system needs to restart.

Though the active control of combustion instability has achieved some success, there are still some basic questions.

1. The common means of exerting force in the active control of combustion instability is to modulate propellant flow rate. However, as has been seen in Section **7.5**, if the combustion

oscillation is caused by chemical reaction processes the effect of changing propellant flow rate on combustion oscillation is relatively small. For example, Neumeier's active control experiment found [38] that pressure and composition show different vibration characteristics. That is, pressure vibrates in three natural frequencies of the combustion chamber, while the components vibrate in forced vibration frequencies that are different from natural frequencies. In addition, when there are many unstable modes, and the frequencies between modes are relatively close or unstable modes switch with time, it is necessary to execute real-time monitoring of unstable modes in the combustion chamber. Gutmark and Parr analyzed the time-changing pressure spectra using a wavelet analysis method [39]. However, this algorithm is relatively slow, so this method is difficult to apply in an actual system before further development of a wavelet fast transform method is achieved. In the study of active control strategies, one perhaps needs to evaluate the ability of new control schemes such as chaos control to inhibit combustion instability.

2. Current control strategies are mostly based on linear control theory. Combustion instability is a kind of nonlinear vibration where the superposition principle is not applicable. Depending on whether there exist combustion oscillations and on the amplitude of oscillations, the characteristics of the transfer function of the system may be very different. Therefore, in the design of control parameters, one cannot directly use the parameters from open-loop experiments.

3. For some combustion instabilities, active control may be powerless. For example, some vibrations caused by combustion instability can destroy the entire engine within several milliseconds; it is difficult for an active controller to respond so rapidly.

Therefore, study of the control scheme of combustion instability from the perspective of an incentive mechanism is still very necessary.

7.6.3 A Third Control Method

To radically inhibit combustion instability one must start from the incentive mechanism of unstable combustion. One way to change the excitation process of unstable combustion is to change the structure and location of the injector. Already, many cases have proven that modifying injectors can inhibit a combustion instability that is difficult to control originally. Many cases in which combustion instability is inhibited by modifying injectors have been compiled in the literature [2]. They show the important effects of injector design on the combustion instability. However, the methods of changing the structure and size of injectors are not commonly used. This is because it is often not very clear how to change the structure of the injector to inhibit combustion instability beforehand, and it is also not clear why changing the structure of the injector can inhibit the combustion instability.

However, if the combustion instability is caused by chemical kinetics of propellants, one can theoretically predict how to improve the injector to inhibit combustion instability. Propellants will pass through a low temperature premixed region after injection into the combustion chamber, which is a sensitive region of combustion instability. Reducing or eliminating the low-temperature premixed combustion region can make the combustion become stable.

A gas–liquid coaxial swirl nozzle uses the impinging effect of air flow to promote liquid propellant atomization. Its inner nozzle is a centrifugal nozzle of liquid, and its outer nozzle is a straight-flow nozzle of gas. Cold tests of injectors have shown that the impinging effect of the peripheral gas flow will reduce the spray angle of the centrifugal nozzle, i.e., the faster the gas flow is the smaller the premixed degree of the two phases. Therefore, it can be inferred that increasing the speed of the airflow can help reduce the premixed degree of the two phases, thus helping to improve the combustion stability of coaxial centrifugal nozzle. Numerous test results related to combustion stability of the United States hydrogen-oxygen rocket engine confirmed that the speed difference between gas hydrogen and liquid oxygen in the coaxial nozzle is the key parameter controlling combustion instability.

A liquid–liquid impinging nozzle uses the impinging effect of two liquid jets to promote liquid propellant atomization. If the two propellant jets are the same species, it is called a self-strike injector, otherwise it is called co-strike injector. Generally, the premixed degree of a co-strike injector is higher than that of a self-strike injector. Thus, the self-strike injector is more stable since it reduces the low-temperature premixed degree when compared with the co-strike injector. Many combustion tests related to hypergolic propellant have confirmed that the self-strike injector is more stable than the co-strike injector.

A liquid–liquid bipropellant centrifugal nozzle uses the centrifugal force to atomize the liquid propellant. To enhance the propellant combustion, the spray angle of the inner nozzle is usually larger than that of the outer nozzle, so that the liquid sheet of fuel and oxidizer intersect near the nozzle. Obviously, the smaller the interaction between the fuel and oxidant, the smaller the degree of the premix. The larger the distance of the interaction point from injection panel, the farther the premixed region is from the injection panel. Therefore, weakening the interaction between the liquid sheets is beneficial in weakening the low-temperature premixed degree, and is conducive to the combustion stability of a bipropellant centrifugal nozzle. Combustion stability tests related to the Russian liquid oxygen/kerosene engine have confirmed that decreasing the retraction ratio to weaken the interaction between the liquid sheets can achieve high stability.

Similarly, for the combustion instability excited by a low-temperature premixed region, the length of baffle should only be the length of the low-temperature premixed region. Several experimental results for the United States Apollo spacecraft F-1 confirmed that the head region of the combustion chamber is mainly divided into three regions: (i) the region closest to the jet plane, including the spray fan zone and all of the process of fuel droplets generation (about 3 inches downstream of the jet plane); in this region, almost all fuel droplets have formed and LOX liquid droplets have almost completely evaporated, (ii) the fuel evaporation region, which basically extends to 10 inches downstream of the jet plane; (iii) the region in which both fuel and oxidant are gas phases, and the combustion is almost completed. The baffle eventually used extends downstream about 3 inches from the jet plane; the baffle actually only covered the first region. The acoustic tune location should be near the low-temperature premixed zone; at present, almost all the tunes are installed in the head of the thrust chamber.

The above inference is consistent with experimental laws, suggesting that the chemical reactions can be used to uniformly interpret these tests. Before, it was considered that the droplet evaporation process is the control unit of unstable combustion, and the main measure to control propellant combustion instability is to control droplet size distribution—the smaller the droplet is, the more concentrated the combustion process is, so large droplets are conducive to combustion stability. But the theory of evaporation process controlling combustion instability contradicts the above experiments. For example, for a gas–liquid coaxial centrifugal nozzle, a high jet velocity of gas will lead to better atomization of liquid propellant; according to the theory of evaporation process controlling combustion instability, a high jet velocity of gas should not be conducive to stability, but combustion tests proved that a high jet velocity of gas favors combustion stability of a gas–liquid coaxial centrifugal nozzle.

Another way to change the excitation process of combustion instability is to change the chemical properties of the propellants. If combustion instability is caused by chemical factors, changing the chemical properties of the propellants may fundamentally eliminate combustion instability. Adding a catalyst to reduce the activation energy in the propellants, changing the chemical reaction channels, and avoiding autocatalytic chain reactions can reduce the possibility of combustion instability.

References

[1] Zhu N C et al. Combustion Instability in Liquid Propellant Rocket Engine. Beijing: National Defense Industry Press (in Chinese), 1980.

[2] Wren G P, Coffee T P. Pressure oscillations in regenerative liquid propellant guns. Journal of Propellants, Explosives and Pyrotechnics, 1995, 20(1):225–231.

[3] Yang V, Anderson W. (eds) (1995) Liquid Rocket Engine Combustion Instability. Progress in Astronautics and Aeronautics, Volume 169, American Institute of Aeronautics and Astronautics, Washington, DC.

[4] Harrje D T, Reardon F H. Liquid Propellant Rocket Instability. Russian ed. Mir, Moscow, 1975.

[5] Liu G Q. Principle of Liquid Rocket Engine. Beijing: The Aerospace Press (in Chinese), 1993, pp. 251–302.

[6] Crocco. L, Cheng X Y. Combustion Instability Theory of Liquid Rocket Engine. National Defense Industry Press (in Chinese), 1965.

[7] Priem R J, Habiballah D C. (1968) Combustion Instability Limits Determined by a Nonlinear Theory and a One-Dimensional Model. NASA-CR-920.

[8] Zhuang F C, Zhao W T, Nie W S. Numerical Simulation of Application Baffles to Suppression of Combustion Instabilities in Liquid Rocket Engine. J. Institute Command Technol. (in Chinese), 1997, 8(1).

[9] Zhuang F C, Zhao W T, Liu W D, et al. Liquid Rocket Combustion Instability Analysis Methodology by CFD J. Combust. Sci. Technol. (in Chinese), 2001, 7(1): 16–20.

[10] Liu W D, Wang Z G, Zhou J. (1996) Numerical Simulation of Unsteady Flow in Liquid Propellant Rocket Engine With PISO Algorithm. AIAA Paper 96–3124.

[11] Liu W D, Wang Z G, Zhou J, et al. Numerical Analysis Model of Radial Combustion Instability in Liquid Propellant Rocket Engine. J. Propulsion Technol. (in Chinese), 1997, 18(6):5–9.

[12] Liu W D, Wang Z G, Zhou J, et al. A Numerical Analysis Model of Tangential Combustion Instability in Liquid Propellant Engine. Journal of Propulsion Technology (in Chinese), 1998, 19(1):16–19.

[13] Zhao W T. Numerical Simulation of the Nonlinear Combustion Stability of Rocket Engine. PhD Thesis, National University of Defense Technology (in Chinese), 1997.

[14] Hong X. Study on High-Frequency Combustion Instability in Liquid Propellant Rocket Engine. PhD Thesis, Shanghai Jiaotong University (in Chinese), 1998.

[15] Nie W S. Study on Combustion Instability of Hypergolic Propellant Rocket Engine. PhD Thesis, National University of Defense Technology (in Chinese), 1998.

[16] Huang Y H, Wang Z G, Zhou J. An Experimental Study on Acoustic Characteristics of Gas-Liquid Coaxial Injector of Liquid Rocket Engine. J. Propulsion Technology (in Chinese), 1996, 17(4):37–41.

[17] Zhou J, Huang Y H. (1996) Flowrate and Acoustics Characteristics of Coaxial Swirling Injector of Hydrogen/Oxygen Rocket Engine. 32nd AIAA/ASME/SAE/ASEE Joint Propulsion Conference and Exhibit, AIAA Paper 96–3135.

[18] Huang Y H, Wang Z G. (1997) Acoustic Model for the Self-oscillation of Coaxial Swirl Injector. 33rd AIAA/ASME/SAE/ASEE Joint Propulsion Conference and Exhibit, AIAA Paper 97–3323.

[19] Huang Y H, Wang Z G, Zhou J. An Experimental Study on Acoustic Characteristics of Gas-Liquid Coaxial Injector of Liquid Rocket Engine. Journal of Propulsion Technology (in Chinese), 1996, 17(4):37–41.

[20] Huang Y H, Wang Z G, Zhou J. Experiment Acoustic Model for the Self-oscillation of Coaxial Swirl Injector and its Influence to Combustion of Liquid Rocket Engine. Chin. J. Acoustics, 1998, 17(2):163–170.

[21] Huang Y H, Wang Z G, Zhou J. (1999) The Characteristics of Nonlinear Acoustic Waves in Engine. AIAA Paper 99–2916.
[22] Wang Z G, Huang Y H. (2000) A Nonlinear Analytical Model of High-Frequency Combustion Instability in Rocket Engine. AIAA Paper 2000–3299.
[23] Huang Y H, Wang Z G, Zhou J. (2000) Numeric Simulation for Combustion Instability of Liquid Propellant Rocket Engine. AIAA Paper 2000–3297.
[24] Huang Y H, Wang Z G, Zhou J. (2000) Nonlinear Thermodynamic Analyses of Combustion Instability in Rocket Engine. AIAA Paper 2000–3298.
[25] Huang Y H, Wang Z G. Nonlinear Analysis for High-Frequency Combustion Instability in Rocket Engines. *Journal of Propulsion Technology* (in Chinese), 2000, **21**(5).
[26] Huang Y H, Wang Z G. (2001) Theoretical Model of Liquid Rocket Engine Combustion Instability. Sino-France Joint Propulsion Conference, 2001, Beijing.
[27] Huang Y H, Wang Z G, Zhou J. (2002) Global Model of Liquid Rocket Engine Combustion Instability Based on Chemistry Dynamics. AIAA Paper 2002–3992.
[28] Huang Y H, Wang Z G, Zhou J. Nonlinear Theory of Combustion Stability in Liquid Rocket Engine Based on Chemistry Dynamics. *Sci. Chin., Ser*. B, 2002, **45**(4): 373–382.
[29] Huang Y H, Wang Z G, Zhou J. Numerical Simulation of Combustion Stability of Liquid Rocket Engine Based on Chemistry Dynamics. *Sci. Chin., Ser*. B, 2002,**45**(5): 551–560.
[30] Huang Y H, Wang Z G. Nonlinear Acoustic Wave in Combustors. *Journal of Propulsion Technology* (in Chinese), 2002, **23**(6):492–495.
[31] Huang Y H, Wang Z G, Zhou J. Experimental Investigation on Combustion Instability of Tripropellant Rocket Engine. *Journal of Propulsion Technology (in Chinese)*, 2003, **24**(1):71–73.
[32] Huang Y H. (2001) Theoretical, Numerical Simulation and Experimental Investigations of Combustion Instability in Liquid Rocket Engine. PhD Thesis, National University of Defense Technology (in Chinese).
[33] Ma D Y. Theory and nonlinearity of thermoacoustics. *Journal of Acoustics (in Chinese)*, 1999, **24**(4):337–350.
[34] Sun W S. *Combustion Instability in Solid Rocket Motors*. Beijing Institute of Technology Press (in Chinese), 1988.
[35] Sirignano W A. Fluid Dynamics of Sprays. *Fluids Eng*., 1993, **115**: 345–378.
[36] Mayer W H O, Schik A H A, Vielle B, *et al*. Atomization and Breakup of Cryogenic Propellants Under High-Pressure Subcritical and Supercritical Conditions. *Journal of Propulsion and Power*, 1998, **14**(5): 835–842.
[37] Cavanagh J, Cox R A, Olson G. Computer Modeling of Cool Flames and Ignition of Acetaldehyde. *Combust. Flame*, 1990, **82**(1): 15–39.
[38] Neumeier Y, Nabi A, Zinn B T. (1996) Investigation of the Open Loop Performance of Active Control System Utilizing a Fuel Injector Actuator. AIAA Paper 96–2757.
[39] Gutmark E, Parr T P. (1990) Use of Chemiluminescence and Neural Networks in Active Combustion Control. Presented at the Twenty-third Symposium (International) on Combustion, Orléans, 22–27 July 1990.
[40] Harrje D J, Reardon F H (Ed.) (1972) Liquid Propellant Rocket Instability, NASA SP-194.

8

Numerical Method and Simulations of Liquid Rocket Engine Combustion Process

The working process in liquid rocket engines is very complex, mainly including the liquid propellant jet atomization, propellant droplet evaporation, mixing and burning of propellant components, accelerated discharging of high temperature gas, and heat transfer between gas and combustion chamber wall and so on. It is very important to understand the physical processes and to establish suitable physical models to describe them for numerical simulations of the liquid rocket engine combustion process. This chapter presents numerical simulation models of the spray combustion flow process in liquid rocket engines, introduces the SIMPLE algorithm and the PISO algorithm applied to the combustion process, and finally gives several calculation examples of the liquid rocket engine combustion process.

8.1 Governing Equations of Two-Phase Multicomponent Reaction Flows

From the three laws of mass conservation, momentum conservation, and energy conservation, the governing equations of the spray combustion two-phase multicomponent chemical reaction flow in liquid rocket engines are derived. This includes gas governing equations and liquid phase governing equations. The coupling between two phases can be described by the source term interacting between gas and liquid; a Eulerian coordinate system is used to describe the gas phase equations, while a particle track method under Lagrangian coordinates is used to simulate the movement of tracked droplets.

Most of the formulae given in this section have been presented in previous chapters. They are repeated here to give the reader a complete understanding and to enhance the systematism of numerical simulations of combustion process in a liquid rocket engine. The mathematical physical model listed here is a simplified version of models described in previous chapters, focusing

Internal Combustion Processes of Liquid Rocket Engines: Modeling and Numerical Simulations,
First Edition. Zhen-Guo Wang.
© 2016 National Defense Industry Press. Published 2016 by John Wiley & Sons Singapore Pte Ltd.

on the model of the propellant spray combustion flow process. The heat transfer model, the turbulent combustion model, and the instable combustion model are not included in this chapter.

8.1.1 Gas Phase Governing Equation

The Favre average Navier–Stokes equations of a three-dimensional unsteady multicomponent chemical reaction flow can be expressed in the following conservation form [1]:

$$\frac{\partial Q}{\partial t} + \frac{\partial (E - E_v)}{\partial x} + \frac{\partial (F - F_v)}{\partial y} + \frac{\partial (G - G_v)}{\partial z} = H \tag{8.1}$$

Among them:

$$
Q = \begin{bmatrix} \bar{\rho} \\ \bar{\rho}\tilde{u} \\ \bar{\rho}\tilde{v} \\ \bar{\rho}\tilde{w} \\ \bar{\rho}\tilde{e} \\ \bar{\rho}\tilde{Y}_i \end{bmatrix}
\quad
E = \begin{bmatrix} \bar{\rho}\tilde{u} \\ \bar{\rho}\tilde{u}\tilde{u} + \bar{p} \\ \bar{\rho}\tilde{u}\tilde{v} \\ \bar{\rho}\tilde{u}\tilde{w} \\ \tilde{u}(\bar{\rho}\tilde{e} + \bar{p}) \\ \bar{\rho}\tilde{u}\tilde{Y}_i \end{bmatrix}
\quad
F = \begin{bmatrix} \bar{\rho}\tilde{v} \\ \bar{\rho}\tilde{v}\tilde{u} \\ \bar{\rho}\tilde{v}\tilde{v} + \bar{p} \\ \bar{\rho}\tilde{v}\tilde{w} \\ \tilde{v}(\bar{\rho}\tilde{e} + \bar{p}) \\ \bar{\rho}\tilde{v}\tilde{Y}_i \end{bmatrix}
\quad
G = \begin{bmatrix} \bar{\rho}\tilde{w} \\ \bar{\rho}\tilde{w}\tilde{u} \\ \bar{\rho}\tilde{w}\tilde{v} \\ \bar{\rho}\tilde{w}\tilde{w} + \bar{p} \\ \tilde{w}(\bar{\rho}\tilde{e} + \bar{p}) \\ \bar{\rho}\tilde{w}\tilde{Y}_i \end{bmatrix}
$$

$$
E_v = \begin{bmatrix} 0 \\ \bar{\tau}_{xx} \\ \bar{\tau}_{xy} \\ \bar{\tau}_{xz} \\ \tilde{u}\bar{\tau}_{xx} + \tilde{v}\bar{\tau}_{xy} + \tilde{w}\bar{\tau}_{xz} - \bar{q}_x \\ \bar{\rho}_i \bar{D}_{imi}\partial \tilde{Y}_i/\partial x \end{bmatrix}
\quad
F_v = \begin{bmatrix} 0 \\ \bar{\tau}_{yx} \\ \bar{\tau}_{yy} \\ \bar{\tau}_{yz} \\ \tilde{u}\bar{\tau}_{xy} + \tilde{v}\bar{\tau}_{yy} + \tilde{w}\bar{\tau}_{yz} - \bar{q}_y \\ \bar{\rho}_i \bar{D}_{imi}\partial \tilde{Y}_i/\partial y \end{bmatrix}
$$

$$
G_v = \begin{bmatrix} 0 \\ \bar{\tau}_{zx} \\ \bar{\tau}_{zy} \\ \bar{\tau}_{zz} \\ \tilde{u}\bar{\tau}_{zx} + \tilde{v}\bar{\tau}_{zy} + \tilde{w}\bar{\tau}_{zz} - \bar{q}_z \\ \bar{\rho}_i \bar{D}_{imi}\partial \tilde{Y}_i/\partial z \end{bmatrix}
\quad
H = \begin{bmatrix} S_{d,m} \\ S_{d,u} \\ S_{d,v} \\ S_{d,w} \\ S_{d,h} \\ \bar{\omega}_i \end{bmatrix} \tag{8.2}
$$

In the formula:

$i = 1, 2, \ldots, N_s - 1$, N_s is the total component fraction;
$\bar{\rho}_i$ is the density of each component,
$\bar{\rho}$ is the density of mixed gas;
$\tilde{u}, \tilde{v}, \tilde{w}$ are the velocity components in the direction x, y, z;
$\bar{p}$ is the pressure;
$\tilde{Y}_i$ is the mass fraction of component i;
$\bar{\omega}_i$ is the production rate of the mass of component i;
$S_{d,m}, S_{d,u}, S_{d,v}, S_{d,w}, S_{d,h}$ are the gas/liquid two-phase interaction and chemical reaction source terms;

τ_{ij} is the viscous stress component:

$$\bar{\tau}_{xx} = -\frac{2}{3}\mu(\nabla \cdot V) + 2\mu\frac{\partial \tilde{u}}{\partial x}$$

$$\bar{\tau}_{yy} = -\frac{2}{3}\mu(\nabla \cdot V) + 2\mu\frac{\partial \tilde{u}}{\partial y}$$

$$\bar{\tau}_{zz} = -\frac{2}{3}\mu(\nabla \cdot V) + 2\mu\frac{\partial \tilde{u}}{\partial z}$$

$$\bar{\tau}_{xy} = \bar{\tau}_{yx} = \mu\left(\frac{\partial \tilde{u}}{\partial y} + \frac{\partial \tilde{v}}{\partial x}\right)$$

$$\bar{\tau}_{yz} = \bar{\tau}_{zy} = \mu\left(\frac{\partial \tilde{w}}{\partial y} + \frac{\partial \tilde{v}}{\partial z}\right)$$

$$\bar{\tau}_{xz} = \bar{\tau}_{zx} = \mu\left(\frac{\partial \tilde{w}}{\partial x} + \frac{\partial \tilde{u}}{\partial z}\right)$$

$\bar{q}_x, \bar{q}_y, \bar{q}_z$ is the energy flux caused by heat conduction and component diffusion:

$$\bar{q}_x = -k\frac{\partial \tilde{T}}{\partial x} - \bar{\rho}\sum_{i=1}^{N_s}\bar{D}_{im}\bar{h}_i\frac{\partial \tilde{Y}_i}{\partial x}$$

$$\bar{q}_y = -k\frac{\partial \tilde{T}}{\partial y} - \bar{\rho}\sum_{i=1}^{N_s}\bar{D}_{im}\bar{h}_i\frac{\partial \tilde{Y}_i}{\partial y}$$

$$\bar{q}_z = -k\frac{\partial \tilde{T}}{\partial z} - \bar{\rho}\sum_{i=1}^{N_s}\bar{D}_{im}\bar{h}_i\frac{\partial \tilde{Y}_i}{\partial z}$$

$\bar{D}_{im}$ is the mass diffusion coefficient of the mixture component i:

$$\bar{D}_{im} = (1 - \tilde{X}_i)/\sum_{i,j\neq i}(\tilde{X}_j/\bar{D}_{ij}) \tag{8.3}$$

At the middle and low pressure, the interaction diffusion coefficient of the two-component mixed gas is:

$$\widetilde{D}_{ij} = 1.883 \times 10^{-2} \sqrt{\widetilde{T}^3 \left(M_i + M_j\right) / M_i M_j} / \overline{p} \sigma_{ij}^2 \Omega_D \tag{8.4}$$

In the formula:

$\widetilde{X}_i$ is the mole fraction of component i;
M_i, M_j are the molecular weights of the gas components i, j;
σ_{ij} is the characteristic length;
Ω_D is the collision integral.

Internal energy:

$$\widetilde{e} = \sum_{i=1}^{N_s} \widetilde{Y}_i \overline{h}_i + \frac{1}{2} \left(\widetilde{u}^2 + \widetilde{v}^2 + \widetilde{w}^2\right) - \frac{\overline{p}}{\overline{\rho}} \tag{8.5}$$

Enthalpy of components:

$$\overline{h}_i = h_i^0 + \int_{T_{\text{ref}}}^{T} c_{pi} \, dT \tag{8.6}$$

The specific heat of each composition at constant pressure using a polynomial fitting formula is:

$$c_{pi} = a_{1,i} + a_{2,i} T + a_{3,i} T^2 + a_{4,i} T^3 + a_{5,i} T^4 \tag{8.7}$$

Coefficients for each component are given in the literature [2,3].
In addition, assume that the multicomponent gas phase mixture complies with the ideal gas state equation and the local thermodynamic equilibrium hypothesis:

$$\overline{p} = R\widetilde{T} \sum_{i=1}^{N_s} \frac{\overline{\rho}_i}{M_i} \tag{8.8}$$

8.1.2 Liquid Particle Trajectory Model

Most of the liquid propellant evaporates and burns after being atomized into small droplets by injection devices. A liquid rocket engine spray combustion process is a typical process of two phase flow. At present, two-phase flow models mainly include the single fluid model, particle trajectory model, and quasi fluid model. The particle trajectory model is the most widely used in numerical calculations of the liquid rocket engine combustion process. Based on the particle track model assumptions, the gas is treated as continuum and the droplet groups are treated as a discrete system. The liquid spray can be divided into typical groups of discrete droplets, and the Lagrangian method is used to track the movement and transport of these discrete drop-lets in the whole flow field. The droplet trajectory is solved from the droplet dynamics equation, and the coupling solution of the mass, momentum, and energy exchange between droplet and gas phase is used to develop parameters such as temperature, the radius of the droplet, and gas phase field.

In the particle trajectory model, the trajectory of discrete phase particles (droplets) is obtained by integrating the differential equation of the particles forces in Lagrangian coordin-ates. The particles balance equations are:

$$\frac{d\vec{x}}{dt} = \vec{V}_p \tag{8.9}$$

$$\frac{d\vec{V}_p}{dt} = F_D\left(\vec{V} - \vec{V}_p\right) + \frac{\vec{g}\left(\rho_p - \rho\right)}{\rho_p} + \vec{F} \tag{8.10}$$

$$F_D = \frac{18\mu}{\rho_p d_p^2}\frac{C_D Re}{24} \tag{8.11}$$

where:

$\vec{V}$ is the gas phase velocity;

$\vec{V}_p$ is the particle velocity;

μ is the viscosity of fluid dynamics;

ρ is fluid density;

ρ_p is the particle density;

d_p is the particle diameter;

$\vec{g}$ is the acceleration due to gravity;

$\vec{F}$ represents all the other forces, including Stefan flow, the pressure gradient force, and other volume forces;

$F_D\left(\vec{V} - \vec{V}_p\right)$ is the resistance of the particle per unit mass;

C_D is the coefficient of the resistance.

Assuming that each droplet is a round ball, C_D can adopt the following expression:

$$C_D = \frac{24}{Re}\left(1 + b_1 Re^{b_2}\right) + \frac{b_3 Re}{b_4 + Re} \tag{8.12}$$

where:

$$\begin{aligned}
b_1 &= \exp(2.3288 - 6.4581 + 2.4486\varphi^2) \\
b_2 &= 0.0964 + 0.5565\varphi \\
b_3 &= \exp(4.905 - 13.8944\varphi + 18.4222\varphi^2 - 10.2599\varphi^3) \\
b_4 &= \exp(1.4681 + 12.2584\varphi - 20.7322\varphi^2 + 15.8855\varphi^3)
\end{aligned} \tag{8.13}$$

In the formula, the shape factor ϕ is defined as follows:

$$\phi = \frac{s}{S} \tag{8.14}$$

where s is the spherical particle surface area, which has the same volume as the actual particle, and S is the actual particle surface area.

Given the development of the gas phase flow field and droplet radius, density, and temperature, the ordinary differential equations can be solved directly to obtain the movement and trajectory of the droplet.

The mass, momentum, and energy exchange between droplet and gas phase not only determines the development of the temperature, speed, size of the fuel droplet, and the evaporation rate of the fuel directly, but also affects the gas phase governing equation directly. Therefore, the source terms:

$$S_{d,m}, S_{d,u}, S_{d,v}, S_{d,w}, S_{d,h}$$

in the gas phase Navier–Stokes equation must be modeled to close gas phase equations.

For any grid cell in the flow field, the source term in the gas phase due to the action of the droplet can be computed as follows:

The mass:

$$S_{d,m} = \frac{\Delta m_d}{m_{d,0}} \dot{m}_{d,0} \tag{8.15}$$

The momentum:

$$\vec{S}_{d,v} = \left(\frac{18\mu C_D Re}{\rho_d d_d^2 24} \left(\vec{V} - \vec{V}_d \right) + \vec{F}_x \right) \dot{m}_d \Delta t \tag{8.16}$$

The energy:

$$S_{d,h} = \left[\frac{\bar{m}_d}{m_{d,0}} c_{drop} \Delta T_d + \frac{\Delta m_d}{m_{d,0}} \left(-h_{fg} + h_{pyrol} + \int_{T_{ref}}^{T_d} C_{p,1} dT \right) \right] \dot{m}_{d,0} \tag{8.17}$$

In the formula:

Δm_d is the change of the unit mass of the droplets through the control unit;

$m_{d,0}$ is the initial mass;

$\dot{m}_{d,0}$ is the initial mass flow rate;

$\dot{m}_d$ is the mass flow rate of the droplet;

Δt is the time step in the calculation;

$\bar{m}_d$ is the average mass of the droplet in a grid cell in calculation;

ΔT_d is the change of droplet temperature in grid cell in calculation;

h_{fg} is the latent heat of vaporization of the fuel droplet;

h_{pyrol} is the high-temperature pyrolysis heat of the volatile components of the fuel droplet; $C_{p,\text{l}}$ is the specific heat at constant pressure of the volatile components of fuel droplet.

8.1.3 Turbulence Model

Based on the Boussinesq hypothesis, in the gas phase governing equations, the viscosity coefficient (μ) can be decomposed into the laminar viscous coefficient (μ_{l}) and the turbulent viscosity coefficient (μ_{t}: $\mu = \mu_{\text{l}} + \mu_{\text{t}}$).

Among them, μ_l is given by the Sutherland formula:

$$\mu_{\text{l}} = \mu_{\text{ref}}(T/T_{\text{ref}})^{3/2}(T_{\text{ref}} + S_{\text{T}})/(T + S_{\text{T}}) \tag{8.18}$$

In the formula, T_{ref} is the reference temperature, S_{T} is the equivalent temperature, and μ_{ref} is the reference viscosity coefficient corresponding to T_{ref}.

In solving RANS, μ_{t} can be given by the compressibility modified $k-\omega$ turbulence model [4]:

$$\frac{\partial}{\partial t}(\bar{\rho}k) + \frac{\partial}{\partial x_i}(\bar{\rho}k\tilde{u}_i) = \frac{\partial}{\partial x_j}\left(\Gamma_k \frac{\partial k}{\partial x_j}\right) + 2\mu_{\text{t}}\bar{S}_{ij}\bar{S}_{ij} - \bar{\rho}\beta^* f_{\beta^*}\omega \tag{8.19}$$

$$\frac{\partial}{\partial t}(\bar{\rho}\omega) + \frac{\partial}{\partial x_i}(\bar{\rho}\omega\tilde{u}_i) = \frac{\partial}{\partial x_j}\left(\Gamma_\omega \frac{\partial \omega}{\partial x_j}\right) + 2\frac{\omega}{k}\mu_{\text{t}}\bar{S}_{ij}\bar{S}_{ij} - \bar{\rho}\beta f_\beta\omega^2 \tag{8.20}$$

In the formula:

$$\Gamma_k = \mu_{\text{l}} + \frac{\mu_{\text{t}}}{\sigma_k} \quad \Gamma_\omega = \mu_{\text{l}} + \frac{\mu_{\text{t}}}{\sigma_\omega} \quad \mu_{\text{t}} = \frac{\bar{\rho}k}{\omega}$$

$$f_{\beta^*} = \begin{cases} 1 & \chi_k \leq 0 \\ \dfrac{1 + 680\chi_k^2}{1 + 400\chi_k^2} & \chi_k > 0 \end{cases} \quad \chi_k \equiv \frac{1}{\omega^3}\frac{\partial k}{\partial x_j}\frac{\partial \omega}{\partial x_j}$$

$$\beta^* = \beta_i^*[1 + \zeta^* F(M_{\text{t}})]$$

$$f_\beta = \frac{1 + 70\chi_\omega}{1 + 80\chi_\omega} \quad \chi_\omega = \left|\frac{\Omega_{ij}\Omega_{jk}S_{ki}}{(\beta_i^*\omega)^3}\right| \quad \Omega_{ij} = \frac{1}{2}\left(\frac{\partial \tilde{u}_i}{\partial x_j} - \frac{\partial \tilde{u}_j}{\partial x_i}\right)$$

$$\beta = \beta_i\left[1 - \frac{\beta_i^*}{\beta_i}\zeta^* F(M_{\text{t}})\right]$$

$$F(M_{\text{t}}) = \begin{cases} 0 & M_{\text{t}} \leq M_{t0} \\ M_{\text{t}}^2 - M_{t0}^2 & M_{\text{t}} > M_{t0} \end{cases} \quad M_{\text{t}}^2 \equiv \frac{2k}{\gamma RT}$$

$$\sigma_k = \sigma_\omega = 2.0, \quad \zeta^* = 1.5, \quad \beta_i^* = 0.09, \quad \beta_i = 0.072, \quad M_{t0}^2 = 0.25$$

In the LES solution, we can use the Smagorinsky–Lilly model to calculate the sub-grid turbulence viscosity:

$$\mu_t = \rho L_s^2 |\bar{S}| \tag{8.21}$$

In the formula:

$$|\bar{S}| = \sqrt{2\bar{S}_{ij}\bar{S}_{ij}}$$

$$\bar{S}_{ij} = \frac{1}{2}\left(\frac{\partial \tilde{u}_i}{\partial x_j} + \frac{\partial \tilde{u}_j}{\partial x_i}\right)$$

L_s is the mixing length of the sub-grid:

$$L_s = \min\left(\kappa d, C_s V^{1/3}\right) \tag{8.22}$$

where:

κ is the von Karman constant,
d is the nearest distance to the wall,
C_s is the Smagorinsky constant,
V is the volume of the cell.

8.1.4   Droplets Atomizing Model

In numerical calculations of the liquid rocket engine combustion process, the spray model as the initial and boundary conditions of the spray combustion calculation has an important influence on the calculation results. The spray model includes spray size distribution model, the flow intensity distribution model, and mixing ratio distribution model, respectively obtaining the mean diameter distribution of spray droplets, the position distribution of droplets, and the initial velocity distribution of droplets. As the atomization process is more complicated, we usually do not consider the details of the liquid atomization process. Instead, for the injector used in the experiments, the distribution characteristics of droplets (including parameters such as droplet size distribution, velocity distribution, and temperature) are directly given based on experimental results in combination with the empirical formula to directly simulate the liquid atomization results. Next, we will introduce in brief several spray models for commonly used injectors [5].

Centrifugal Injector
In engineering applications, the Rosin–Rammler distribution is widely used to describe the droplet spray size distribution of the centrifugal injector:

$$\frac{dR_i}{dd_i} = B_n d_i^{n-1} \exp\left(-Bd_i^n\right) \tag{8.23}$$

where R_i is the percentage of the mass integral of the droplets whose diameter is less than d_i in the total mass of all droplets; B_n are constants for the specific injector and working condition.

Self-Impinging Injector

For self-impinging injectors, the Nukiyama–Tanasawa distribution can be used:

$$\frac{dR_i}{dd_i} = Ad_i^5 \exp(-Bd_i) \tag{8.24}$$

where A is the empirical coefficient associated with the specific injector and working conditions, which can be determined by experiments. For doublet self-impinging injectors, the drop size distribution is:

$$\frac{dR_i}{dd_i} = \left(\frac{3.915}{d_{30}}\right)^6 \frac{d_i^5}{120} \exp(-3.915d_i/d_{30}) \tag{8.25}$$

In the formula, d_{30} is the droplet volume mean diameter defined as:

$$d_{30}^3 = \sum N_i d_i^3 / \sum N_i.$$

According to the N-T distribution, it can be found that $d_{30}^3 = B^{-3}\Gamma(6)/\Gamma(3)$.

Unlike Impinging Injector

Doublet unlike impinging injectors are mainly used for hypergolic propellants. This injector has very little atomization fineness test and needs to make predictions on the droplet size distribution. Rocketdyne uses molten paraffin wax and water as the test propellant, giving the following correlation:

$$d_{m,ox} = \frac{2.37 \times 10^5}{d_{ox}^{0.38} V_{ox}^{0.86} V_f^{1.19}} \tag{8.26}$$

$$d_{m,f} = 2.1 \times 10^4 \frac{d_f^{0.27} d_{ox}^{0.023}}{V_{ox}^{0.33} V_f^{0.74}} \tag{8.27}$$

In the formula, the injector diameter is given in units of cm, the injection speed in m s^{-1}, and d_m in μm; the subscript "ox" means the oxidant and "f" means the fuel.

Coaxial Injector

Coaxial injectors are usually used in oxyhydrogen, liquid oxygen/methane cryogenic propellant rocket engines, and so on. The coaxial injector atomization model is:

$$D_m = k\left(\frac{O}{F}\right)\sqrt{\frac{\mu_g A_g}{R\rho_l T_g}} p_c \tag{8.28}$$

In the formula:

D_m is the mean diameter of the droplet;
O/F is the mixture ratio of oxidizer and fuel;
T_g is the injection temperature of the hydrogen;
p_c is the chamber pressure;
μ_g is the hydrogen molecular mass;
R is the gas constant;
ρ_l is the density of the liquid oxygen;
k is a constant.

After getting the mean droplet diameter at the corresponding working conditions, we can use the Rosin–Rammler distribution to obtain a more accurate droplet size distribution.

8.1.5   Droplet Evaporation Model

The internal flow and heat transfer process of the droplet usually have two limit cases, which are known as the surface model and the uniform temperature model.

The surface model assumes that the coefficient of temperature conductivity of the fuel droplet $\alpha \rightarrow 0$, assuming that a high temperature difference always exists between the surface of a drop and its internal mass. Assuming that the droplet internal temperature is equal to the initial droplet temperature, only the droplet surface temperature reaches the thermodynamic equilibrium temperature, and the droplet can exchange heat and mass with external air flow through the surface. The surface model is used in the treatment of droplet evaporation in still air.

The uniform temperature model is also known as the infinite thermal conductivity model. It assumes that the coefficient of temperature conductivity of the drop $\alpha \rightarrow \infty$, and that the droplet internal circulation and other factors mean that the droplet temperature is always uniform and identical to the surface temperature. This model is applicable to the droplet evaporation of small droplets, strong convection, and when the droplet internal circulation is good.

When we deal with the droplet preheating problem, surface models usually ignore droplet evaporation in the heating period, and the droplet internal heating is calculated according to the unsteady heat conduction. When the droplet surface temperature reaches the equilibrium temperature, the evaporation where the heat and mass transfer are balanced between the two phases is established. The uniform temperature model assumes that the droplet evaporation is determined by the diffusion in the droplet preheating period, which is a non-equilibrium evaporation process; the heat transmitted to the droplet not only heats the droplet but also provides the heat required for droplet evaporation; After reaching the equilibrium temperature, we have a stable evaporation process when the diffusion and heat transfer is in phase equilibrium. The equilibrium temperature of droplet steady evaporation is closely related to the flow conditions, usually growing with increasing flow temperature. When the current temperature is extremely high, the equilibrium temperature will be close to the boiling temperature of liquid drops. Unlike the model of droplet evaporation in static air, a droplet undergoes deformation due to the ambient air flow effect in convection conditions. The evaporation model cannot use the one-dimensional spherically symmetric method, which makes it more difficult to model this process. Consequently, the droplet evaporation in convection conditions must use more of an engineering approximation method, such as the "converting film" theory.

Before a drop reaches the evaporation temperature, the heating mechanism can be described by the formula:

$$m_p c_p \, dT_p/dt = hA_p \left(T_\infty - T_p \right) \tag{8.29}$$

where m_p, c_p, and T_p are the droplet mass, specific heat, and temperature, respectively, A_p is the droplet surface area, T_∞ is the local gas temperature, and h is the heat convection coefficient.

After the droplet temperature reaches the evaporation temperature (before the boiling point temperature), assuming that the droplet evaporates in the convective flow and the physical condition is uniform inside the droplet, the evaporation rate is controlled by the concentration difference between the droplet surface and the gas phase:

$$N_i = k_c \left(C_{i,s} - C_{i,\infty} \right) \tag{8.30}$$

where N_i is the molar mass flow of the vapor and $C_{i,s}$ is the vapor concentration on droplet surface. Assuming that the vapor pressure on the drop surface is equal to the saturated vapor pressure (p_{sat}) at the drop temperature (T_p), the following correlation can be obtained:

$$C_{i,s} = p_{sat} \left(T_p \right) / RT_p$$

$C_{i,\infty}$ is the concentration in the gas phase, given by the transport of the gas component:

$$C_{i,\infty} = X_i p_{op} / RT_\infty \, ;$$

k_c is the mass transfer coefficient, given by the Nusselt equation:

$$\mathrm{Nu} = k_c d_p / D_{i,m} = 2.0 + 0.6 \mathrm{Re}_d^{0.5} \mathrm{Sc}^{0.33} ,$$

where $D_{i,m}$ is the diffusion coefficient of the vapor, Sc is the Schmidt number ($\mu/\rho D_{i,m}$), and d_p is the droplet diameter.

Therefore, the droplet mass in the next time step is:

$$m_p(t + \Delta t) = m_p(t) - N_i A_p M_{w,i} \Delta t \tag{8.31}$$

In addition, the development of the drop temperature is described by the heat balance equation between the droplet and the gas phase:

$$m_p c_p \cdot \frac{dT_p}{dt} = hA_p \left(T_\infty - T_p \right) + \frac{dm_p}{dt} \cdot h_{fg} \tag{8.32}$$

where dm_p/dt is the evaporation rate and h_{fg} is the latent heat of droplet evaporation.

8.1.6   Chemical Reaction Kinetics Model

For a chemical reaction system composed of N_R reactions, the production rate of component i is:

$$\omega_i = M_{w,i} \sum_{r=1}^{N_R} \hat{\omega}_{i,r} \qquad (8.33)$$

where $\hat{\omega}_{i,r}$ is the Arrhenius molecular production rate of component i in the r-th reaction and $M_{w,i}$ is the molecular weight of component i.

The general form of the r-th reaction equation is:

$$\sum_{i=1}^{N} v'_{i,r} M_i \underset{k_{b,r}}{\overset{k_{f,r}}{\rightleftharpoons}} \sum_{i=1}^{N} v''_{i,r} M_i \qquad (8.34)$$

where:

N is the total component number of the reaction;

$v'_{i,r}, v''_{i,r}$ represent the stoichiometric coefficients of the chemical component i in the positive and reverse reactions, respectively;

M_i is the molecular weight of chemical component i;

$k_{f,r}, k_{b,r}$ is the rate constants of the positive and reverse reaction.

Then:

$$\hat{\omega}_{i,r} = \Gamma \left(v''_{i,r} - v'_{i,r} \right) \left(k_{f,r} \prod_{j=1}^{N} \left[C_{j,r} \right]^{\eta'_{j,r}} - k_{b,r} \prod_{j=1}^{N} \left[C_{j,r} \right]^{\eta''_{j,r}} \right) \qquad (8.35)$$

where $C_{j,r}$ is the molar concentration of component j; $\eta'_{i,r}, \eta''_{i,r}$ represent, respectively, the reaction index of the chemical component j in the positive and reverse reaction; Γ represents the third body effect.

The rate constant in a positive reaction $(k_{f,r})$ is given by the Arrhenius equation:

$$k_{f,r} = A_r T^{\beta_r} e^{-E_r/RT} \qquad (8.36)$$

In the quasi equilibrium assumption, the rate constant in the reverse reaction can be given by the chemical equilibrium constant:

$$k_{b,r} = k_{f,r}/K_r \qquad (8.37)$$

The chemical equilibrium constant (K_r) can be expressed as:

$$K_r = \exp\left(\Delta S_r^0 / R - \Delta H_r^0 / RT \right) \left(P_{\text{atm}}/RT \right) \sum_{r=1}^{N_R} \left(v''_{j,r} - v'_{j,r} \right) \qquad (8.38)$$

where:

$$\Delta S_r^0 / R = \sum_{i=1}^{N} \left(v_{i,r}'' - v_{i,r}' \right) S_i^0 / R \tag{8.39}$$

$$\Delta H_r^0 / RT = \sum_{i=1}^{N} \left(v_{i,r}'' - v_{i,r}' \right) \cdot \left(H_i^0 / RT \right) \tag{8.40}$$

In the formula, S_i^0 and H_i^0, respectively, represent the entropy and enthalpy of component i under standard conditions; P_{atm} is the standard atmospheric pressure.

Chemical reaction kinetics models of the chemical reaction are generally divided into four categories: the detailed reaction mechanism, the elementary reaction mechanism, simplified reaction mechanism, and the overall reaction mechanism. In the actual calculation, we usually adopt the simplified reaction mechanism or the overall reaction mechanism due to limitations of computer speed and memory.

Gas phase equations, liquid phase equations, turbulence model, droplet spray and the evaporation model, two phase flow model, chemical kinetics model, and the component equations constitute the whole two-phase multicomponent combustion flow governing equations, which can be solved simultaneously.

8.2 Numerical Methodology

The governing equations system describing the liquid propellant rocket engine spray combustion process has many equations that are strongly coupled. Therefore, they can only be solved using numerical methods. To obtain the numerical solution in accordance with the actual situation, an accurate physical model, an appropriate grid, and a stable and reliable numerical method are required. This section introduces the SIMPLE algorithm and the PISO algorithm, which are widely used in the processes of combustion, heat transfer, and flow.

8.2.1 Overview

The SIMPLE algorithm was presented by D.B. Spalding and S.V. Patankar from Imperial College, London, in the early 1970s (as reported in Reference [6]). The method is the finite difference method essentially, except that the control volume integral method is used in the construction of discrete equations, making it slightly different from the common finite difference method. The main feature of the algorithm is that pressure rather than the commonly-used density is used as the independent variable of the continuity equation, the pressure correction equation is applied to correct the velocity field gradually to make it satisfy the continuity equation. Then, in calculations of low-speed flow (or incompressible flow), problems due to the small variation in density and the difficulty in calculating the pressure field can be solved. Another feature is that a staggered grid technique is used to overcome the numerical oscillation often occurring in the calculation process and guarantees the simulation stability.

The PISO algorithm is an improved algorithm further developed by R.I. Issa in 1986 on the basis of the SIMPLER algorithm. The philosophy here is that one implicit prediction step and two explicit correction steps are used to complete the calculation of each time step. The PISO algorithm uses pressure as the independent variable, as does the SIMPLER algorithm. An accurate and complete pressure equation is derived from the discrete continuity equation and momentum equation in the PISO algorithm, rather than a pressure correction equation as in the SIMPLER algorithm. The algorithm is suitable for all kinds of steady/unsteady, compressible/incompressible, and two-phase combustion problems with various flow velocities. Furthermore, we do not need to iterate at each time step when calculating the unsteady flow, and thus it can be used in numerical analysis of the rocket engine combustion process.

8.2.2 The Commonly-Used Discretization Scheme

From comparison of the mass conservation equation, momentum conservation equation, energy conservation equation, and component mass conservation equations, it is observed that though the dependent variables of the four equations are different the conservation properties of the physical quantity per unit volume per unit time is shown in each equation. If we use ϕ to represent the general variable, the general form of the above governing equations can be expressed as:

$$\frac{\partial(\rho\phi)}{\partial t} + \text{div}(\rho\boldsymbol{u}\phi) = \text{div}(\Gamma\,\text{grad}\phi) + S \tag{8.41}$$

In the formula, ϕ is the general variable, which can represent the solution variables such as u, v, w, T, and so on, Γ is the generalized diffusion coefficient, and S is the generalized source term. In Equation 8.41, each term successively represents the unsteady term, convection term, diffusion term, and source term.

With appropriate mathematical treatment of all the governing equations, the dependent variable, unsteady term, convection term, and diffusion term of the equation can be written in a standard form. The other terms on the right-hand side of the equations are defined as a source term, resulting in a general differential equation. General differential Equation 8.41 can be used to solve different types of flow and heat transfer problems. For different ϕ, by simply calling program repeatedly, giving Γ and S appropriate expressions, and specifying appropriate initial and boundary conditions, it can be solved.

To explain the characteristics of various discretization schemes easily, one-dimensional steady-state convection–diffusion problems with no source term are selected for discussion (as shown in Figure 8.1). Assuming the velocity field as u, we mainly investigate the

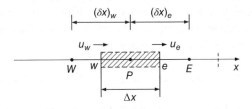

Figure 8.1   The control volume (P) and the interface velocity.

generalized node P, the adjacent nodes E and W, and the interface e and w of the control volume. The following equation can be obtained by integrating transport Equation 8.41 in the control volume P:

$$(\rho u \phi A)_e - (\rho u \phi A)_w = \left(\Gamma A \frac{d\phi}{dx} \right)_e - \left(\Gamma A \frac{d\phi}{dx} \right)_w \tag{8.42}$$

Integral of the continuity equation can produce the following equation:

$$(\rho u A)_e - (\rho u A)_w = 0 \tag{8.43}$$

To get the discrete equations of the convection–diffusion problems, we must carry out some approximate treatment to physical quantities in the interface in Equation 8.42. For convenience, in the following discussion, two new physical quantities F and D are defined. F represents the convective mass flux per unit area through the interface, with the abbreviation of convective mass flow; D represents the diffusion conductivity at the interface:

$$F \equiv \rho u \tag{8.44}$$

$$D \equiv \frac{\Gamma}{\delta x} \tag{8.45}$$

Then, the values of F and D on the control volume interface, respectively, are computed by:

$$F_w = (\rho u)_w, \quad F_e = (\rho u)_e \tag{8.46}$$

$$D_w = \frac{\Gamma_w}{(\delta x)_w}, \quad D_e = \frac{\Gamma_e}{(\delta x)_e} \tag{8.47}$$

The one-dimensional Peclet number (P_e) can be defined by:

$$P_e = \frac{F}{D} = \frac{\rho u}{\Gamma / \delta x} \tag{8.48}$$

P_e represents the intensity ratio of the convection and diffusion. When P_e is 0, the convection–diffusion problem becomes the pure diffusion problem, i.e., there is no flow in the flow field, only diffusion. When $P_e > 0$, the fluid flows in the positive direction of x. When $P_e < 0$, the fluid flows in the negative direction of x. When P_e is very large, the convection–diffusion problem become a pure convection problem, and the effect of diffusion can be neglected.

In addition, another two assumptions are introduced:

1. The interface area of the interfaces e and w of the control volume has the relationship: $A_w = A_e = A$.
2. The diffusion term of the right-hand end of the equation is always expressed using the central difference scheme.

Then, Equation 8.42 can be written as:

$$F_e \phi_e - F_w \phi_w = D_e(\phi_E - \phi_P) - D_w(\phi_P - \phi_W) \qquad (8.49)$$

At the same time, the integral result of the continuous Equation 8.43 is:

$$F_e - F_w = 0 \qquad (8.50)$$

To simplify matters, we assume that the velocity field are known in some way. Then, F_w and F_e are known. To solve Equation 8.49, we need to calculate the solution of the generalized unknown quantity ϕ on the interface e and w. To accomplish this task, we must decide how the interface physical quantity is represented by the interpolation of the node physical quantities, i.e., the discretization scheme discussed in the following.

Central Difference Scheme

The central difference scheme can be described as a method whereby the interface physical quantity is calculated using the linear interpolation formula. For a given uniform grid, the interface physical quantity ϕ of the control volume can be obtained by:

$$\begin{cases} \phi_e = \dfrac{\phi_P + \phi_E}{2} \\[2ex] \phi_w = \dfrac{\phi_P + \phi_W}{2} \end{cases} \qquad (8.51)$$

The above equations are substituted into the convection term of Equation 8.49:

$$F_e \frac{\phi_P + \phi_E}{2} - F_w \frac{\phi_P + \phi_W}{2} = D_e(\phi_E - \phi_P) - D_w(\phi_P - \phi_W) \qquad (8.52)$$

The above equation can be rewritten as follows:

$$\left[\left(D_w - \frac{F_w}{2} \right) + \left(D_e + \frac{F_e}{2} \right) \right] \phi_P = \left(D_w + \frac{F_w}{2} \right) \phi_W + \left(D_e - \frac{F_e}{2} \right) \phi_E \qquad (8.53)$$

The discrete form of the continuity Equation 8.50 is introduced, and the above equation becomes:

$$\left[\left(D_w - \frac{F_w}{2} \right) + \left(D_e + \frac{F_e}{2} \right) + (F_e - F_w) \right] \phi_P = \left(D_w + \frac{F_w}{2} \right) \phi_W + \left(D_e - \frac{F_e}{2} \right) \phi_E \qquad (8.54)$$

Using a_P, a_E, and a_W to denote the coefficients preceding ϕ_P, ϕ_W, and ϕ_E, we can obtain the discrete equation of the convection–diffusion equation in the central difference scheme:

$$a_P \phi_P = a_W \phi_W + a_E \phi_E \qquad (8.55)$$

In the equation:

$$
\left.
\begin{aligned}
a_W &= D_w + \frac{F_w}{2} \\[2mm]
a_E &= D_e - \frac{F_e}{2} \\[2mm]
a_P &= a_E + a_W + (F_e - F_w)
\end{aligned}
\right\}
\tag{8.56}
$$

Discrete equations that have the form of Equation 8.55 can be given on all nodes (center of control volume), forming linear algebraic equations. The unknown quantity ϕ of the equations are the node values such as ϕ_P, ϕ_W and ϕ_E in Equation 8.55. By solving these equations, the spatial distribution of ϕ can be obtained.

Equation 8.55 is the result of the diffusion term and the convection term discretized by the central difference scheme. Coefficients a_E and a_W include the influence of diffusion and convection. The terms D_e and D_w in the coefficients are due to the central difference of the diffusion term, representing the influence of the diffusion process. The part of the coefficients relating to F_e and F_w is due to implementation of the piecewise linear interpolation method at the interface on a uniform grid, representing the effect of convection.

It has been proved that when $P_e < 2$ the calculation results of the central difference nearly coincide with the exact solution. However, when $P_e > 2$, the calculation results of the central difference lose all physical meanings. It can be observed from the coefficient of the discrete equations that $a_E < 0$ when $P_e > 2$. The coefficients a_E and a_W represent the effect of the physical quantity of neighboring points E and W on the point P by the convection and diffusion. When the discrete equation is written in the form of Equation 8.55, a_E, a_W, and a_P must be larger than zero as negative coefficients can lead to an unphysical solution. The requirement of a positive coefficient comes from consideration of the iterative solving of the equations. Generally, iteration methods are used to solve Equation 8.55, and the sufficient condition for the convergence of the iteration methods is that $(\Sigma|a_{nb}|)/|a_p'| \leq 1$ at all nodes with $(\Sigma|a_{nb}|)/|a_p'| < 1$ at one node at least. The a_p' here is the main coefficient of the equations without the source term $(a_p' = a_p - S_P)$, the subscript nb represents all the adjacent nodes around the node P.

Note that P_e of the control volume defined by Equation 8.48 is the combination of the following parameters: fluid characteristics (ρ and Γ), flow characteristics (u), and computational grid characteristics (δx). Then, for the given ρ and Γ, we must ensure that the velocity (u) is very slow (corresponding to the low Reynolds flow controlled by convection) or the grid space is very small in order to meet $P_e < 2$. Due to the constraint, the central difference scheme cannot be used as the discretization scheme for the general flow problems. Other more appropriate discretization scheme must be developed.

Second Order Upwind Scheme

As with the first order upwind scheme, the second order upwind scheme uses the physical quantity at upstream nodes to calculate the interface physical quantity of the control volume. However, the second order upwind scheme not only needs to use the value at the nearest upstream node, but also needs the value at the second upstream node.

When the flow moves in the positive direction, namely, $u_w > 0$, $u_e > 0$ ($F_w > 0$, $F_e > 0$), the following approximation is used in the second order upwind scheme:

$$\phi_w = 1.5\phi_W - 0.5\phi_{WW}, \quad \phi_e = 1.5\phi_P - 0.5\phi_W \tag{8.57}$$

Then, discrete Equation 8.49 becomes (notice that the central difference scheme is still used here to discretize the diffusion term):

$$F_e(1.5\phi_P - 0.5\phi_W) - F_w(1.5\phi_W - 0.5\phi_{WW}) = D_e(\phi_E - \phi_P) - D_w(\phi_P - \phi_W) \tag{8.58}$$

After rearrangement:

$$\left(\frac{3}{2}F_e + D_e + D_w\right)\phi_P = \left(\frac{3}{2}F_w + \frac{1}{2}F_e + D_w\right)\phi_W + D_e\phi_E - \frac{1}{2}F_w\phi_{WW} \tag{8.59}$$

When the flow moves along the negative direction, namely, $u_w < 0$, $u_e < 0$ ($F_w < 0$, $F_e < 0$), the following approximation is used in the second order upwind scheme:

$$\phi_w = 1.5\phi_P - 0.5\phi_E, \quad \phi_e = 1.5\phi_E - 0.5\phi_{EE} \tag{8.60}$$

Then, discrete Equation 8.49 becomes:

$$F_e(1.5\phi_E - 0.5\phi_{EE}) - F_w(1.5\phi_P - 0.5\phi_E) = D_e(\phi_E - \phi_P) - D_w(\phi_P - \phi_W) \tag{8.61}$$

After rearrangement:

$$\left(D_e - \frac{3}{2}F_w + D_w\right)\phi_P = D_w\phi_W + \left(D_e - \frac{3}{2}F_e - \frac{1}{2}F_w\right)\phi_E + \frac{1}{2}F_e\phi_{EE} \tag{8.62}$$

Combining Equations 8.59 and 8.62, and using a_P, a_W, a_{WW}, a_E, and a_{EE} to represent the coefficients preceding ϕ_P, ϕ_W, ϕ_{WW}, ϕ_E, ϕ_{EE} in the equation, the discrete equation of the convection–diffusion equation using the second order upwind scheme can be obtained:

$$a_P\phi_P = a_W\phi_W + a_{WW}\phi_{WW} + a_E\phi_E + a_{EE}\phi_{EE} \tag{8.63}$$

where:

$$\left. \begin{array}{l} a_P = a_E + a_W + a_{EE} + a_{WW} + (F_e - F_w) \\[2mm] a_W = \left(D_e + \frac{3}{2}\alpha F_w + \frac{1}{2}\alpha F_e\right) \\[2mm] a_E = \left(D_e - \frac{3}{2}(1-\alpha)F_e - \frac{1}{2}(1-\alpha)F_w\right) \\[2mm] a_{WW} = -\frac{1}{2}\alpha F_w \\[2mm] a_{EE} = \frac{1}{2}(1-\alpha)F_e \end{array} \right\} \tag{8.64}$$

When the flow moves in the positive direction (i.e., $F_w > 0$ and $F_e > 0$) $\alpha = 1$; when the flow moves in the negative direction (i.e., $F_w < 0$ and $F_e < 0$) $\alpha = 0$.

The second order upwind scheme is based on the first order upwind scheme, considering the effect of curvature of the physical quantity at the distribution curve among the nodes. In the second order upwind scheme, actually, only the convection term uses the second order upwind scheme, while the diffusion term still use the central difference scheme. It is easy to prove that discrete equations using the second order upwind scheme have the truncation error of second order precision. In addition, a significant characteristic of the second order upwind scheme is that the discrete equation not only includes the unknown quantity at the neighboring nodes but also requires the quantity at other nodes adjacent to the neighboring nodes, meaning that it is no longer a three diagonal equation.

8.2.3   Discrete Equations

The grid shown in Figure 8.2 is used to divide the whole computational domain. The intersections of the solid line in the grid are the computational nodes, and the shaded region enclosed by the dotted line is a control volume. The interface of the control volume is located at the middle of two nodes, and thus every node is inside a control volume.

P is used to represent a generalized node. Its adjacent nodes in the east and west are represented by E and W, and its adjacent nodes in the south and north are represented by S and N. The control volume corresponding to each node uses the relevant character to represent. The four blocks at node P in Figure 8.2 are the control volume P. The letters e, w, s, and n are used to represent the four interfaces in the East, South, West and North of the control volume. The width of the control volume in the x and y directions is denoted by Δx and Δy, the volume of the control volume is $\Delta V = \Delta x \times \Delta y$. The distance from node P to E, W, S, and N is, respectively, denoted by $(\delta x)_e$, $(\delta x)_w$, $(\delta x)_s$, $(\delta x)_n$.

For the grid shown in Figure 8.2, integral of governing Equation 8.41 in the control volume (P) and time period Δt (from t to $t + \Delta t$) gives:

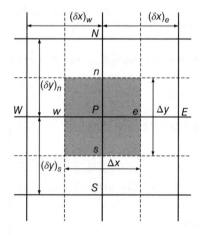

Figure 8.2   Grid and control volume in a two-dimensional problem.

$$\int_t^{t+\Delta t} \int_{\Delta V} \frac{\partial(\rho\phi)}{\partial t} dV dt + \int_t^{t+\Delta t} \int_{\Delta V} \text{div}(\rho u\phi) dV dt$$

$$= \int_t^{t+\Delta t} \int_{\Delta V} \text{div}(\Gamma \text{grad}\phi) dV dt + \int_t^{t+\Delta t} \int_{\Delta V} S dV dt \qquad (8.65)$$

The computational methods for integral of the unsteady term and the source term in the above equation are the same as in the one-dimensional problem. For the integral of the convection term and diffusion term, some special consideration is needed.

To obtain the volume integral of the convection term and the diffusion term in the above equation, the Gauss divergence theorem is introduced:

$$\int_{\Delta V} \text{div}(a) dV = \int_{\Delta S} v \cdot a \, dS = \int_{\Delta S} v_i a_i \, dS \qquad (8.66)$$

where:

ΔV is the three-dimensional integral domain;
ΔS is the closed boundary corresponding to ΔV;
a is an arbitrary vector;
v is the unit normal vector of the surface element dS;
a_i and v_i are the components of vector a and v, respectively.

The above equation obeys the index summation convention of the tensor.
The above equation can be written in a conventional form:

$$\int_{\Delta s} \left(\frac{\partial a_x}{\partial x} + \frac{\partial a_y}{\partial y} + \frac{\partial a_z}{\partial z} \right) dV = \int_{\Delta S} \left(a_x v_x + a_y v_y + a_z v_z \right) dS \qquad (8.67)$$

Now we will explain how to conduct an integral calculation for each term in Equation 8.65.

Unsteady Term
When dealing with the unsteady term, the value of the physical quantity ϕ is assumed to be ϕ_P in the whole control volume (P). Assuming that the change of density ρ in time period of Δt is negligible, the unsteady term can be written as:

$$\int_t^{t+\Delta t} \int_{\Delta V} \frac{\partial(\rho\phi)}{\partial t} dV dt = \int_{\Delta V} \left[\int_t^{t+\Delta t} \frac{\partial(\rho\phi)}{\partial t} dt \right] dV = \rho_P^0 \left(\phi_P - \phi_P^0 \right) \Delta V \qquad (8.68)$$

where the physical quantity with superscript "0" is the quantity value at time t, while the physical quantity at time $t + \Delta t$ does not have a superscript. The physical quantity with subscript "P" is the quantity value at the node point of the control volume (P).

Source Term

$$\int_{t}^{t+\Delta t}\int_{\Delta V} S\,\mathrm{d}V\,\mathrm{d}t = \int_{t}^{t+\Delta t} S\Delta V\,\mathrm{d}t = \int_{t}^{t+\Delta t} (S_C + S_P\phi_P\Delta V)\,\mathrm{d}t \tag{8.69}$$

Note that the linearization is introduced in the above equation.

Convection Term

On the basis of the Gauss divergence theorem, after the conversion from volume into area of the integral, we have:

$$\int_{t}^{t+\Delta t}\int_{\Delta V} \mathrm{div}(\rho u\phi)\,\mathrm{d}V\,\mathrm{d}t$$

$$= \int_{t}^{t+\Delta t} \left[(\rho u\phi A)_e - (\rho u\phi A)_w + (\rho v\phi A)_n - (\rho v\phi A)_s\right]\mathrm{d}t \tag{8.70}$$

$$= \int_{t}^{t+\Delta t} \left[(\rho u)_e\phi_e A_e - (\rho u)_w\phi_w A_w + (\rho v)_n\phi_n A_n - (\rho v)_s\phi_s A_s\right]\mathrm{d}t$$

where A is the area of the control volume interface.

Diffusion Term

Similarly, on the basis of Gauss divergence theorem, after the conversion from volume to area of the integral, we have:

$$\int_{t}^{t+\Delta t}\int_{\Delta V} \mathrm{div}(\Gamma\,\mathrm{grad}\phi)\,\mathrm{d}V\,\mathrm{d}t$$

$$= \int_{t}^{t+\Delta t} \left[\left(\Gamma\frac{\partial\phi}{\partial x}A\right)_e - \left(\Gamma\frac{\partial\phi}{\partial x}A\right)_w + \left(\Gamma\frac{\partial\phi}{\partial x}A\right)_n - \left(\Gamma\frac{\partial\phi}{\partial x}A\right)_s\right]\mathrm{d}t \tag{8.71}$$

$$= \int_{t}^{t+\Delta t} \left[\Gamma_e A_e\frac{\phi_E - \phi_P}{(\delta x)_e} - \Gamma_w A_w\frac{\phi_P - \phi_w}{(\delta x)_w} + \Gamma_n A_n\frac{\phi_N - \phi_P}{(\delta y)_n} - \Gamma_s A_s\frac{\phi_P - \phi_s}{(\delta y)_s}\right]$$

Note that Equation 8.71 used the central difference scheme to disperse the value ϕ of the interface. This is consistent practice in the finite volume method. In the previous derivation of a one-dimensional problem's discrete equation, no matter what discretization scheme was used for the convection term, the diffusion term always uses the central difference scheme for dispersion.

After obtaining the equation of each individual expression, the following two aspects are addressed.

First, a specific discretization scheme is needed in the convection term to express the interface physical quantities ϕ_e, ϕ_w, ϕ_n, and ϕ_s using the quantity at the node point, e.g. the first-order up-wind scheme can be used.

Second, in the convection term, the diffusion term and the source term introduce the fully implicit time integration scheme, such as:

$$\int_{t}^{t+\Delta t} \phi_P \, dt = \phi_P \Delta t$$

Hence, the equation is:

$$a_P \phi_P = a_W \phi_W + a_E \phi_E + a_S \phi_S + a_N \phi_N + b \tag{8.72}$$

This is a discrete equation of a two-dimensional instantaneous convection–diffusion problem obtained by using the fully implicit time integration scheme. Coefficients $a_W, a_E, a_S,$ and a_N are dependent on the specific discretization scheme for the convection term. If we use the first order upwind scheme, we have:

$$\left.\begin{aligned}
a_W &= D_w + \max(0, F_w) \\
a_E &= D_e + \max(0, -F_e) \\
a_S &= D_s + \max(0, F_s) \\
a_N &= D_n + \max(0, -F_n) \\
a_P &= a_W + a_E + a_S + a_N + b + (F_e - F_w) + (F_n - F_s) + a_P^0 - S_P \Delta V \\
b &= S_C \Delta V + a_P^0 \phi_P^0 \\
a_P^0 &= \frac{\rho_P^0 \Delta V}{\Delta t}
\end{aligned}\right\} \tag{8.73}$$

If we use other discretization schemes, the formula of coefficients $a_W, a_E, a_S,$ and a_N can also be derived by the same method.

When extending from 2D to 3D, the third coordinate z will be added. Accordingly, the control volume will change from a rectangle as shown above to a cube. The interfaces in the up and down directions will be added, denoted by t (top) and b (bottom), respectively, and the two neighboring nodes are denoted by T and B.

The discrete equation of three-dimensional instantaneous convection–diffusion problem obtained by using the fully implicit time integration scheme is:

$$a_P \phi_P = a_W \phi_W + a_E \phi_E + a_S \phi_S + a_N \phi_N + a_B \phi_B + a_T \phi_T + b \tag{8.74}$$

8.2.4 Discretization of the Momentum Equation Based on the Staggered Grid

In a staggered grid the velocity components and pressure are located in different grid systems. A staggered grid is used to solve the numerical problem caused by the zigzag pressure distribution that is produced when discretizing the governing equation in the ordinary grid. The staggered grid is also the basis of the SIMPLE algorithm.

To present a direct demonstration, only the momentum equation of steady state problem is first discussed. The solid line shows the original calculation grid line, and the solid points are computational notes (the center of the master control volume). The dotted line shows the interface of the master control volume. Here, the gird lines represented by solid lines are denoted by capital letters. The solid vertical lines in the x direction are, respectively, numbered ..., $I-1$, I, $I+1$,...; the solid horizontal lines in the y direction are, respectively, numbered ..., $J-1$, J $J+1$,.... The dotted lines used to represent the interfaces of the control volume are denoted by lowercase letters: the dotted vertical lines in the x direction are, respectively, numbered ..., $i-1$, i, $i+1$,...; the dotted horizontal lines in the y direction are respectively numbered ..., $j-1$, j, $j+1$,... (Figure 8.3).

The variable arrangement system shown in Figure 8.3 can accurately denote the positions of grid nodes and interfaces of the control volume. The node used to store the scalar is, in this book called the scalar node, and it is the intersection of two grid lines (solid line), shown by two capital letters, e.g., point P is denoted by (I, J). Pressure value $P_{I,J}$ is defined and stored at scalar node (I, J). The rectangular area surrounding the scalar node (I, J) is a scalar control volume. The velocity (u) is stored at the e and w interfaces. Intersections of the scalar control volume's interface lines and grid lines are called u velocity nodes, or velocity node for short, which is denoted by the combination of a lowercase and an uppercase, e.g., the w interface is defined by (i, J). The rectangular area surrounding the velocity node (i, J) is the u control volume. Similarly, the position used to store the v speed is called the v velocity node, which is denoted by the combination of an uppercase and a lowercase, e.g., the s interface is defined by (I, j). The rectangular area surrounding the velocity node (I, j) is the v control volume.

We can use the velocity grid with the forward or backward dislocation. Here a uniform grid with backward velocity grid dislocation is used. The distance from the i position of the u velocity $u_{i,J}$ to the scalar node (I, J) is $-\frac{1}{2}\delta x_u$; the distance from the j position of v velocity $v_{I,j}$ to the scalar node (I, J) is $-\frac{1}{2}\delta x_v$.

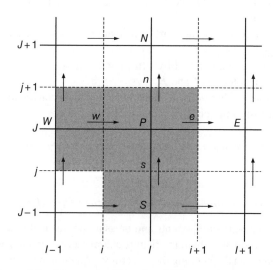

Figure 8.3 Staggered grids and variable arrangement system.

The method and process of producing discrete equations on the staggered grid are exactly the same as on the regular grid, but we need to pay attention to the change of the control volume used. Since all the scalars (such as, pressure, temperature, density, and so on) are still stored in the main control volume on the staggered grid, the discretization process and result of the transport equations with these scalars as dependent variable are exactly the same as previously. When deriving the u and v discrete equations on a staggered grid, the main difference is that the control volume used in the integral is no longer the original primary control volume but, instead, is the respective control volume of u and v, and at the same time the pressure gradient is separated from the source term. For example, for u control volume, the integral for this term is:

$$\int_{y_j}^{y_{j+1}} \int_{x_{I-1}}^{x_I} \left(-\frac{\partial p}{\partial x} \right) dxdy \cong (p_{I-1,J} - p_{I,J})A_{i,J}$$

Thus, according to the method and process of deriving the discrete equation (for the steady state problem, ignore the temporal integration), and considering the momentum equation using u control volume in the u direction, we can write the discrete form of the momentum equations for velocity $u_{i,J}$ at position (i, J):

$$a_{i,J} u_{i,J} = \sum a_{nb} u_{nb} + (p_{I-1,J} - p_{I,J})A_{i,J} + b_{i,J} \tag{8.75}$$

In Equation 8.75, $A_{i,J}$ is the area of the east or west interface of the u control volume; it is actually Δy in the two-dimensional problem:

$$A_{i,J} = \Delta y = y_{j+1} - y_j \tag{8.76}$$

The term b is the source term part of the u momentum equation (not including pressure). For steady state problems, we have:

$$b_{i,J} - S_{uC} \Delta V_u \tag{8.77}$$

In Equation 8.77, S_{uC} is the constant part of the linear decomposition $S_u = S_{uC} + S_{uP} u_P$ of the source term S_u. If S_u do not change with velocity u, $S_{uP} = 0$. Also in Equation 8.77, ΔV_u is the volume of the u control volume. The pressure gradient term has been discretized by linear interpolation, which uses the pressure difference between two nodes on the border of the u control volume.

The summation notation $\sum a_{nb} u_{nb}$ contains four neighboring points E, W, N, and S, and they are $(i-1,J)$, $(i+1,J)$, $(i,J+1)$, and $(i,J-1)$.

In the same way, the discrete momentum equation for velocity $v_{I,j}$ at position (I,j) can be derived:

$$a_{I,j} v_{I,j} = \sum a_{nb} v_{nb} + (p_{I,J-1} - p_{I,J})A_{I,j} + b_{I,j} \tag{8.78}$$

8.2.5   The SIMPLE Algorithm of Flow Field Computing

The SIMPLE (semi-implicit method for pressure-linked equations) method was proposed by Patankar and Spalding in 1972. It is a kind of numerical method used to solve both incompressible flows and compressible flows. The core of SIMPLE is the processing of "prediction–correction". It solves the pressure field on a staggered mesh in order to solve the momentum equation.

The philosophy behind the SIMPLE algorithm can be described as follows: when the pressure field is given (which can be either an assumed value or the result from the last calculation iteration), the velocity field can be obtained by solving the discrete momentum equation. Since the pressure field is assumed or inaccurate, the velocity field generally will not satisfy the continuity equation. It is, therefore, necessary to correct the given pressure field. The principle of correction is that the velocity that corresponds to the corrected pressure field should satisfy the continuity equation at this iteration level. Based on this principle, the correlation between pressure and velocity, which is prescribed by discretized momentum equation, is substituted into the discretized continuity equation, resulting in the corrected pressure equation, from which the corrected pressure can be solved. Then a new velocity field can be obtained according to the corrected pressure field. Whether the velocity field converges is checked. If the velocity field does not converge, the corrected pressure field is set to be the given field and the next level's calculation is carried out until a converged velocity field is obtained.

In the above calculation process, two key issues of SIMPLE are how to get the corrected pressure field (i.e., derivation of pressure correction equation) and how to get the "corrected velocity" from the corrected pressure field (derivation of velocity correction equation). Therefore, these two problems are first solved, as follows, and then the calculation steps of SIMPLE are presented.

Velocity Correction Equation

A problem of two-dimensional stable-state laminar flow is solved here in the Cartesian coordinate system. Given an assumed initial pressure field (p^*), the discretized momentum equations can be solved with the pressure field, and then the corresponding velocity components u^* and v^* can be obtained.

The following correlations can be obtained from the discretized momentum equations:

$$a_{i,J}u^*_{i,J} = \sum a_{nb}u^*_{nb} + \left(p^*_{I-1,J} - p^*_{I,J}\right)A_{i,J} + b_{i,J} \tag{8.79}$$

$$a_{I,j}v^*_{I,j} = \sum a_{nb}v^*_{nb} + \left(p^*_{I,J-1} - p^*_{I,J}\right)A_{I,j} + b_{I,j} \tag{8.80}$$

The difference between the correct pressure field (p) and the assumed pressure field (p^*) is defined as p', and the following correlation can be given:

$$p = p^* + p' \tag{8.81}$$

Similarly, the corrected velocity components u' and v' are defined to correlate the correct velocity field (u, v) with the assumed velocity field (u^*, v^*). The correct velocity field (u, v)

can be obtained by substituting the corrected pressure field (p) into the discretized momentum equation. Assuming that the source term b is fixed, the following correlations can be obtained by adding the relation between the corrected pressure and the corrected velocity:

$$a_{i,J}u'_{i,J} = \sum a_{nb}u'_{nb} + \left(p'_{I-1,J} - p'_{I,J}\right)A_{i,J} \tag{8.82}$$

$$a_{I,j}v'_{I,j} = \sum a_{nb}v'_{nb} + \left(p'_{I-1,J} - p'_{I,J}\right)A_{I,j} \tag{8.83}$$

It is observed that we can obtain the corrected velocity (u', v') via the pressure field (p'). The above equations also show that the corrected velocity of any points consists of two parts: one is the difference between two adjacent nodes' corrected pressure values on the same direction of the velocity, which is the direct driving force for producing velocity correction; the other is brought in from the corrected value of the neighboring velocity, which can also be seen as the indirect influence of the corrected value of ambient pressure on the velocity under discussion.

To simplify the solution process, we can introduce an approximate treatment by omitting $\sum a_{nb}u'_{nb}$ and $\sum a_{nb}v'_{nb}$ related to corrected velocity in the equation. This treatment is an important feature of SIMPLE. Then we can get:

$$u'_{i,J} = d_{i,J}\left(p'_{I-1,J} - p'_{I,J}\right) \tag{8.84}$$

$$v'_{I,j} = d_{I,j}\left(p'_{I-1,J} - p'_{I,J}\right) \tag{8.85}$$

where:

$$d_{i,J} = \frac{A_{i,J}}{a_{i,J}}, \quad d_{I,j} = \frac{A_{I,j}}{a_{I,j}}$$

Substituting these equations into the velocity correction equation, we can obtain:

$$u_{i,J} = u^*_{i,J} + d_{i,J}\left(p'_{I-1,J} - p'_{I,J}\right) \tag{8.86}$$

$$v_{I,j} = v^*_{I,j} + d_{I,j}\left(p'_{I-1,J} - p'_{I,J}\right) \tag{8.87}$$

It can be deduced from the above equations that, if the pressure correction (p') is given, we can carry out the corresponding velocity correction to the assumed velocity field (u^*, v^*) and find the correct velocity field (u, v).

Similarly, equations for $u_{i+1,J}$ and $v_{I,j+1}$ are:

$$u_{i+1,J} = u^*_{i+1,J} + d_{i+1,J}\left(p'_{I,J} - p'_{I+1,J}\right) \tag{8.88}$$

$$v_{I,j+1} = v^*_{I,j+1} + d_{I,j+1}\left(p'_{I,J} - p'_{I,J+1}\right) \tag{8.89}$$

where:

$$d_{i+1,J} = \frac{A_{i+1,J}}{a_{i+1,J}}, \quad d_{I,j+1} = \frac{A_{I,j+1}}{a_{I,j+1}}$$

Pressure Correction Equation

In the above subsection, we only considered momentum equations. However, as mentioned earlier, the velocity field is also constrained by the continuity equation. For two-dimensional steady problems, the continuity equation can be written as:

$$\frac{\partial(\rho u)}{\partial x} + \frac{\partial(\rho v)}{\partial y} = 0 \tag{8.90}$$

Discretization of the equation in the scalar control volume can give:

$$\left[(\rho u A)_{i+1,J} - (\rho u A)_{i,J}\right] + \left[(\rho u A)_{I,j+1} - (\rho u A)_{I,j}\right] = 0 \tag{8.91}$$

Substitution of the corrected velocity into the above equation results in the following correlation:

$$
\begin{cases}
\rho_{i+1,J} A_{i+1,J} \left[u^*_{i+1,J} + d_{i+1,J} \left(p'_{I,J} - p'_{I+1,J} \right) \right] \\[2mm]
-\rho_{i,J} A_{i,J} \left[u^*_{i,J} + d_{i,J} \left(p'_{I-1,J} - p'_{I,J} \right) \right]
\end{cases}
$$
$$
+
\begin{cases}
\rho_{I,j+1} A_{I,j+1} \left[v^*_{I,j+1} + d_{I,j+1} \left(p'_{I,J} - p'_{I,J+1} \right) \right] \\[2mm]
-\rho_{I,j} A_{I,j} \left[v^*_{I,j} + d_{I,j} \left(p'_{I-1,J} - p'_{I,J} \right) \right]
\end{cases}
= 0
\tag{8.92}
$$

After rearrangement, we can find:

$$\left[(\rho dA)_{i+1,J} + (\rho dA)_{i,J} + (\rho dA)_{I,j} + (\rho dA)_{I,j}\right] p'_{I,J}$$
$$= (\rho dA)_{i+1,J} p'_{I+1,J} + (\rho dA)_{i,J} p'_{I-1,J} + (\rho dA)_{I,j+1} p'_{I,J+1} + (\rho dA)_{I,j} p'_{I,J-1} \tag{8.93}$$
$$+ \left[(\rho u^* A)_{i,J} - (\rho u^* A)_{i+1,J} + (\rho v^* A)_{I,j} - (\rho v^* A)_{I,j+1}\right]$$

This equation can be simplified as:

$$a_{I,J} p'_{I,J} = a_{I+1,J} p'_{I+1,J} + a_{I-1,J} p'_{I-1,J} + a_{I,J+1} p'_{I,J+1} + a_{I,J-1} p'_{I,J-1} + b'_{I,J} \tag{8.94}$$

where:

$$a_{I+1,J} = (\rho dA)_{i+1,J}$$
$$a_{I-1,J} = (\rho dA)_{i,J}$$
$$a_{I,J+1} = (\rho dA)_{I,j}$$
$$a_{I,J-1} = (\rho dA)_{I,j}$$
$$a_{I,J} = a_{I+1,J} + a_{I-1,J} + a_{I,J+1} + a_{I,J-1}$$
$$b'_{I,J} = (\rho u^* A)_{i,J} - (\rho u^* A)_{i+1,J} + (\rho v^* A)_{I,j} - (\rho v^* A)_{I,j+1}$$

(8.95)

The source term b' of the equation is the "continuity" imbalance resulting from incorrect velocity field (u^*, v^*). By solving the equation, we can determine the corrected pressure (p') at any point in space.

The term ρ is the density at the interface of the scalar control volume. Since ρ is defined and stored at the nodes of the scalar control volume (the center of control volume), it should be obtained by interpolation. There is no value that can be directly used at the interface of the scalar control volume. No matter which interpolation method is adopted, ρ should always be the same at the interface of two control volumes.

Finally, the calculation algorithm of SIMPLE according to the philosophy of SIMPLE is shown in Figure 8.4.

8.2.6   PISO Algorithm

The PISO (pressure implicit with splitting of operators) method was proposed by Issa in 1986. It is a kind of pressure/velocity calculation program used for non-iterative calculation of unsteady compressible flows. Later, it was also widely used for iterative calculation of steady problems.

The differences between PISO and SIMPLE lie in that SIMPLE is a two-step algorithm, which consists of a prediction step and a correction step, while PISO consists of a prediction step and two correction steps [7, 8]. After the first correction step, PISO carries out a second correction to satisfy better the momentum equation and continuity equation simultaneously. Since PISO has adopted three steps of assumption–correction–recorrection, it can speed up the convergence rate in a single iteration.

Step of Assumption
As with SIMPLE, PISO obtains velocity components u^* and v^* by solving momentum equations via the assumed pressure field (p^*).

First Step of Correction
The velocity field (u^*, v^*) generally does not satisfy the continuity equation unless p^* is correct. To make it satisfy the continuity equation, we should introduce the first correction step of SIMPLE, which gives a velocity field (u^{**}, v^{**}). This correction equation is exactly the same as

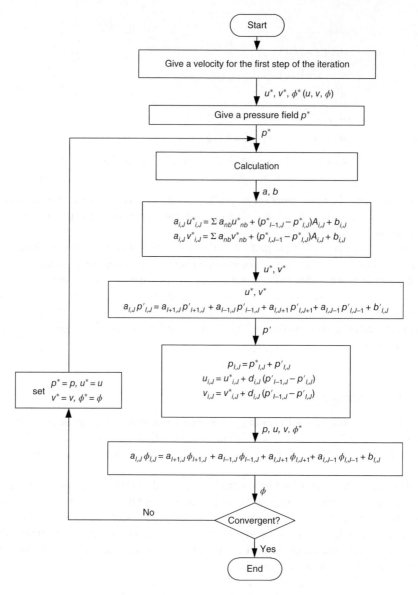

Figure 8.4   SIMPLE calculation algorithm.

SIMPLE. However, considering there is a second correction step in PISO, we use a different symbol:

$$p^{**} = p^* + p', \quad u^{**} = u^* + u', \quad v^{**} = v^* + v' \tag{8.96}$$

This set of equations is used to define the corrected velocity u^{**} and v^{**}:

$$u_{i,J}^{**} = u_{i,J}^* + d_{i,J}\left(p_{I-1,J}' - p_{I,J}'\right) \tag{8.97}$$

$$v^{**}_{I,j} = v^{*}_{I,j} + d_{I,j}\left(p'_{I-1,J} - p'_{I,J}\right) \tag{8.98}$$

Second Correction Step

To improve the SIMPLE calculations, PISO requires a second correction step. The momentum equations of u^{**} and v^{**} are as follows:

$$a_{i,J}u^{**}_{i,J} = \sum_{nb} a_{nb}u^{*}_{nb} + \left(p^{**}_{I-1,J} - p^{**}_{I,J}\right)A_{i,J} + b_{i,J} \tag{8.99}$$

$$a_{I,j}v^{**}_{I,j} = \sum_{nb} a_{nb}v^{*}_{nb} + \left(p^{**}_{I,J-1} - p^{**}_{I,J}\right)A_{I,j} + b_{I,j} \tag{8.100}$$

By solving the momentum equation again, we can get a recorrected velocity field (u^{***}, v^{***}):

$$a_{i,J}u^{***}_{i,J} = \sum_{nb} a_{nb}u^{**}_{nb} + \left(p^{***}_{I-1,J} - p^{***}_{I,J}\right)A_{i,J} + b_{i,J} \tag{8.101}$$

$$a_{I,j}v^{***}_{I,j} = \sum_{nb} a_{nb}v^{**}_{nb} + \left(p^{***}_{I,J-1} - p^{***}_{I,J}\right)A_{I,j} + b_{I,j} \tag{8.102}$$

Notably, the sum in the correction step should be calculated via u^{**} and v^{**}.

In this step, we should subtract Equation 8.66 from Equation 8.68 and subtract Equation 8.67 from Equation 8.69, and get:

$$u^{***}_{i,J} = u^{**}_{i,J} + \frac{\sum a_{nb}\left(u^{**}_{nb} - u^{*}_{nb}\right)}{a_{i,J}} + d_{i,J}\left(p''_{I-1,J} - p''_{I,J}\right) \tag{8.103}$$

$$v^{***}_{I,j} = v^{**}_{I,j} + \frac{\sum a_{nb}\left(v^{**}_{nb} - v^{*}_{nb}\right)}{a_{I,j}} + d_{I,j}\left(p''_{I-1,J} - p''_{I,J}\right) \tag{8.104}$$

where p'' is the recorrection of pressure; p^{***} can be expressed as:

$$p^{***} = p^{**} + p'' \tag{8.105}$$

By substituting the equation of u^{***} and v^{***} into the continuity equation, we can obtain the second correction pressure equation:

$$a_{I,J}p''_{I,J} = a_{I+1,J}p''_{I+1,J} + a_{I-1,J}p''_{I-1,J} + a_{I,J+1}p''_{I,J+1} + a_{I,J-1}p''_{I,J-1} + b''_{I,J} \tag{8.106}$$

The coefficients are as follows:

$$a_{I,J} = a_{I+1,J} + a_{I-1,J} + a_{I,J+1} + a_{I,J-1}$$

$$a_{I+1,J} = (\rho dA)_{i+1,J}$$

$$a_{I-1,J} = (\rho dA)_{i,J}$$

$$a_{I,J+1} = (\rho dA)_{I,j}$$

$$a_{I,J-1} = (\rho dA)_{I,j} \tag{8.107}$$

$$b''_{I,J} = \left(\frac{\rho A}{a}\right)_{i,J} \sum a_{nb}\left(u^{**}_{nb} - u^{*}_{nb}\right) - \left(\frac{\rho A}{a}\right)_{i+1,J} \sum a_{nb}\left(u^{**}_{nb} - u^{*}_{nb}\right)$$

$$+ \left(\frac{\rho A}{a}\right)_{I,j} \sum a_{nb}\left(v^{**}_{nb} - v^{*}_{nb}\right) - \left(\frac{\rho A}{a}\right)_{I,j+1} \sum a_{nb}\left(v^{**}_{nb} - v^{*}_{nb}\right)$$

It could have been an equation similar to Equation 8.62:

$$\left[(\rho u^{**}A)_{i,J} - (\rho u^{**}A)_{i+1,J} + (\rho v^{**}A)_{I,j} - (\rho v^{**}A)_{I,j+1}\right]$$

in $b''_{I,J}$. However, since u^{**} and v^{**} satisfy the continuity equation, the equation:

$$\left[(\rho u^{**}A)_{i,J} - (\rho u^{**}A)_{i+1,J} + (\rho v^{**}A)_{I,j} - (\rho v^{**}A)_{I,j+1}\right]$$

is equal to 0.

Now we can get second correction pressure (p'') by solving Equation 8.73 and determine the second corrected pressure field via:

$$p^{***} = p'' + p^{**} = p' + p'' + p^*.$$

Finally, by solving Equations 8.70 and 8.71 we can obtain a second corrected velocity field. In the non-iterative calculation of unsteady problems, the pressure field p^{***} and velocity field (u^{***}, v^{***}) are believed to be correct.

Since PISO requires the pressure correction equation to be solved twice, it needs extra storage space to calculate the source term of second pressure correction equation. Though this method involves a lot of calculation, it has a comparatively high calculation speed and, thus, a high efficiency.

The algorithm of PISO is shown in Figure 8.5:

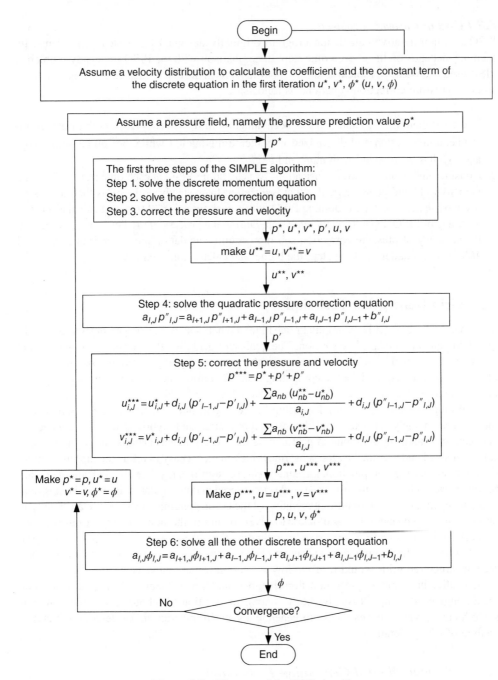

Figure 8.5   Diagram of the PISO algorithm.

PISO Used in Unsteady Problems

PISO is a non-iterative calculation program originally designed for unsteady problems. Its accuracy depends on the chosen time step. In comparison with the PISO for steady problems, PISO used in unsteady problems requires modifications in the discretized momentum equations and two pressure correction equations as follows:

1. The coefficient a_P ($a_{i,J}$ and $a_{I,j}$) has been added with $a_p^0 = \rho_P^0 \Delta V / \Delta t$ in both discretized u momentum equation and discretized v momentum equation while the source term b ($b_{i,J}$ and $b_{I,j}$) has been added with $a_P^0 u_P^0$ and $a_P^0 v_P^0$.
2. Considering the above two modifications, we can adopt PISO at each time step to calculate velocity field and pressure field. What is different from steady problems is that there is no need to use iteration when adopting PISO within each time step of transient calculation. The accuracy of PISO depends on the time step. During the process of prediction and correction, the accuracy of pressure correction and momentum calculation are of order of $3(\Delta t^3)$ and $4(\Delta t^4)$. The smaller the time step, the higher the calculation accuracy.

8.3   Grid Generation Techniques

The precondition of CFD (Computational Fluid Dynamics) is the proper design and high-qualified generation of grid generation. The quality of grid directly affects the computational accuracy of a numerical solution. Meanwhile, the process of grid generation takes up 60–80% of the whole time required for the computing task. Therefore, the development of grid generation techniques has an important impact on the advance of CFD.

According to the connection relationship between grid nodes, grids can be classified into three types, namely, structured grid, unstructured grid, and hybrid grid. The connection between grid nodes in a structured grid is ordered and regulated. Data is stored orderly and can be indexed and searched by the subscript, which makes it easy to refine the grid in a certain direction so that the computational accuracy can be significantly improved. The connection between grid nodes in an unstructured grid is unordered and unregulated. Each of them is relatively independent, which requires generation of the corresponding data structure artificially in order to index and search the grid data. However, owing to the homogeneity of unstructured grid cells, it is hard to refine the grid in a certain direction. Thus, normally it requires more grid cells to improve the computational accuracy. For viscous flow calculations, adoption of complete unstructured grids would result in low resolution in the boundary layer.

Generally, the shapes of precombustion chambers and thrust chambers in a liquid-propellant rocket engine are simple (usually rotary shape). In addition, the boundary is a continuous smooth curve, which makes it easy to generate a grid. In this section we describe the classic method of grid generation.

8.3.1   Structured Grid Generation Technology

Traditional ways of structured grid generation are the algebraic approach, elliptic partial differential equations approach, and hyperbolic differential equations approach.

The algebraic approach is to combine algebra, triangle, analytic geometry, and other mathematical approaches to generate the grid. This approach means changing the physically unregulated areas into regulated areas in the computational domain through some algebraic equations.

The algebraic approach is quick. Since it does not need an iterative process, the calculation cost is small. Thus it is widely used. However, the algebraic approach makes it hard to control the grid quality. Generally, the generated grids are non-orthogonal, and thus grids around an object plane and corners often cannot meet the requirement of calculation accuracy.

There are various algebraic grid generation methods, such as directly drawing, coordinate transformation, double boundary method, and so on. Currently, the mature and efficient algebraic approach is to use the known boundary point coordinates to interpolate the unknown grids, that is, the infinite interpolation method. Specifically, applied interpolation approaches are Lagrange interpolation, Hermite interpolation, and so on. The infinite interpolation method is an interpolation theory based on multi-directions or multivariables. The function values of interpolation function on a boundary face match the given value. Thus, interpolation can be defined as "infinite".

The approach employing elliptical equations generates grids by solving elliptical equations. The smoothing feature of elliptical equations is a frequently used differential equation approach in grid generation. The earliest representative of this kind of approaches is the renowned TTM, which enable us to get uniform grids in a computational domain by solving a planar Laplace equation. When the right-hand terms (which can also be called source terms) are added into the Laplace equation, we can get Poisson's equation. Then we can control the density and orthogonality of grids by adjusting the source term's value. However, the disadvantages of elliptical equations are: (i) it is hard to determine the source term and (ii) there use takes a long time. Generally, the grid lines around the object surface are supposed to orthogonal to object surface and the distance between the first gridline on object surface and object surface is controllable.

Currently, the ways of controlling source terms can be summarized in the following steps:

1. Assume the source term of entry at the border—the easiest way to do this is to make the source term 0.
2. Change the value of the source term at the border, i.e., so-called "source entry controlling", which can make it as close as possible so as to generate the ideal grid mesh.
3. Insert the source term value around the object surface and the outer boundary into the grid points within flow field.
4. Use the iterative solution to solve the Poisson equation which is used when the source term is added.
5. Repeat steps 2–4 until the ideal grid is worked out.

Step 4 is called "inner iterations" while steps 2–4 are called "outer iterations". Generally, the iterative source term value is changed only in outer iterations, and the source term value remains unchanged in inner iterations.

Currently, source term methods can be divided into two categories: one is to derive expressions of source term directly according to the orthogonal grid spacing requirements; the other is get the ideal grid mesh by adopting an "artificial" means of term source value according to the changes of source term during the process of iteration. The former method requires a lot of calculation and the convergence is slow. Moreover, the orthogonal grid spacing and the distance of the first layer cannot be directly controlled at the same time. Therefore, currently the latter method is more commonly used. The specific methods of elliptical rotary grid generation are as follows.

For the two-dimensional problem, conformal mapping provides a way of grid generation method that meets the requirement of orthogonality and smoothness. The physical plane and the grid point for calculation should be expressed as complex numbers:

$$Z = x + iy \tag{8.108}$$

$$\varsigma = \xi + i\eta \tag{8.109}$$

If $\varsigma = \varsigma(Z)$ is an analytic function, then the transformation from (x, y) into (ξ, η) is called conformal transformation. The superiority of conformal transformation lies in that the angle between the ξ, η grid lines is equal to the angle between the x, y grid lines (i.e., if the grids on (ξ, η) plane are orthogonal, then the grids on (x, y) are also orthogonal), and the transformation is sufficiently smooth. The relation $\varsigma = \varsigma(Z)$ is the condition of analytic function, which is the so-called Cauchy–Riemann condition:

$$\frac{\partial \xi}{\partial x} = \frac{\partial \eta}{\partial y}, \frac{\partial \xi}{\partial y} = \frac{\partial \eta}{\partial x} \tag{8.110}$$

The only disadvantage of the conformal mapping method is that the structure of conformal mapping becomes very complex for general shape of regions, and it only suits two-dimensional problems. To solve these problems, it is noted that by Equation 8.110 we can get:

$$\begin{cases} \dfrac{\partial^2 \xi}{\partial x^2} + \dfrac{\partial^2 \xi}{\partial y^2} = 0 \\[3mm] \dfrac{\partial^2 \eta}{\partial x^2} + \dfrac{\partial^2 \eta}{\partial y^2} = 0 \end{cases} \tag{8.111}$$

Therefore, we can set this Laplace equation (Equation 8.111) as the initial grid generation equation. Obviously, in general, Equations 8.111 and 8.110 are not equivalent; so the generated grid meshes are not always orthogonal. However, since Equation 8.111 is an elliptic equation, the transformations between (x, y) and (ξ, η) are smooth and corresponding, respectively.

However, it is hard to solve Equation 8.111 directly since the independent variable of the equation is (x, y), and a mesh should be generated by solving the equation on a non-rectangular area. The way to solve this problem is to transform Equation 8.111 by using Equation 8.110, resulting in:

$$\left.\begin{array}{c} \alpha x_{\xi\xi} - 2\beta x_{\xi\eta} + \gamma x_{\eta\eta} = 0 \\ \alpha y_{\xi\xi} - 2\beta y_{\xi\eta} + \gamma y_{\eta\eta} = 0 \end{array}\right\} \tag{8.112}$$

where:

$$\left.\begin{array}{l} \alpha = \left(x_\eta\right)^2 + \left(y_\eta\right)^2 \\ \beta = x_\xi x_\eta + y_\xi y_\eta \\ \gamma = \left(x_\xi\right)^2 + \left(y_\xi\right)^2 \end{array}\right\} \tag{8.113}$$

Since the solution domain of (ξ, η) is a rectangular area, we can solve Equation 8.112 by using the finite difference method under a rectangular grid, and then we can find $x_{i,j}, y_{i,j}$ which corresponds to ξ_i, η_j.

The grid distributions are supposed to be more clustered in a flow field where the changes of parameters are drastic, so that the real flow field can be reproduced more accurately. However, the grids on physical plane that are worked out through the aforementioned Laplace equation transform method cannot be refined where it is needed, and the distributions of grids are substantially uniform. Consequently, Thompson and others proposed, in 1977, a method using the Poisson equation with source terms as the transfer function. When controlling the density distribution of the grid lines through the source term $P(\xi, \eta)$ and $Q(\xi, \eta)$ in the equation, the transformation equation becomes:

$$\frac{\partial^2 \xi}{\partial x^2} + \frac{\partial^2 \xi}{\partial y^2} = P(\xi, \eta) \tag{8.114}$$

$$\frac{\partial^2 \eta}{\partial x^2} + \frac{\partial^2 \eta}{\partial y^2} = Q(\xi, \eta) \tag{8.115}$$

After the inverse transformation:

$$\alpha x_{\xi\xi} - 2\beta x_{\xi\eta} + \gamma x_{\eta\eta} = -J^2 \left[x_\xi P(\xi, \eta) + x_\eta Q(\xi, \eta) \right] \tag{8.116}$$

$$\alpha y x_{\xi\xi} - 2\beta y_{\xi\eta} + \gamma y_{\eta\eta} = -J^2 \left[y_\xi P(\xi, \eta) + y_\eta Q(\xi, \eta) \right] \tag{8.117}$$

In the formula, the Jacobian determinant is $J = x_\xi y_\eta - x_\eta y_\xi$. If we want the ξ lines to be focused around $\xi = \xi_i$, then we can get:

$$P(\xi, \eta) = -a \text{Sign}(\xi - \xi_i) \exp[-c(\xi - \xi_i)] \tag{8.118}$$

If we want the ξ lines to be focused around (ξ_i, η_i), then we can get:

$$Q(\xi, \eta) = -b \text{Sign}(\xi - \xi_i) \exp\left[-d\sqrt{(\xi - \xi_i)^2 + (\eta - \eta_i)^2} \right] \tag{8.119}$$

where:

$$\text{Sign}(x) = \begin{cases} 1 & x > 0 \\ 0 & x = 0 \\ -1 & x < 0 \end{cases} \tag{8.120}$$

Then, the general form of source terms $P(\xi, \eta)$ and $Q(\xi, \eta)$ is:

$$\begin{aligned} P(\xi, \eta) = &-\sum_{i=1}^{n} a_i \text{Sign}(\xi - \xi_i) \exp[-c_i(\xi - \xi_i)] \\ &-\sum_{i=1}^{m} b_j \text{Sign}(\xi - \xi_j) \exp\left[-d_j \sqrt{(\xi - \xi_j)^2 + (\eta - \eta_j)^2} \right] \end{aligned} \tag{8.121}$$

$$Q(\xi,\eta) = -\sum_{i=1}^{n} a_i \mathrm{Sign}(\eta-\eta_i)\exp[-c_i(\eta-\eta_i)]$$

$$\hspace{3cm} -\sum_{i=1}^{m} b_j \mathrm{Sign}(\eta-\eta_j)\exp\left[-d_j\sqrt{(\eta-\eta_j)^2+(\xi-\xi_j)^2}\right] \tag{8.122}$$

In the formulas, Sign is a signed function; a_i and b_i stand for amplitude that affects the size of the volume adjustment; c_i and d_i are the attenuation factors reflecting the degree of closeness of the grid lines—these coefficients in these two formulas may be various and the specific numerical values need to be determined by the numerical calculation. Notably, the first terms on the right-hand side of Equations 8.121 and 8.122 mean that grid lines are getting close to grid lines of different $\xi_i =$ constants or $\eta_i =$ constant, and the second terms mean that each grid point is getting close to (ξ_i, η_i).

The distribution of the mesh of a flow field can also be controlled by adjusting the distributions of mesh nodes at the boundary, but it will not change the grid nodes at fixed boundary and it does not structure the source term. When the iteration is convergent, it is hard to ensure orthogonality of the grid lines at the boundary. Moreover, there is no source term in the equations, and thus the mesh distribution in the inner region is hard to control.

The quality of mesh generated by solving partial differential equations is high, and the solved equations are in a unified form, which makes it easy to construct a universal program. However, it requires massive calculation and takes a long time to generate mesh. Currently, the most widely applied method is what which combines the algebraic method and elliptical numerical grid generation method, i.e., the elliptical numerical grid generation method is a smooth process of meeting a certain condition of grid distributions. It needs some grids contributions at the border and an initial distribution of grid points, which are generally generated by the algebraic method.

8.3.2   Unstructured Mesh Generation Techniques

Unstructured mesh generation technology has mainly formed three basic methods: advancing front method, Delaunay method, and quadtree/octtree method. In the last two decades, to adapt to more and more complicated flow field calculation for projects, many new grid generation technologies have been developed based on traditional ones such as: moving boundary grids, self-adaptive mesh, multi-grid, and so on. Currently in computational fluid dynamics, numerical grid technology is a very active research area, where many new methods are emerging, but there are still many issues remaining to be resolved.

Since the structured grid is equivalent in topology to a uniform contributed grid in the rectangular region, whose nodes are defined in each layer of grid line, and the node number in each layer is the same, it is hard to generate body-fitted grid for complex shapes. An unstructured mesh does not have a regular topological structure or the concept of "layer" and its distribution of grid nodes is arbitrary, so it has flexibility. Since the early 1990s, the unstructured grid has undergone rapid development.

Triangular or tetrahedral elements adopted by the unstructured mesh can easily fill the entire calculation area without the need to consider the distribution of the constitutive property or orthogonality of the connection of grids. The contribution of nodes and cells has good controllability, which has a strong adaptation to domain geometry; grid generation for a complex shape is relatively easy and time-saving. The data of an unstructured grid is randomly stored, which is conducive to control of the grid density and realizing self-adaption so that it can improve the calculation accuracy. Since the mesh nodes can be quickly added and deleted, it is comparatively easier in terms of handling border issues. In applications such as free surface flows, multi-material flows, and relative motion of multi-body, unstructured grids have advantages that structured grids cannot match.

However, the unstructured grid also has some drawbacks. Compared with a structured grid, it requires a lot more physical structure for data storage and it takes more time to address during the calculation. In dealing with the viscous problem of high Reynolds number, the highly stretched triangle/tetrahedral grids cannot accurately simulate gradient terms of the flow parameters and if grids in each direction are densely located the calculation cost will be enormous.

In this section, the grid generation process is briefly introduced with a two-dimensional (triangular element) advancing front method as an example. The main steps are: generating background grid; forming the initial front by splitting the inner and outer boundaries; introducing new nodes advancing front and generating mesh; smoothing and optimizing the generated mesh:

1. *Background grid:* This refers to a set of sparse meshes that cover the entire computational domain. Its nodes store the controlling parameters required in grid generation, such as the basic length of cells, the stretching ratio of mesh and stretching direction of mesh. For simplicity, here only its basic length is used. The background grid is not the final part of mesh, and thus it does not need to be consistent with boundary. However, any point within the calculation area can be obtained from the parameters used in its linear interpolation of the three vertices of the mesh, which has ensured the continuous distribution of grid cells. The most basic background grid is formed by inputting data manually; the new and improved measure is to automatically generate it using the Delaunay method.
2. *Formation of the initial front:* For a two-dimensional n grid, the front is made of sides (vectors) and with the initial front can be determined by separately subdividing control parameters provided by background, object surface boundary, and far field boundary. The sides (vectors) of the outer boundary of the control body is moves in a counterclockwise direction; the inner boundaries are closed loops with directions, but each side (vector) of it is arranged in the counterclockwise direction, which will eventually turn into the sides of triangles within the triangular structure. Generally, the initial front of the double diversification is the sum of all the inner borders.
3. *Advancing front and generating mesh:* After the initial front is formed, the control parameters provided by background grid can be introduced into the internal nodes and then generate the mesh. This process involves connecting new nodes or front nodes that have already existed to form triangular cells and refine the front.
4. *Optimizing the mesh:* To further improve the quality of the grid, the grid generated above needs to be optimized. The most common method is to use Laplacian iteration.

Compared with a structured mesh, an unstructured mesh has problems such requiring more memory and longer CPU time. The method of accelerating convergence used by structured grids cannot be used in unstructured grids. The generation of unstructured grids in viscous flow calculation needs further study. Thus, a mixed grid combining structured and unstructured grids has been examined.

The rectangular grid mesh is the simplest kind of structured grid meshes that does not have to calculate the Jacobian matrix, and it is simpler and more efficient than the body-fitted grid. However, it is not fit for dealing with a complex boundary. Therefore, the simplest hybrid meshes consist of the close-to-surface unstructured grid and faraway rectangular grid. This kind of hybrid mesh is effective for non-viscous flow calculation.

Since the beginning of the 1990s, hybrid meshes combining the advantages of structured and unstructured grids have begun to be developed. The idea is to use the structured grid mesh in places with high gradient such as walls and then use the unstructured grid to fill up the areas between the structured grids. This can make full use of the high accuracy of a structured grid in viscous flow calculation and the small workload of unstructured grid generation, making it suitable for calculation of viscous flows with a complex shape. In comparison with the structured grid, the hybrid grid generation is easier. In comparison with the unstructured grid, the hybrid grid is more suitable for the simulation of viscous flows, and requires less calculation cost. In recent years, it has become a new hot research area in CFD.

Both structured and unstructured grids have adopted the finite volume method. The flow variables are stored at the cell center, and the data exchanges all happen in the cell center. At the interface, both triangular and quadrilateral elements are outside the virtual layer, and the center flow variables of a virtual triangle is determined by the center of quadrilateral elements, and vice versa. Then, the data exchanges between structured and unstructured grid meshes can be achieved. Each iteration step during the calculation will be followed by the update of the flow variables at the interface between structured grid and unstructured grid. The calculation on a hybrid grid first resolves on the structured and unstructured grids, respectively, according to the given initial and boundary conditions, and then exchange solution information at the interface. If this does not meet the convergence conditions, continue solving and exchanging data, and repeat the previous steps until the convergence conditions are met and output results.

8.4   Simulations of Combustion in Liquid Rocket Engines and Results Analysis

8.4.1   Numerical Analysis of Dual-States Hydrogen Engine Combustion and Heat Transfer Processes

A dual-states hydrogen engine can realize single stage to orbit by the method of changing the mixing ratio, which can reduce the costs of carrier launch and improve reliability. The engine adopts high-altitude oxygen-rich conditions at low attitude and has a high density of specific impulse. When the engine reaches a certain height, it will switch to a fuel-rich working condition that is close to equivalence ratio to maintain a high specific impulse. However, a dual-states hydrogen engine closes some parts of propellant flow during state transition, which will inevitably lead to fierce changes of combustion flow field in the thrust chamber and affects both the temperature and heat flow distribution at the combustion chamber wall.

This example adopts a three-dimensional model to perform numerical modeling on the combustion flow field of an engine before and after working conditions transfer. Under the given temperature of combustion outer wall, a coupling calculation is carried out on the heat transfer between the gas and the combustion chamber wall; we then compare the combustion flow field and efficiency when using a coaxial centrifugal injector with that obtained when using a coaxial injector.

8.4.1.1 Physical Model

It is assumed that the hydrogen has been gasified before getting into the combustion chamber. The three-dimensional Reynolds averaged Navier–Stokes equations are selected as the gas-phase controlling equation. The standard $\kappa-\varepsilon$ two-equation model is used as the turbulence model. The hydroxide single-step and total package chemical reaction model is selected as the chemical reaction model, considering the forced convection of gas to wall and the heat conduction effect of the wall; the discrete phase model is used to describe the random orbit oxygen particle Lagrangian coordinate system for droplet motion. Drops are grouped according to the Rosin–Rammler distribution function, and the propellant atomization model can be expressed as:

$$q_i = 1 - \exp\left[-0.639\left(\frac{d_i}{D}\right)^N\right] \tag{8.123}$$

where q_i is the ratio of the mass of the drops with a diameter smaller than d_i, D is the diameter of the intermediate droplet mass, and N is the size of the distribution parameter. Both D and N are measured experimentally.

The evaporation rate of liquid oxygen is calculated using the droplet atmospheric evaporation model.

The finite rate chemistry model is used to simulate single-step hydrogen reaction. The reaction rate is given by the Arrhenius equation:

$$R = C_A \bar{\gamma}_{\text{fu}} \bar{\gamma}_{\text{ox}} \exp\left(-\frac{E}{RT}\right) \tag{8.124}$$

The temperature and heat flux distribution at the combustion chamber wall are unknown, which requires the forced convection of the gas to the wall and heat conduction of the wall to be solved together.

The Gas forced convection heat transfer expression is:

$$q_k = \alpha_{\text{aw}}\left(T_{\text{aw}} - T_{\text{wg}}\right) \tag{8.125}$$

The gas convection heat transfer coefficient adopts the following formula:

$$\alpha_{\text{kg}} \approx 74.3 \times C_{\text{pg}} \times \eta_g^{0.18} \times G_g^{0.82}/D^{1.82} \times \left(T_g/T_{\text{wg}}\right)^{0.35} \tag{8.126}$$

where C_{pg}, η_g, and G_g are the specific heat, average dynamic viscosity, and total mass flow rate of the close-to-wall layer gas, respectively; The parameter D is the inner diameter of the combustion chamber at cross section; T_g and T_{wg} are the temperatures of gas and wall contacted with gas, respectively.

The combustion chamber wall heat conduction equation is given by:

$$q_k' = -\lambda \frac{dT}{dL}\bigg|_w \tag{8.127}$$

where λ stands for the thermal conductivity of the solid wall.

The following conditions are valid at the coupling interface:

1. continuous temperature:

$$T_{wg}\bigg|_g = T_{wg}'\bigg|_w \tag{8.128}$$

2. continuous heat flux:

$$q_k\big|_g = q_k'\big|_w \tag{8.129}$$

where subscripts "g" and "w" denote the physical quantity close to the gas and solid wall on the gas–solid surface.

8.4.1.2   Numerical Methods

The time correlation method is used to solve the hydrogen combustion reaction flow field and the controlling volume method is chosen to discretize the equations. In the space domain, the second-order upwind scheme is used to solve the momentum equation, continuity equation, energy equation, and components equations; in time domain, the explicit Runge–Kutta method is applied for the iterative solution.

While solving the reaction flow field via SIMPLE, the heat conduction of gas and wall is also coupled to obtain the solution. The gas area based on the starting temperature distribution of the coupled boundaries is solved and part of the heat flux and temperature gradient of coupling boundaries can be obtained; then solve the solid wall region, obtain a new temperature distribution on coupled boundaries and use it as the input of gas area, repeating the process until convergence.

8.4.1.3   Grid Mesh for Calculation

The modeled surface of model engine is an axially symmetric structure composed of a double arc. Considering that there are 12 outer and 6 inner injectors, evenly distributed, the calculation region along the symmetry plane takes 1/12. The surface grid distribution is shown in Figure 8.6.

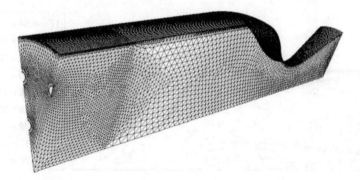

Figure 8.6 Mesh for combustion and heat transfer analysis.

8.4.1.4 Boundary Conditions

The entrance boundary conditions are given by the flow rate, temperature, and inlet velocity, and the parameter distribution is given by the atomization model of centrifugal injectors. The export conditions are given by a linear extrapolation. The gradient of variable in the plane of symmetry equals to zero on the direction of X and Y. The speed of the inner wall meets the no-slip conditions of a solid wall. The heat transfer boundary conditions meet the coupling interface conditions; the outer wall is considered as isothermal.

8.4.1.5 Calculation Conditions

Calculations are run at two working conditions: (i) oxygen-rich conditions: hydrogen flow is 0.08307 kg s^{-1}, inlet temperature is 150 K; oxygen flow 1.0799 kg s^{-1}, inlet temperature of 93 K; excess oxygen coefficient of 1.5; (ii) rich burning conditions: hydrogen flow rate is 0.08307 kg s^{-1}, inlet temperature of 150 K; oxygen flow 0.4984 kg s^{-1}, inlet temperature of 93 K; excess oxygen coefficient is 0.75. The outer wall of the combustion chamber walls and other solid wall temperature is set as $T = 500$ K.

For both oxygen-rich and fuel-rich conditions, one certain component should be completely consumed in chemical reaction. The ratio of burnt amount of this propellant to the total flux is defined as the combustion efficiency, and is used to evaluate atomization and combustion of different injectors.

The wall thrust chamber is made of zirconium copper, the thermal conductivity of which can be measured by experiment. The physical parameters of hydrogen gas, oxygen gas, oxygen, and water vapor can be found in References [2,9].

8.4.1.6 Analysis of the Results

Flow Field Analysis
Figures 8.7 and 8.8 show the temperature distributions on one symmetry plane under working conditions of oxygen-rich and fuel-rich for centrifugal injectors, respectively. As can be seen, the maximum flame temperature can be reach up to 4000 K, which is higher than the actual flame adiabatic temperature of the hydrogen reaction. This is mainly because of the simplified

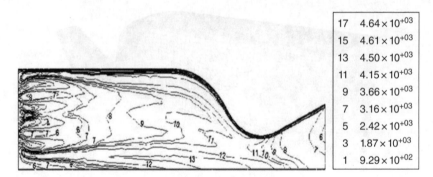

17	$4.64 \times 10^{+03}$
15	$4.61 \times 10^{+03}$
13	$4.50 \times 10^{+03}$
11	$4.15 \times 10^{+03}$
9	$3.66 \times 10^{+03}$
7	$3.16 \times 10^{+03}$
5	$2.42 \times 10^{+03}$
3	$1.87 \times 10^{+03}$
1	$9.29 \times 10^{+02}$

Figure 8.7   Temperature profile for oxygen-rich conditions.

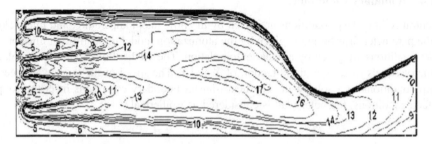

Figure 8.8   Temperature profile for fuel-rich conditions.

calculation in which a single step reaction hydroxide is used without considering the dissociation reaction; however, in the real physical hydrogen combustion process, when the temperature reaches a certain value many complex dissociation reactions occur that need to absorb large amount of heat and, thus, the temperature of flame is lower than the calculated one. However, the general temperature distribution is reasonable.

There also exists a partial higher temperature distribution area at the inlet side of the jet; it can be seen by combining the combustion products of the centrifugal injector binding profile (Figures 8.9 and 8.11) that these areas have a higher water vapor content. This is mainly due to the use of centrifugal injectors. Liquid oxygen will be discharged from the injector at an angle off-axis injection. After gasification, it generates high-speed combustion which results in complex recirculation zones at the jet expansion. While increasing the effect of mixing, these recirculation zones bring large amount of high temperature vapor products of combustion to the inlet face, which cause partially high temperature areas at the inlet face.

Comparing the distribution of products of straight-flow combustion injector (Figures 8.10 and 8.12), we can see that the water vapor content of the inlet face is lower than that with the centrifugal injector. This is because for the straight-flow injector the inlet droplet spray angle is small, such that entry into the combustion chamber is almost parallel to the axial direction; on the other hand, the effect of atomizing mixing of a straight-flow injector is inferior to that of a centrifugal injector, the combustion products in the inlet section are fewer, and the hydrogen components that that recirculation zone brings are mainly of low temperature. This

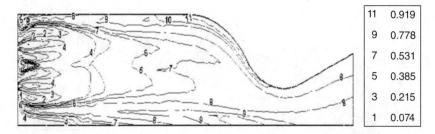

Figure 8.9   Mol. coefficient profile of H_2O for oxidant-rich condition, centrifugal injector.

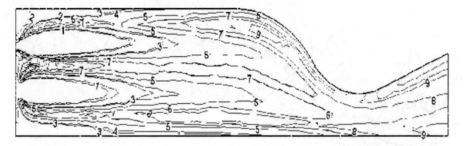

Figure 8.10   Mol. coefficient profile of H_2O for oxidant-rich condition, straight-flow combustion injector.

causes the temperature distribution in the inlet interface of the centrifugal injector to be higher than for the straight-flow injector. Therefore, in engineering applications, if centrifugal injectors are applied, we should pay more attention to cooling the inject surface.

In addition, from the results of the centrifugal injector, we can also see that in the central area there is also a higher temperature distribution, and that these regions also have higher water vapor distribution. This is due to the presence of a piece of a hollow region in the center jet area of a centrifugal injector, so that part of the hydrogen gas will pass through the liquid oxygen into the region and react. For straight-flow injector, the water vapor content in this area is low.

By comparing Figures 8.7 and 8.8 we can also conclude that overall the temperature distribution under the oxygen-rich conditions is lower than that under fuel-rich conditions, which is mainly because under the oxygen-rich conditions there is plenty of low-temperature oxygen excess.

Both flow fields under these two kinds of injectors have obvious stratification, showing the diffusion flame characteristics. In addition, when using a DC injector, this stratification is more serious than for centrifugal injectors. This shows that the DC mixing injector atomization injector is worse than the centrifugal injector. Figures 8.11 and 8.12 show the results of water vapor molar concentration distribution under fuel-rich conditions. Obviously, when using the

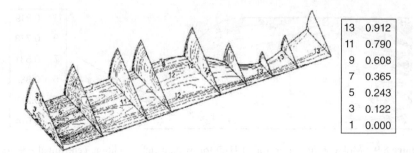

13	0.912
11	0.790
9	0.608
7	0.365
5	0.243
3	0.122
1	0.000

Figure 8.11   Mol. coefficient profile of H_2O for fuel-rich condition, centrifugal injector.

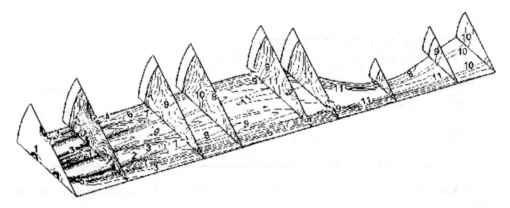

Figure 8.12   Mol. coefficient profile of H_2O for fuel-rich condition, straight-flow combustion injector.

DC injector, the concentration of the combustion products close to the wall surface and the inlet is lower, and the combustion stratification is more profound.

Heat Transfer Analysis

Figures 8.13 and 8.14 compare the combustion chamber wall heat flux and temperature distribution along the axial direction in the centrifugal injector under, respectively, fuel-rich and oxygen-rich conditions. It is found that under oxygen-rich conditions, the wall temperature and heat flux is larger than under fuel-rich conditions. This can be seen from the analysis of temperature distribution in Figures 8.7 and 8.8 where high temperature zone in oxygen rich condition is closer to the wall than that in fuel-rich condition. This was mainly due to the excess hydrogen in the fuel-rich conditions, because the hydrogen is in the outer ring of the centrifugal injector a large part of the excess hydrogen flow was along the wall, separating hot combustion products from the wall. The temperature and highest values of heat flux can also be obtained, which both appeared in the vicinity of the throat ($x = -0.05$); the maximum heat flux can be up to 107 W m^{-2}. Consequently, in the design, special attention should be paid to thermal protection for oxygen-rich conditions.

Comparison of the Combustion Efficiency

Figures 8.15 and 8.16 compare the combustion efficiency of two kinds of injectors under oxygen-rich and fuel-rich conditions, respectively. By analysis we can draw the following conclusions: (i) for oxygen-rich conditions, the combustion efficiency of the centrifugal

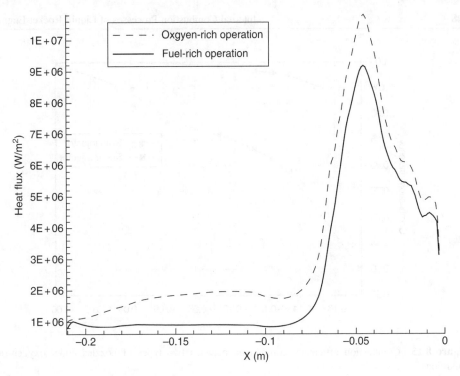

Figure 8.13 Surface heat flux comparison between oxygen-rich operation and fuel-rich operation, centrifugal injector.

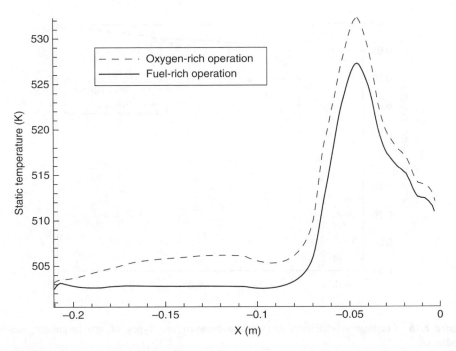

Figure 8.14 Surface temperature comparison between oxygen-rich operation and fuel-rich operation, centrifugal injector.

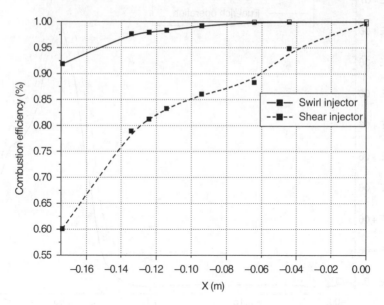

Figure 8.15 Combustion efficiency comparison between two types of nozzles under oxygen-rich conditions.

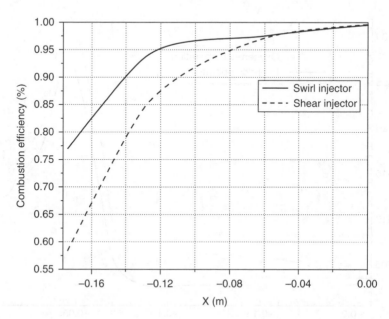

Figure 8.16 Combustion efficiency comparison between two types of nozzles under fuel-rich conditions.

injector is higher than for the DC injector. (ii) Under fuel-rich conditions, combustion efficiency in the section centrifugal injector combustion was higher than in the DC type injector, but the combustion efficiencies in the throat are similar. (iii) For both injectors the reaction is completed at the exit.

8.4.2 Numerical Heat Transfer Simulation of a Three-Component Thrust Chamber [10]

The heat transfer process of a three-component thrust chamber is complex, mainly including the propellant droplet phase-change heat transfer, heat transfer among gases, convection and radiation heat transfer between gas and the thrust chamber wall surface, heat conduction within cooling groove fins, coolant and cooling slot convection heat transfer, and so on. Of course, the process becomes more complicated when coupling the interactions, which is confronted in the numerical simulation.

8.4.2.1 Physical Model

The case study given here is for a three-element model of an engine. In the model of the engine thrust chamber injector panel is evenly arranged on a ring of five injectors, with an injector structure of centrifugal external mixing type injectors or DC external mixing type injectors. These two kinds of external mixing type injector consist of a center hole and two concentric ring slits where the oxygen is jetted from the inner hole, kerosene is jetted out from the middle circular seam spray, and gas and hydrogen are injected by the outer loop slot jet. Liquid atomization is realized by using the difference between scour collision and the shear stress caused by the gas–liquid velocity difference. The liquid atomized droplet distribution has a Rosin–Rammler distribution function, the distribution index of which is 2.1, and the diameter distribution of kerosene and liquid oxygen are as shown in Table 8.1.

After evaporation of the droplets atomization, the chemical reaction is calculated by the Arrhenius model, where the single overall reaction model of kerosene and oxygen and kerosene and hydrogen is selected.

A three-component rocket engine shows high temperature and high chamber pressure, which will cause massive convective and radiative heat transfer towards the walls. Especially in the near injector throat, the maximum heat flux can be as high as 10–160 MW m^{-2}. In the model of the engine thrust chamber with head inlet, injector outlet coolant cooling method, the liquid water is at room temperature, the cooling water tank pressure is 5.0 MPa, the cooling water flow rate is 8.9 kg s^{-1}, and the cooling groove is milled. The thrust chamber is made of

Table 8.1 Simulation results of droplet disperse diameter.

	($10^5\times$) LO$_2$	($10^5\times$) Kerosene
d_{max}(m)	7	8
d_{min}(m)	1	1
d_{avg}(m)	3	5

zirconium copper, the length of the combustion chamber is changeable, and there are holes in the connecting pipe to sustain cooling water flow. The thrust chamber wall surface heat flux is used as the boundary condition; one-dimensional calculation and evaluation of the turbulent flow and convection heat transfer can be conducted, the result of which shows that the limit of the wall temperature does not exceed that of zirconium copper, thus the cooling scheme is reliable.

8.4.2.2 Grid and Boundary Conditions

A hybrid grid combining structured grid and unstructured grid is used; an unstructured grid is used for the head region of the complex thrust chamber and the structured grid is used in the rest. Since the engine thrust chamber and cooling channels have a rotationally symmetric structure, in order to save the number of grid computations, the distribution of the thrust chamber body part (Figure 8.17) can be calculated; 1/10 area of the original thrust chamber is selected according to the symmetry of the injection injector structure. The combination of the symmetry of the structure to simplify the calculation does not result in deviation of the calculation results. The structure is shown in Figure 8.18. The combustion section grid is divided into two regions, meshes are connected at the region connection location: because droplet injection, atomization, evaporation, vapor diffusion, mixing, combustion, and other complex physical and chemical reactions will be take place in the thrust chamber head region, the grid is relatively denser. In addition, because of the complex structure of the injector, only the unstructured grid is suitable for the specific region. In the combustion chamber and the cooling channel in the boundary layer near the wall, due to the effects of the laminar boundary layer, the grids near the wall should be intensified. Moreover, due to the presence of a large temperature gradient in the zirconium copper surface, mesh intensification should also be performed to meet the precision requirement.

Figure 8.17   Jet panel, cooling, and thrust chamber.

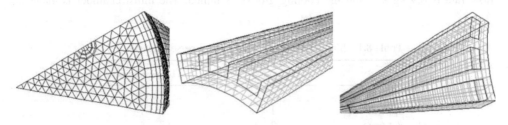

Figure 8.18   Schematic diagram of grid computing.

The boundary conditions are given as follows:

At the combustion chamber entrance: the flow rate of gaseous hydrogen is 1.601 kg s^{-1}, and the temperature is 302.2 K.

The combustion chamber outlet: there is no back flow in supersonic flow, therefore no boundary condition can be given; numerical boundary conditions are obtained by first order extrapolation supplement; in the subsonic region, such as in the boundary layer, parameters are given according to the atmosphere.

Cooling passage inlet: the flow rate of cooling water is 9 kg s^{-1}, and the temperature is 302.2 K.

Cooling passage outlet: the outlet pressure is given.

Wall: no seepage, no slip wall, i.e., $u = v = w = 0$, considering the insulation wall condition. For a discrete phase, when the liquid drop hits the wall surface, certain a reflecting angle is given as the boundary condition.

Plane of symmetry: gradient of all variables is equal to zero in both X, Y directions.

8.4.2.3 Analysis of the Simulation Results

Analysis of Combustion and Flow

The three elements are generally liquid oxygen rocket propellant, kerosene, and hydrogen, so the injector used is usually a gas–liquid coaxial injector. The present simulation case is for the three-component engine DC external mixing type injector and centrifugal external mixing type injector. The DC type injector has the advantages of a simple structure and easy processing. The structure of a centrifugal injector is complex, with generally larger flow. However, because the liquid forms a high-speed rotary motion inside the injector, it turns a film under the action of centrifugal force after leaving the injector, which is conducive to propellant atomization.

According to Figures 8.19–8.24, in the three-element model of other engines operating under the same conditions, a change of injectors has no obvious impact on the engine combustion

Figure 8.19 Static pressure profile in the thrust chamber, straight-flow combustion injector.

Figure 8.20 Static pressure profile in the thrust chamber, centrifugal injector.

Figure 8.21 Static temperature profile in the thrust chamber, straight-flow combustion injector.

Figure 8.22   Static temperature profile in thrust chamber, centrifugal injector.

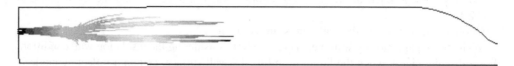

Figure 8.23   Kerosene droplet, straight-flow combustion injector.

Figure 8.24   Kerosene droplet, centrifugal injector.

chamber pressure. The engine room pressure is determined mainly by the propellant flow rate and thrust chamber structure. Calculation shows that the pressure DC injector combustion chamber is stable at 3.43 MPa, and the combustion efficiency is estimated to be 90.75%. For a centrifugal injector, the pressure chamber combustion is stable at 3.52 MPa, slightly higher than for the DC type injector under the same conditions at the combustion chamber pressure, and the combustion efficiency is estimated to be 91.5%. Obviously, the centrifugal injector is helpful, to a certain extent, in improving the performance of the engine.

The pressure of a DC type injector combustion chamber hot commissioning is stable at 3.3 MPa; the corresponding combustion efficiency is 87.5%. The pressure of a centrifugal injector combustor hot commissioning is stable at 3.38 MPa; the combustion efficiency is estimated to be 89.5%. Comparing the test results with the simulation results of combustion chamber pressure, we find that the test results are slightly lower than the simulation results, indicating that the simulation model of the example used is applicable, and the simulation results are reliable. Meanwhile, the discrepancies are caused by the uncertainties from the propellant flow meter, pressure sensors that were used to determine the corresponding physical quantity and so on. Meanwhile, the numerical simulation of the thrust chamber hot test process uses many hypothesis and simplified models, which also brings uncertainties to the results.

In the scale model the engine heat test of combustion efficiency is not too high, which may be because the heat inside the engine is partly dissipated by the cooling water. The engine temperature decreased, resulting in low combustion efficiency. Consequently, simulation of the engine combustion process at the wall under thermal insulation condition was taken, the results of which show that the pressure for the DC type injector thrust chamber is stable at 3.75 MPa, and the combustion efficiency is estimated to be 93.75%, which is slightly higher than the combustion efficiency of combustion chamber with cooling. The pressure for a centrifugal injector combustion chamber was stable at 3.77 MPa, which slightly higher than for the DC injector chamber pressure; the combustion efficiency is estimated to be 94.25%. The visible emission cooling reduces the total temperature of the combustion chamber. This leads to a decrease in combustion efficiency and a reduction in the ventricular pressure. Additionally, comprehensive analysis of the pressure data of centrifugal and DC type injectors show that the centrifugal injector helps liquid propellant atomization, thus intensifying the process of combustion and enhancing the combustion efficiency.

Analysis of the Discrete Phase Model

Figures 8.21 and 8.22 show that the combustion chamber temperature can be static; although for different injectors the highest temperature inside the engine does not have much impact, the low temperature zone in DC injector engine is apparently larger than that in the centrifugal injector, the minimum temperature is lower by about 50 K. Comparing the combustion temperature map of the chamber head we can see, immediately after liquid propellant erupting out of the centrifugal injector, a strong chemical reaction takes place, so that the temperature in this region increased significantly. The DC type injector of the propellant mixes and burns intensely after a long distance, which produces a short low temperature zone in the head of a thrust chamber. From the analysis we can see that the thrust chamber temperature centrifugal injector favors the evaporation, mixing combustion for liquid propellant.

Analyzing Figures 8.23 and 8.24, we can see that the droplet ejected from a centrifugal injector evaporates much faster than for the DC type. For the ejected liquid oxygen, liquid oxygen in the centrifugal injector (model of engine cylinder length is 244.7 mm) is completely evaporated at a distance of 48 mm from the jet plane, while for the DC type injector, the distance is 81 mm; for ejected kerosene, the centrifugal injector kerosene droplet can be completely evaporated 96 mm away from the jet plane, while with the use of a DC injector the data gives a distance of 148 mm. This is because the centrifugal injector droplet velocity in the engine axial component is smaller compared with the DC type; moreover, in the atomization performance of a centrifugal injector, which makes the low engine head rather short, high temperature gas accelerates the evaporation of the droplet. It can also be seen from the figure that the oxygen angle of atomized kerosene droplet atomization inside a centrifugal injector hole is about 15° larger than the middle circular seam spray angle; this leads to erosion between films, which exacerbates the hybrid collision between the fuel and oxidizer droplets, which is beneficial to the atomization, evaporation, and mixing. The DC type injector does have this feature. The spray-hole oxygen angle and the central ring seam atomized kerosene droplet angle are basically the same, without the scour mixing. Obviously, the centrifugal injector greatly enhances the spray evaporation. It is also because the atomization centrifugal injector angle is large that low temperature gas hydrogen is washed into the combustion chamber wall, and moves along the wall a long distance, forming a low temperature hydrogen gas protective film on the wall of the combustion chamber. This exerts a very good protective effect on the combustion chamber wall. However, the performance of the DC type injector is relatively poor.

For the thrust chamber that adopts a centrifugal injector, the good atomization and mixing of its propellant promotes the thrust chamber combustion; as a result, the low temperature zone in the head becomes shorter. This is also the direct reason why the pressure and combustion efficiency of a centrifugal injector thrust chamber are both slightly higher than for the DC type injector combustion chamber.

Heat Transfer Analysis

The results show that the thrust chamber wall heat flux is dense, the heat flux varies greatly along the gas flow direction, and especially the heat flow density changes sharply in the injector area. Along the circumferential direction, the thrust chamber heat flux density distribution is not uniform; the closer to the injection surface, the greater the density distribution differs, which is mainly due to the uneven mixture ratio of the propellant near the wall, with the process of combustion. Distribution of the posterior circumferential heat flux gradually tends to be uniform in the combustion chamber, which is profound in the DC type injector thrust chamber wall. Placing the thrust chamber wall heat load in the throat area of the injector, the maximum heat flux approaches 2×10^7 W m^{-2}.

For the scale model engine cooling, we use the discharge of cooling water on the outside; water injected from the thrust chamber head to the cooling channel, exhausted from the engine injector end, and at the entrance of water is at normal temperature. According to the one-dimensional calculation results, due to the huge water flow as well as the large heat capacity, the outlet water temperature is less than 350 K; experiment has also proved that the outlet water temperature is below its boiling point. Thus, in the simulation of a discharge type cooling channel, a single-phase flow governing equation is used to deal with the flow.

From Figures 8.25 and 8.26 we can see that the different types of injectors do not have much impact on the overall thrust chamber wall surface temperature, but there is some difference between the wall temperatures of the thrust chamber at the head of the two types of injector. This is mainly because the DC type injector performs more poorly than the centrifugal injector in atomizing. There is a small low temperature district in the thrust chamber head where the heat flux on the wall of the gas is very low, and the cooling water flows in from the head of the thrust chamber through cooling. Therefore, there is a relatively low temperature in the wall of the DC injector thrust chamber surface, the length of which is about 50 mm. Beyond the low temperature zone, there is generally no temperature difference between the two types of injector thrust chamber fin (Figures 8.27 and 8.28).

From the temperature of the thrust chamber wall surface we can see that the temperature of the thrust chamber wall adopting a centrifugal injector is far more uniform than that using the

Figure 8.25   Wall temperature profile in the thrust chamber, straight-flow combustion injector.

Figure 8.26   Wall temperature profile in the thrust chamber, centrifugal injector.

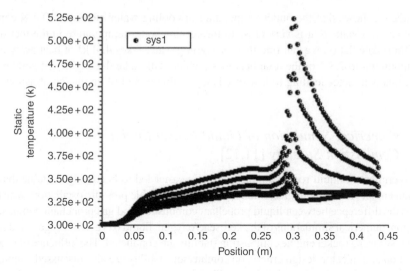

Figure 8.27   Fin surface temperature profile in the thrust chamber, straight-flow combustion injector.

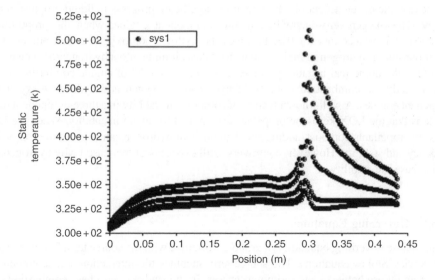

Figure 8.28   Fin surface temperature profile in the thrust chamber, centrifugal injector.

DC injector, which further illustrates that the uniform mixture of the propellants lead to combustion and heat transfer uniformity. As can be seen from the figures, the room wall temperature climaxes at the throat; the wall temperature in the throat zone near gas side is less than 520 K, which is 350 K lower than the thermal fatigue temperature of zirconium copper material, which shows that the cooling scheme is reliable and thus the thrust chamber wall is safe. The wall temperature curve in the combustion chamber is gentle, but in the throat of the injector wall and convergent section the temperature rises dramatically.

Calculation shows that the outlet temperature of cooling water increases 10.2 K compared with the entry, a value that is very close to the cooling water temperature in the thermal test. After simple calculation we can find that the cooling water takes 420 kJ of heat away per second, (amounting to 7.8% of the total heat loss), which will make the engine temperature drop. Clearly, this is a factor leading to low efficiency of this model of engine combustion.

8.4.3   Numerical Simulation of Liquid Rocket Engine Combustion Stability [11,12]

In an experiment, a liquid rocket engine is usually easier led to becoming unstable than a gas propellant engine, resulting in greater amplitude of the unstable pressure oscillation, which demonstrates the difference between liquid propellant combustion and gas propellant combustion. In this section, numerical simulations of the kerosene/oxygen, liquid oxygen/hydrogen, and oxygen/hydrogen/kerosene rocket engine combustion stability are conducted. The influence of propellant combination and injector design criteria on combustion stability are also discussed, emphasizing the incentive mechanism of combustion oscillation.

The reasons for using the three-component engine as a numerical study are as follows: the oxygen/kerosene/hydrogen three-component rocket engine is a future RLV candidate propulsion device, the development of which requires significant progress in liquid propulsion technology. The oxygen/kerosene/hydrogen three-component engine, when the proportion of hydrogen in fuel accounts for 0%, becomes the hydrocarbon/oxygen rocket engine; when the proportion of hydrogen in fuel accounts for 100%, it turns into a hydrogen/oxygen engine. Therefore, the numerical instability of the three-element model of engine combustion can be used to study the combustion stability of hydrocarbon/oxygen and hydrogen/oxygen two-component engines. A well-known feature of combustion stability in a three-component rocket engine is that the LOX/kerosene propellants often lead to severe instability. When the LOX/kerosene propellants combust, adding small amounts of hydrogen can increase the combustion efficiency and stability. However, controversy still exists about the reason why hydrogenation can increase the stability.

8.4.3.1   Governing Equation

The gas phase governing equation is a group of formulae with the source term Favre time averaged Navier–Stokes equations. The source term includes the interaction source combustion sources and those between gas and liquid phases. In the study of gas phase combustion only, terms between the gas phase and liquid phase interaction source are set to zero. The high Reynolds number is adapted in turbulence model equations.

The liquid phase is divided into a series of droplet group. Each group of droplets has the same diameter, position, velocity, temperature, and composition properties. When the droplet travels through the combustion chamber, for a time they remain in their initial droplet group. There is no mass and momentum transfer among the groups. Each droplet group is described with the unsteady Lagrangian particle tracing method.

A complete chemical dynamics system of a three-component reaction is very complicated. Numerical study of combustion instability is time consuming, so in this reaction system we use the most simplified case. A simplified reaction system may differ from a real oscillation process, but is permissible in mechanism research for a qualitative perspective. If the chemical reaction is completed by three-step reactions, kerosene is first oxidized into two kinds of products (CO and H_2), which are then oxidized:

$$C_{12}H_{24} + 6O_2 \rightarrow 12CO + 12H_2 \tag{8.130}$$

$$CO + 0.5O_2 \rightarrow CO_2 \tag{8.131}$$

$$H_2 + 0.5O_2 \rightarrow H_2O \tag{8.132}$$

The reaction rate of each of the above three reactions is expressed in Arrhenius law form:

$$\omega_i = -A_i T^{\beta_i} \prod_m \left(\frac{\rho m_m}{M_m} \right)^{v_m} e^{\left(-\frac{E_i}{RT} \right)} \; i = 1, 2, 3 \tag{8.133}$$

The reaction product is set as 1, the product is set as 0. Three reactions were set to be ½. The hydrogen oxidation reaction activation energy is much lower than that of the other two reactions.

8.4.3.2   Numerical Method and Boundary Conditions

The governing equations are solved using implicit PISO methods. To save CPU time, only the two-dimensional problem is considered. The radius of the combustion chamber is 25 cm, and the distance from injection surface to the chamber throat is 125 cm. Since the combustion chamber length is much larger than its radius, the longitudinal vibration mode in the combustion chamber probably be the most unstable mode. Using the embedded grid, the cylindrical segment of combustion chamber is divided into a grid. The combustion chamber and injector convergent section are divided into a body fitted grid.

Hydrogen, oxygen, and kerosene are injected through a central coaxial injector in jet plane into the combustion chamber. Flow rate, temperature, and composition are regulated according to the injector entrance. In addition, the propellant pressure pushes back upstream. At the exit, all variables push from the upstream. Unlike some other numerical study, no special numerical methods facilitate the instability of combustion. The initial field of the numerical study starts from the combustion chamber pressure, and the time step is fixed. A group of droplet are injected into the combustion chamber every ten time-steps. All groups of droplets are endowed with the same initial size (50 μm) and the same initial temperature, but their initial position and velocity vector are randomly distributed in a certain range.

Simulation of the three groups of the gaseous combustion process was performed. Then studies of two-phase combustion processes under five different conditions (Table 8.2) were conducted. The first case uses the LOX/kerosene propellant. Cases 2–4 use the kerosene/oxygen/hydrogen propellant, and the mass percentage of gaseous hydrogen in the fuel is, respectively, 10%, 20%, and 30%. The last case uses liquid oxygen/hydrogen propellant. The designed pressure of the former four conditions in the combustion chamber is 4 MPa, the designed pressure of the combustion chamber in last case is 2.6 MPa. Finally, for the

Table 8.2 Mass flow rate in the simulated five cases of two-phase combustion.

Case No.	1	2	3	4	5
Propellant	LO$_2$/kerosene	GO$_2$/kerosene/ GH$_2$	GO$_2$/kerosene/ GH$_2$	GO$_2$/kerosene/ GH$_2$	LO$_2$/GH$_2$
m_{H2} (g s^{-1})	0.00	12.52	22.08	29.63	39.10
$m_{kerosene}$ (g s^{-1})	149.80	112.72	88.30	69.14	0.00
m_{O2} (g s^{-1})	410.00	388.08	381.9	377.40	234.6

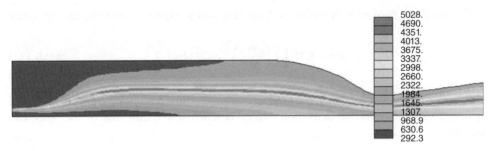

```
5028.
4690.
4351.
4013.
3675.
3337.
2998.
2660.
2322.
1984.
1645.
1307.
968.9
630.6
292.3
```

Figure 8.29 Temperature distribution of the gas-phase diffusion flame in the combustion chamber.

three-component conditions, the gas rate of oxygen is increased and the reaction activation energy is changed.

8.4.3.3 Numerical Simulation Results

Gas Phase Combustion Simulation Results Analysis
The three-element phase diffusion flame and premixed flame were studied. For diffusion flames, a fuel mixture of kerosene vapor and hydrogen gas spray from the inner injector coaxial injector were used, and gas oxygen sprays from the outer injector coaxial injector. The entire flow field of the combustion chamber gas flame can be divided into three areas: fuel, oxidant, and product area. The product district is sandwiched between the fuel and oxidizer zone (Figure 8.29). The topological structure is similar to the flow field division; the temperature in the product area is very high, the fuel and oxidant area at low temperature. The apparent diffusion process is controlled by the gas diffusion flame. Combustion oscillations are not observed under the various research conditions.

For the premixed flame, uniform mixture of kerosene vapor, hydrogen, and oxygen is injected into the combustion chamber via the outer nozzle of the coaxial injector. Unlike the diffusion flame, the whole flow field can be divided into only two areas: combustion area and the product area. The chemical-kinetics-determined reaction rate is so fast that the propellants are all burnt as soon as they enter the combustion chamber.

However, when the steady field was set as the initial field and artificially increases the activation energy in the third reactions, oscillation appeared in the premixed flame. Moreover, the non-combustion zone appeared to be much larger than before (Figure 8.30). The frequency of pressure oscillation is about the first order longitudinal mode of the combustion chamber (Figures 8.31 and 8.32). This oscillation shows that the premixed flame is more instable than the diffusion flame; a high activation energy makes the stability of combustion much more likely.

As can be seen from Figure 8.31, although the initial amplitude of pressure oscillation in the combustion chamber is large, as time increases, non-firing zones will gradually increase and finally flameout. Because the gas phase reaction in the combustion chamber is generally insulation, there are only two states of vapor propellant combustion zone: the ignition and flameout. It is hard for a large and long-lasting amplitude of pressure oscillations to appear in the combustion chamber. In the following study of the two-phase numerical simulation of combustion stability, we will see the differences between two-phase and gas-phase combustion stability.

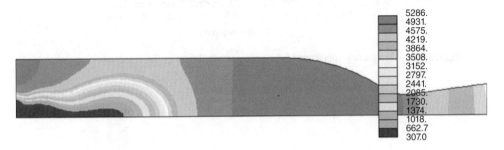

Figure 8.30   Temperature distribution of the premixed flame in the combustion chamber.

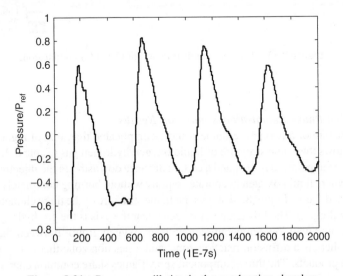

Figure 8.31   Pressure oscillation in the combustion chamber.

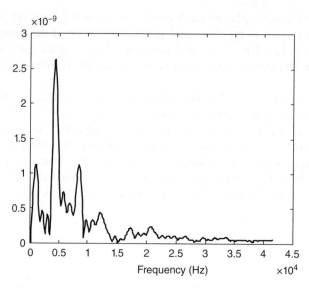

Figure 8.32 Frequency spectrum of the pressure oscillation in the combustion chamber.

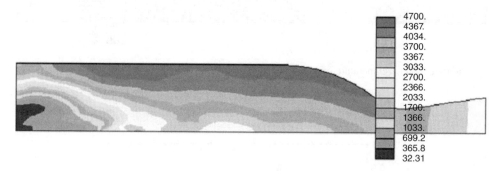

Figure 8.33 Temperature distribution of GH_2/LO_2 combustion.

Analysis of Two Phase Combustion Simulation Results

The flame structure of a two phase flame is more complicated than that of a gas phase flame. Figure 8.33 shows the flame structure of liquid oxygen/hydrogen gas (Table 8.2, case 5). Near the injector, the temperature is low, and a certain distance downstream the injector a flame peak is presented, where liquid oxygen evaporates rapidly without burning completely. The flame is like the premixed flame. Figure 8.34 shows the flame structure of the three elements (Table 8.2, case 2, 10% hydrogen). The low temperature zone near the axis is the gas hydrogen group partition, and the near wall low temperature zone is an oxygen group. Between the oxygen and hydrogen gas there is a diffusion reaction zone, and a reaction zone that is mainly composed of the reaction products. The three-component spray flames share common characteristics with a gas phase diffusion flame and a premixed flame.

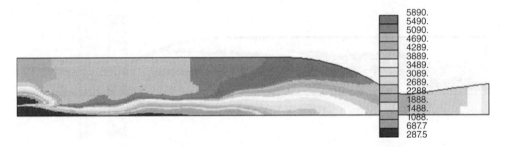

Figure 8.34 Temperature distribution of three-component combustion (10% H$_2$).

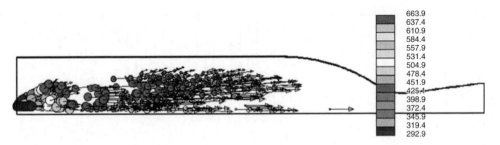

Figure 8.35 Kerosene droplet distribution of three-component combustion (10% H$_2$).

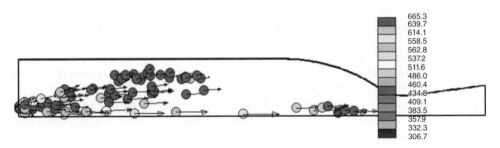

Figure 8.36 Kerosene droplet distribution of three-component combustion (20% H$_2$).

Figure 8.35 shows the distribution of particle kerosene in case 2. Particle color, size, and the vector represent the particle temperature, diameter, and velocity, respectively. Kerosene rapidly evaporates under heating of the reaction zone near the injector. Figure 8.36 shows the distribution of droplets in case 3 (Table 8.2). Since the hydrogen ratio increases, the temperature in the combustion zone near the injector increases, and the mass ratio of kerosene is relatively reduced. Therefore, the kerosene droplet evaporates in a shorter period of time. Figure 8.37 is the distribution of the droplet after oxygen accelerates the inlet speed in case 3. Because the gas oxygen speed increases, a high temperature zone in the head of the combustion chamber fails to form, and only a very small amount of kerosene particle head evaporated. Instead, it went straight through the gas oxygen group partition into the large recirculation zone, close

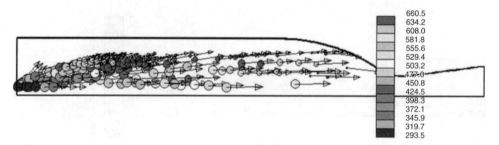

Figure 8.37 Distribution of the three-element particle kerosene (20% h); the oxygen inlet velocity is doubled to 140 m s^{-1}.

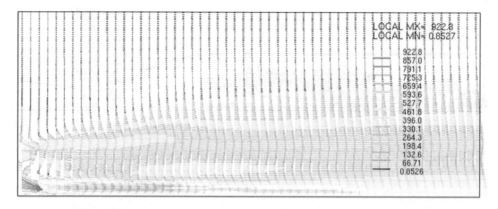

Figure 8.38 Velocity vector in the three-component head (20% H$_2$).

to the wall, where the temperature is high; there the kerosene droplets evaporation completes, and then begin the diffusion with oxygen gas.

Figure 8.38 shows the status of combustion chamber head velocity vector diagram in case 3. There exists a recirculation zone between the hydrogen and oxygen with a high temperature, which promotes the evaporation of passing kerosene. Figure 8.39 shows the increasing jet velocity after gas oxygen combustion chamber head velocity vector in case 3. Obviously, the recirculation zone becomes smaller, and the temperature near the head decreases.

In addition to the liquid oxygen/hydrogen circumstances, the self-irritated oscillation appears massively in all other cases while the oscillation is sustained up to a certain magnitude after the formation of stable limit cycles. Figure 8.40 gives the pressure versus time near the jet surface and the outer surface of a grid in a three-element (10% hydrogen) combustion chamber. In the diagram the pressure and time are relative values, the reference pressure is the local average pressure of the chamber, and the reference time is 1 ms. The spectrum analysis of Figure 8.40 is shown in Figure 8.41. The unstable mode of this condition is a first order longitudinal (1L) and transverse (1R). Figure 8.42 shows the case 4 (Table 8.2, 30% hydrogen) pressure oscillation combustion chamber head case. Figure 8.43 shows the corresponding spectrum transform. Unstable modes of one order, two order, and three order longitudinal mode are presented. Figure 8.44 shows the variation of temperature oscillation corresponding to

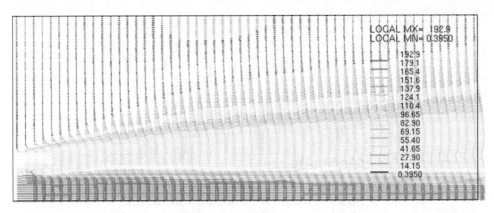

Figure 8.39 Velocity vector in three-element head (20% h); the oxygen inlet velocity is doubled to 140 m s^{-1}.

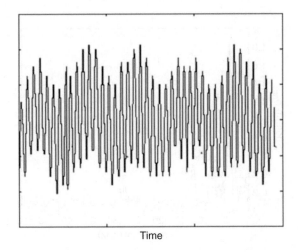

Figure 8.40 Pressure oscillation of three-component combustion (10% H$_2$).

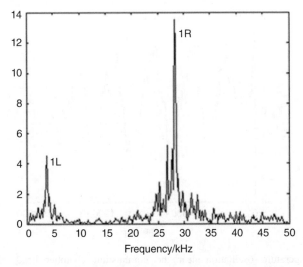

Figure 8.41 Frequency spectrum of pressure oscillation of three-component combustion (10% H$_2$).

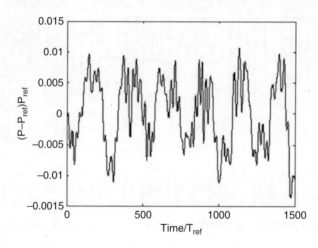

Figure 8.42 Pressure oscillation of three-component combustion (30% H_2).

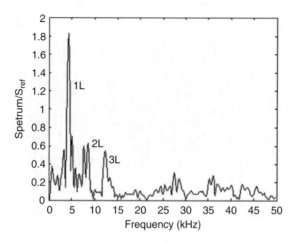

Figure 8.43 Frequency spectrum of pressure oscillation of three-component combustion (30% H_2).

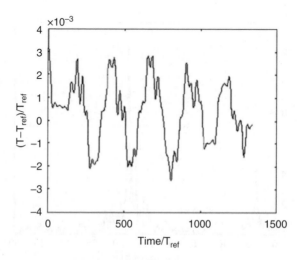

Figure 8.44 Temperature oscillation near the combustion chamber head of three-component combustion (30% H_2).

Figure 8.42 for the case of time. The relative amplitude of the temperature oscillation is much smaller than the amplitude of pressure oscillation. Figure 8.45 shows that the pressure oscillation amplitude changes as the hydrogen content increases in the five cases. The impact of the additional hydrogen in the hydrocarbon/oxygen combustion inhibition is quite obvious.

The numerical results of reducing the kerosene vapor oxidation and CO oxidation reaction activation energy are shown in Figure 8.46. Figure 8.47 shows the corresponding FFT results.

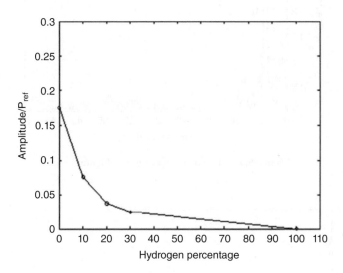

Figure 8.45   Combustion stability versus hydrogen percentage.

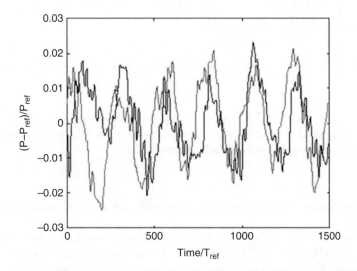

Figure 8.46   Effect of activation energy on combustion stability (case 4, Table 8.2).

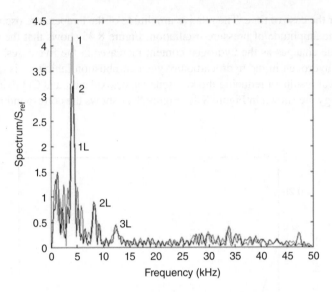

Figure 8.47   Frequency spectrum of Figure 8.46.

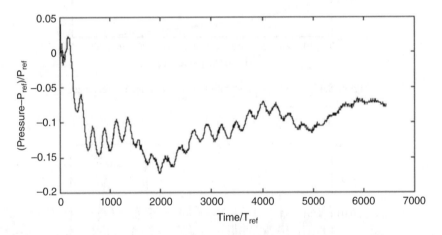

Figure 8.48   Calculation of case 3 as the initial state, reducing the entrance area of oxygen; while increasing the oxygen injection velocity the average pressure of the combustion chamber decreases, and the chamber pressure pulsation is gradually reduced to zero.

The calculation results show a lower combustion oscillation amplitude after reducing the activation energy.

When the jet velocity of oxygen is doubled, combustion oscillation gradually weakened, approaching zero (Figure 8.48). The jet velocity of oxygen is beneficial in improving the combustion stability, but at the same time the average pressure of the combustion chamber decreases to 3.7 MPa, which indicates that the oxygen injection velocity enhances the combustion performance. In the hydrogen/oxygen bipropellant mode, the change of entrance temperature of hydrogen shows no significant effect on the combustion stability.

If the combustion chamber parameters spatial distribution in different time periods are animated, the space oscillation characteristics of combustion can be revealed (Figures 8.49 and 8.50): a sinusoidal injector pressure near the time pulse is not standard, and it is not very smooth, but there are some pressure spikes. Sine and cosine the pressure distributions are not standard in the form of horizontal and vertical space. The pressure vibration amplitude in the combustion chamber head injector is near maximum, while in the injector entrance part, the spatial distribution is much smaller. Since the temperature fluctuation is more complicated than the pressure fluctuation, the temperature not only presents the wavy characteristics but also shows the flow characteristics.

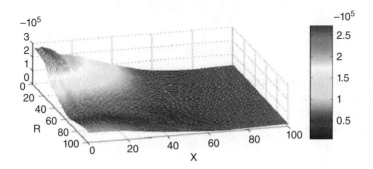

Figure 8.49 Two-dimensional space distribution of cylindrical pressure fluctuation amplitude in case 2.

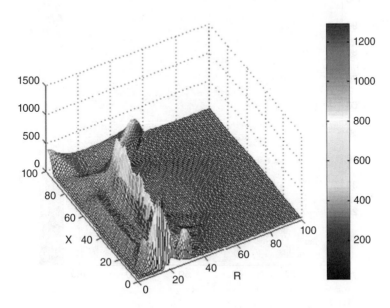

Figure 8.50 Two-dimensional space distribution of cylindrical temperature fluctuation amplitude in case 2.

8.4.3.4 Analysis of the Combustion Oscillation

Numerical results of the above numerical model for the study of the oscillation mechanism of combustion contains some indications, but the determination of what kind of physical mechanisms can lead to combustion oscillation still need further analysis (Figures 8.51–8.62).

The Burning Sensitive Area

The instability sensitive region in the combustion chamber is crucial to determining the combustion instability mechanism, and can be determined by Rayleigh criteria. First, we define the following coefficient:

$$G = \int_T \int_{dV} p(\vec{x},t) Q(\vec{x},t)\, dV\, dt$$

where:

p is the pressure of the combustion chamber;
Q is fluctuating heat release;
T is the pulsation period;
V is the control volume.

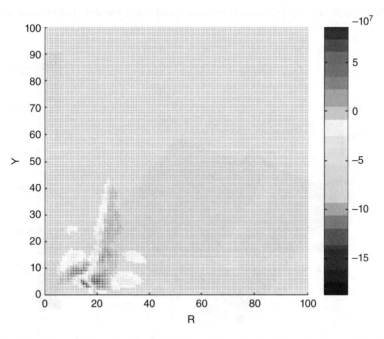

Figure 8.51 Case 2, two-dimensional distribution of the combustion chamber to the perturbation sensitivity of the acoustics.

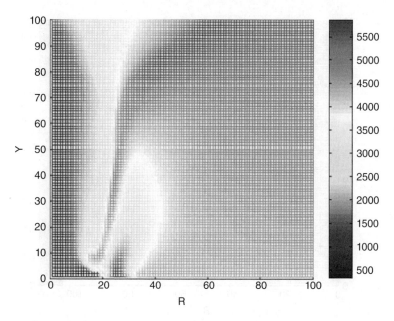

Figure 8.52   Case 2, two-dimensional distribution of the temperature.

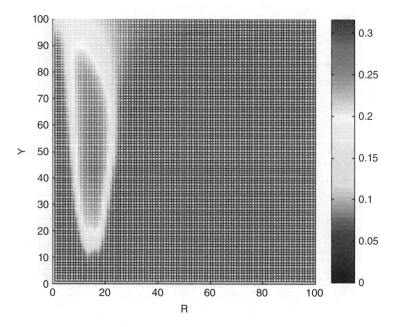

Figure 8.53   Case 2, the two-dimensional distribution of the component of the kerosene in the combustion chamber.

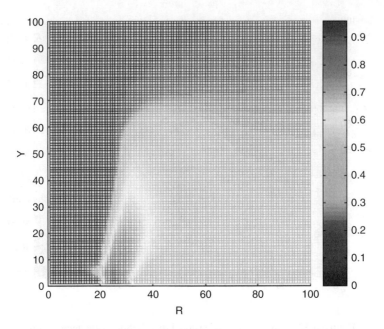

Figure 8.54   Case 2, the two-dimensional distribution of the component of the oxygen of the combustion chamber.

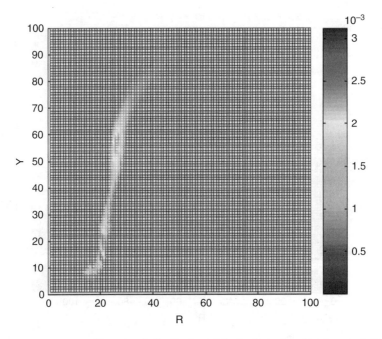

Figure 8.55   Case 2, the two-dimensional distribution of the product of the kerosene and oxygen in the combustion chamber.

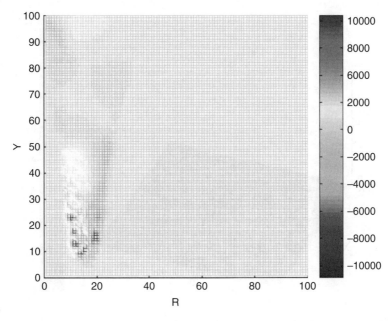

Figure 8.56   Case 2, integral of the kerosene vapor and the product of pressure pulsation over time.

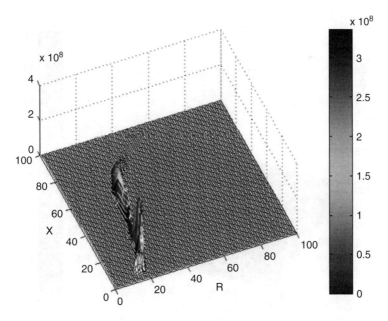

Figure 8.57   Case 2 (10% hydrogen), spatial distribution of the reaction rate of the oxidative decomposition of the kerosene vapor in the combustion chamber.

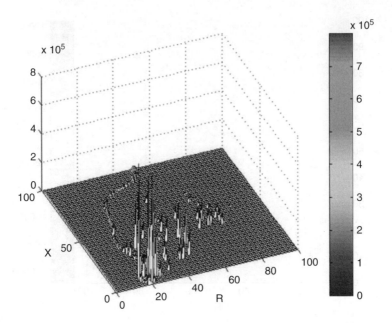

Figure 8.58 Case 3 (20% hydrogen), spatial distribution of the reaction rate of the oxidative decomposition of the kerosene vapor in the combustion chamber.

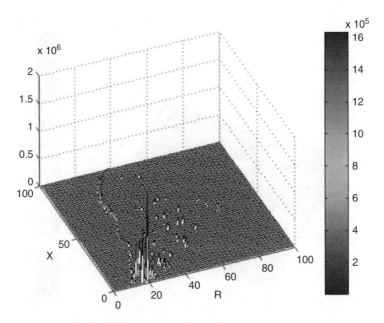

Figure 8.59 Case 4 (30% hydrogen), spatial distribution of the reaction rate of the oxidative decomposition of the kerosene vapor in the combustion chamber.

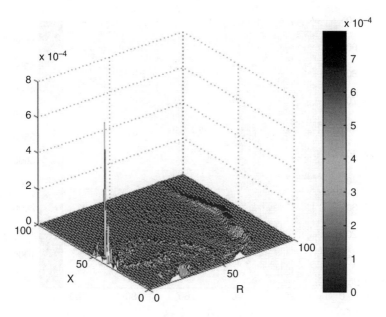

Figure 8.60 Spatial distribution of the chemical reaction rate of the hydrogen/oxygen.

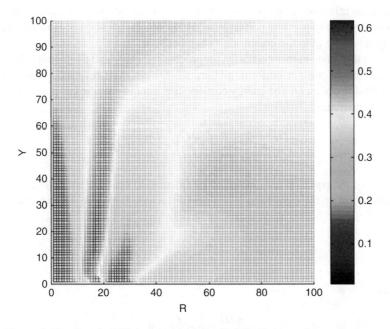

Figure 8.61 Case 3 (20% hydrogen), the spatial distribution of the temperature.

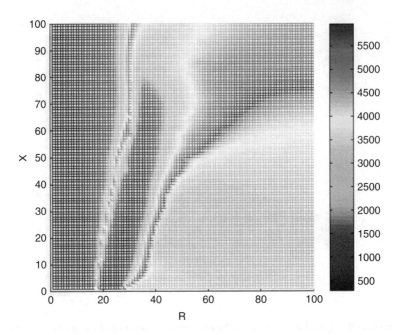

Figure 8.62   Case 3 (20% hydrogen), two-dimensional distribution of the temperature field after the jet velocity of the oxygen is increased.

If G is positive, the local contribution to the pressure oscillation is positive, otherwise the local dissipation effect on the pressure oscillation can be found. As G increases, the local pressure is more sensitive to pressure perturbation. However, the question is: How can we measure the local heat release Q in the combustion chamber?

As found in the above numerical results for processing, phase and exothermic chemical composition fluctuations are not uniform, and are sometimes even reversed. Therefore, we use another method to obtain Q. The energy conservation equation of the gas phase can be written in terms of temperature:

$$\frac{DT}{Dt} = \frac{1}{\rho c_p}\left[\frac{\partial}{\partial x_j}\left(\lambda\frac{\partial T}{\partial x_j}\right) + \frac{Dp}{Dt} - \dot{w}_s Q_s\right] \tag{8.134}$$

Both sides of the equation are multiplied by fluctuation pressure and integrated over time. The diffusion term on the right hand is neglected. Because the phase difference of Dp/Dt and p is 90°, so:

$$\int\frac{Dp}{Dt}p\,dt \approx 0 \tag{8.135}$$

Therefore:

$$\int\frac{DT}{Dt}p\,dt \propto \int Qp\,dt \tag{8.136}$$

Figure 8.51 shows the spatial distribution in case 2 (10% h). The 1–14 radial grids of the combustion chamber head ($X = 0$) are of gas hydrogen injector, and 18–31 grids are for gas oxygen injector, while kerosene droplet mainly focuses in the 15th radial grid and the second longitudinal grid for injection. It can be seen that G is positive only near the injector, and that it is close to zero in most of the other regions. In addition, the impact on the pressure oscillation is relatively insignificant, which in fact should be described as a minor dissipation. There is a small area of negative G zone of kerosene droplet in the vicinity of the concentration injection grid, in which the disturbance to the combustion chamber pressure is negative.

Figure 8.52 shows the temperature distribution of the combustion chamber. Perturbation positive districts of the gaseous hydrogen and oxygen gas are located near the injector. Figure 8.57 is the kerosene component distribution, and Figure 8.54 is the distribution of oxygen component space. Figure 8.59 is the product distribution of the kerosene component and the oxygen component space. It is found that, since the droplet can pass through the gas phase, in certain temperature regions the kerosene fraction and oxygen components can indeed coexist. Moreover, according to the amplitude distribution in Figure 8.49, the disturbance on the sensitive area can be determined based on the following conditions: the intersection of the set temperature is moderate; the fuel and oxidizer component coexist, as well as the existence of the pressure amplitude normal acoustic modes. In the area where the absolute velocity of the chemical reactions is not too fast, a certain amount of fuel and oxidizer components coexist, and the chemical reaction is not particularly slow, thus they are not sensitive to any disturbance.

The evaporation process can be divided by the fluctuating pressure effect on acoustic disturbances with kerosene group, and integrated against time. Figure 8.56 shows the spatial distribution of the kerosene. The kerosene vapor sensitive area of pressure fluctuation is a V shape; in the right-hand part of the V, the kerosene vapor pressure is out of phase with the vapor. This corresponds to a peak of hydrogen and oxygen diffusion flame position where the temperature is very high; For the left-hand half of the V, kerosene vapor and pressure pulsation are in-phase, corresponding to the hydrogen region where the temperature is very low. Additionally, in other regions of the combustion chamber, especially in the cryogenic and propellant coexistence region, the evaporation contribution to pressure oscillation is almost zero. Compared with Figure 8.51, kerosene vapor phase distribution of a sensitive area of pressure fluctuation and combustion heat sensitive area of pressure fluctuation space are found, showing that the simulation of the combustion oscillation is not driven by the kerosene evaporation process.

The spatial distribution of the chemical reaction rate is also quite useful. According to the traditional point of view, in the high temperature and high pressure environment of a liquid rocket engine the chemical reaction rate is very high. Therefore, there will not be a distribution of chemical reaction of finite area in the combustion chamber. However, the simulation result shows the opposite case. Figures 8.57–8.60 are the spatial distributions of the chemical reaction rate under some working conditions. With increasing oxygen content, the chemical reaction distribution of the head of the combustion chamber becomes weaker. For the oxy-hydrogen working condition, there is almost no chemical reaction distribution.

Figures 8.61 and 8.62 are the temperature distributions of the combustion chamber in working case 3 (20% hydrogen). We can see from the figure that, after increasing the jet velocity of oxygen in the tripropellant working condition, the oxygen area propagates forward. After a large number of droplets enter into the high temperature recirculation zone near the wall, evaporation takes place, and the diffusion flame outside the oxygen area is formed. Inside the oxygen area, oxygen and hydrogen form a very thin diffusion reaction line, which shows that the increase of oxygen injection velocity makes the flame become more like the diffusion flame, thus stabilizing the combustion.

Based on the above analysis, the combustion stability is closely related to the chemical reaction spatial distribution in the combustion chamber. As the chemical reaction distribution spreads widely, the combustion stability deteriorates. Therefore, the combustion oscillation phenomenon is the result of the interaction of the chemical reaction and the acoustic process.

References

[1] Yang T, Fang D Y, Tang Q G. *The combustion theory of the rocket engine*. Press of National University of Defense Technology, 2008 (in Chinese).
[2] Bommie J M. (1963) Thermodynamic properties 6000 K for 210 substances involving the first 18 elements. NASA SP–3001.
[3] Tomoaki K., Atsushi M., Koichi M (2002) Ignition and flame-holding characteristics of methane and hydrogen by plasma torch. AIAA Paper 2002–5210.
[4] Pan Y. (2007) Research of the combustion and the flow process of the multi cavity combustor of the scramjet engine. PhD Thesis, Graduate School of National University of Defense Technology (in Chinese).
[5] Nie W S, Feng S J. *The combustion dynamic model and the numerical calculation of the liquid rocket engine*. Press of National Defense Industry, 2011 (in Chinese).
[6] Yan C. *The method and application of the computational fluid mechanics*. Press of Beijing University of Aeronautics and Astronautics, 2006 (in Chinese).
[7] Wang C Y, Wang Z H, Yang X H. *The computational fluid mechanics and the parallel algorithm of it*. Press of National University of Defense Technology, 2000 (in Chinese).
[8] Liu W D, Wang Z G, Zhou J. The research of the PISO algorithm of the second upwind scheme. *J. Aerospace Dynamics*, 1998, **13** (1):81–86 (in Chinese).
[9] Ma Q F. *The practical handbook of the thermal physical properties*. Press of the Chinese Agricultural Machinery, 1986 (in Chinese).
[10] Wu H Y. (2004) The simulation research of the thrust chamber combustion and regenerative cooling of the tripropellant liquid rocket engine. Master's Degree Thesis, Graduate School of National University of Defense Technology (in Chinese).
[11] Huang Y H, Wang Z G, Zhou J. (2002) Numerical simulation of combustion stability of liquid rocket engine based on chemistry dynamics, *Sci. Chin. Ser. B*, **45** (5), 551–560.
[12] Huang Y H. (2001) Research of the theory of the combustion stability, the numerical simulation and the experiment of the liquid rocket engine. PhD Thesis, Graduate School of National University of Defense Technology (in Chinese).

Index

absolute mass concentration, 193
absorption, 238
acoustic energy density, 263
acoustic energy loss, 267
acoustic frequency, 119
acoustic natural, 297
acoustic oscillations, 117, 255
acoustic quality factor, 297
acoustic speed, 118
acoustic vibration density, 262
acoustic vibration rate, 262
acoustic wave, 261
activation energy, 193, 287
active control, 297
activity coefficient, 124
actuators, 297
aerodynamic force, 35
aircraft, 1
alcohol, 132
algebraic stress model, 204
ambient gas, 139
the amplification factor, 280
the amplitude of sound pressure, 262
angular, 240
angular frequency, 262
anisotropic, 167
anti-resonant methods, 297

atmospheric environment, 42
atmospheric pressure, 97
atomization, 9, 26, 283
atomization modeling, 59
attenuation, 14
autocatalytic process, 283

baffles, 295
bag breakup, 46
baseline, 159
Bernoulli equation, 61
binary system, 117
blackbody, 239
blackness, 242
bluff body jet flames, 214
boiling point, 135
boundary condition, 104
boundary layer, 74
boundary-layer structure, 259
box filtering function, 215
breakup morphology, 77
breakup position, 91
breakup time, 77

catastrophe breakup, 48
centrifugal injector, 60
chamber, 2

characteristic, 1
characteristic diameter, 53
the characteristic time, 272
chemical kinetics, 285
chemical process, 192
chemical reaction, 283
the chemical reaction process, 269
chemical reaction source term, 194
Clausius–Clapeyron equation, 142
coaxial shear injector, 70
collision frequency, 193
combustion chamber pressure oscillations, 273
combustion characteristic time, 206
combustion efficiency, 26
combustion instability, 71, 255
combustion preparation process, 68
combustion processes, 255
combustion stability, 26
combustion state, 289
combustion time delay, 258
compatibility, 13
complete combustion, 105
complex amplitude of sound pressure, 262
component conservation, 104
component diffusivity, 137
component distribution, 124
composition domain, 222
compressible, 152
conductivity, 237
configuration parameter, 63
considering, 165
constantly, 244
consumption term, 217
continuous function, 135
continuous stirred tank reactor, 287
continuous thermodynamics theory, 136
convection, 228
convective environment, 108
convergence, 9
co-saturation coefficients, 292
coupled level set and VOF method, 86
critical pressure, 109
cryogenic, 234
curvature, 231
cylindroid cigar shape, 45

damping of nozzle, 268
damping process, 255
deformation, 79
delta function, 222

density-weighted average, 212
the deterministic model, 211
deterministic trajectory model, 81
developed, 159
differential, 166
difficult to control, 260
diffusion coefficient, 195
diffusion flamelet model, 205
dilatation, 162
dimensionless, 12
dimensionless area, 131
dimensionless diameter, 119
dioxide, 245
direct numerical simulation (DNS), 164
discrete multicomponent model, 135
displacement sensitivity index, 277
dissipation rate, 207
distortion, 79
distribution function, 51
disturbance value, 270
divergent, 9
droplet combustion model, 117
droplet damping, 268
droplet drag, 282
droplet evaporation, 97
the droplet evaporation rate, 273
droplet group combustion model, 146
droplet group evaporation, 145
droplet vapor, 99

eddy, 153
eddy diffusion tensor, 211
the eddy scale, 225
eddy viscosity, 220
effect, 157
efficiency, 154
electromagnetic, 238
emission, 232
empirical, 158, 233
energy conservation, 101
energy diffusion, 129
energy loss, 261
ensemble average, 209
equilibrium evaporation model, 130
equilibrium thermodynamics, 285
evaporation, 229
evaporation constant, 101
evaporation entropy, 142
the evaporation process, 269
evaporation rate, 273

evaporation time, 101
exothermic reactions, 255
extinguishing state, 289

fast chemistry reaction model, 194
favre average, 197
favre filtering, 214
Fick's diffusion law, 294
field oscillator model, 285
the filtered flame front curvature, 221
filtered mass density function, 222
filtering characteristic length, 222
filtering function, 214
finite-amplitude oscillations, 287
finite difference, 87
finite diffusion rate, 129
first-order radial vibration mode, 258
fixed grid, 89
flame curvature, 221
flame front, 105
flamelet model, 194
flamelet stretch rate, 221
flamelet wrinkling, 221
flame propagation speed, 257
flame stability characteristics, 257
the flame thickness, 206
flow characteristic time, 206
flow coefficient, 61
flow instability, 221
flow oscillation, 271
flow-rate, 8
flow rate oscillation, 270
the fluctuating pressure field, 283
fluctuation, 165
fluctuation of heat release rate, 261
fluid inertia, 258
fluid property, 34
forward reaction rate constant, 193
framework, 163
free flow, 128
free flow border, 108
freezing evaporation model, 129
frequency of mixing, 222

gas density, 133
gas dissolution, 124
gasification process, 273
gas–liquid coaxial swirl nozzle, 298
gas–liquid mass ratio, 74
Gaussian distribution, 200

generalization, 170
generation term, 217
generator, 2
Gibbs energy, 116
gradient transport model, 211

half-cone angle, 63
half oscillation cycle, 280
harmonic number, 262
harmonic wave, 262
heat conduction coefficient, 108
heat equilibrium, 104
high-frequency instability, 258
high-frequency oscillations, 255
high-frequency pressure oscillations, 287
high frequency pressure wave, 118
high-speed photography, 26
hybrid algorithm, 213
hydraulic jump, 271
hydrocarbon/oxygen flame, 285
hysteretic/viscous losses, 267

ignition area, 133
ignition delay time, 257
ignition quality, 257
impinging-stream Injector, 64
inertial force, 84
infinitesimal, 240
infrared, 243
initial phase angle, 262
injection speed, 257
injector, 26
the injector pressure drop, 258
injector vibrations, 271
integral length scale, 207
intensity, 254
interface equilibrium, 120
interface reconstruction, 86
interface shrinkage, 110
interface tracking method, 85
intermediate-frequency instability, 259
internal loss, 267
internal temperature, 120
irregular shape, 45
isobaric specific heat capacity, 215

Kelvin–Helmholtz instability, 36
kinetic energy, 48
Kolmogorov eddy, 204

Lagrange method, 212
laminar airflow, 44
laminar flame speed, 207
laminar flame thickness, 207
laminar viscous coefficient, 215
large-eddy simulation (LES), 88, 151
Lewis number, 98, 195
lifted jet flames, 214
limit model, 129
linear control theory, 298
linear eddy model, 223
linear function, 131
linear instability, 117, 276
liquid column, 47
liquid film, 26
liquid jet structure, 91
liquid–liquid bipropellant centrifugal nozzle, 299
liquid–liquid impinging nozzle, 299
liquid membrane, 67
liquid oxygen, 109
liquid propellant, 28
liquid ribbon, 37
liquid rocket engine, 26
liquid sheet, 26
liquid speed, 73
local heat release rate, 287
local LEM time scale, 224
Logarithm–normal, 52
longitudinal vibration mode, 266
longitudinal wave, 261
low-frequency instability, 257

mach diamond region, 257
macro-gradient, 294
macro-material flow, 110
Malvern principle, 55
manifolds, 8
mass conservation, 65
mass fraction, 99
mass medial diameter, 54
mean diameter, 53
mean free path, 294
mixing, 283
mixing process, 192, 269
mixture fraction, 196
model coefficient, 222
modeled sub-grid stress tensor, 216
modes interaction model, 285
molar fraction, 136
molar mass, 123

molecular diffusion coefficient, 215
molecular heat conduction, 107
molecular viscosity, 192
monopropellant, 12
Monte Carlo method, 212
multicomponent effect, 123
multicomponent mixture, 123

nonlinear dynamics, 285
nonlinear instability, 276
nonlinear vibration, 283
normal pressure fluctuation, 256
nozzle burr, 26
nozzle vibration, 26
Nukiyama–Tanasawa, 52

one-dimensional longitudinal mode, 263
optical energy, 56
oscillation, 33, 359, 363
oscillation amplitude, 120
oscillation cycle, 280
oscillation evaporation, 118
oscillation frequency, 120
oxidizer, 4, 26

particle interaction model, 211
particle method, 212
particle trajectory model, 81
passive control, 295
penetration, 239
perpendicular, 235
perturbation, 26
phase Doppler Anemometry, 57
phase equilibrium, 104
physical property, 125
pipe, 17
Poisson equation, 213
positive feedback mechanism, 283
Prandtl number, 108, 215
prechamber, 6
pre-exponential factor, 193
premixed flamelet model, 206
premixed flames, 201
pressure disturbance, 119
pressure fluctuation, 261
pressure oscillation, 256
primary breakup, 75
probability density function, 136
propellant, 1
propellant feed system, 256

propellant mixture ratio, 259
the propellant supply system, 269
pulsation, 3, 371
punching, 10
pure evaporation, 109

quantity, 246
quasi-equilibrium, 235
quasi-fluid model, 80
quasi-steady evaporation, 97
quasi-steady-state evaporation theory, 294

radial vibration mode, 267
radiation, 12, 229
random mixture fraction, 197
ray, 251
Rayleigh breakup, 31
Rayleigh criterion, 260
reaction, 16, 303
reaction source term, 215
real amplitude of acoustic pressure, 262
rearrangement, 319
recirculation, 344
reflection, 239
regenerative, 10
regulator, 2
reliability, 232
remarkable, 164
renormalization group (RNG), 158
requirements, 3
residence, 15
resolvable-scale fluctuations, 214
resolvable velocity, 223
response characteristics, 269
response function, 274
Reynolds average method (RANS), 152
Reynolds number, 192
Reynolds stress, 170
R-K state equation, 115
rocket, 1
Rosin–Rammler, 53

sample probability density function, 209
saturated vapor, 113
scalar turbulence diffusion coefficient, 225
scattering, 248
secondary atomization, 77
secondary breakup, 43
second-order, 168
second-order longitudinal vibration mode, 258

second order moment, 138
self-catalytic mechanisms, 290
self-excited oscillations, 260
self-excited system, 267
self-saturation coefficients, 292
sensitive time delay model, 273
sensitivity, 159
sensors, 297
shadowgraph, 90
shaft, 5
shear-stress transport (SST), 160
shutdown, 13
single-fluid model, 80
sinusoidal, 367
small-scale mixing model, 211
solid ball, 108
solvable sub-grid fluctuations, 223
specific heat capacity, 99
specific heat ratio, 287
spectrum, 244
spherical droplet, 98
spherical membrane, 107
splash, 7
spliced process, 225
spray, 8
spray atomization process, 269
spray size distribution, 54
stagnation, 230
start, 13
static environment, 107
static medium, 103
steady-state combustion, 256
steady-state value, 270
Stefan flow, 98
stochastic trajectory model, 81
stoichiometric coefficients, 192
straight-flow, 52
strain, 153
stream, 8
stress, 166
stretching breakup, 48
strong coupling case, 293
structural damping, 267
structure, 1
sub-grid scale fluctuations, 214
sub-grid scale models, 214
sub-grid stress tensor, 216
sub-grid turbulent kinetic energy, 216
sub-layer, 158
supercritical, 16

surface tension, 26
swirler, 7
swirling chamber, 63
symmetric process, 124
system fluid-dynamic processes, 255

tangential vibration mode, 267
tank, 2
temperature field, 283
temperature sensitive process, 283
temperature-sensitivity, 290
tensor, 167
thereafter redistricting process, 226
thermal diffusion coefficient, 102
thermal property, 135
thermoacoustic oscillations, 261
thermodynamic equilibrium, 113
the volume of the combustion chamber, 258
third-order tangential vibration mode, 258
thrust, 2
time-averaged pressure, 119
time delay, 273
total heat release, 290
total number of components, 215
trajectory, 306
transfer, 10
transform, 337
transient heat transfer, 143
transpiration, 228
transported PDF, 194
transport-equation model, 204
transverse mode, 265
transverse wave, 261
triplet mapping method, 224
Trouton's law, 142
turbine, 4
turbo-pump, 2
turbulence, 26, 151, 309
turbulence–chemistry interactions, 194
turbulent combustion, 192
turbulent diffusion coefficient, 207

turbulent flame propagation velocity, 221
turbulent flame speed, 207
turbulent flame thickness, 221
turbulent flame zone, 149
turbulent fluctuations, 197
turbulent kinetic energy, 207
turbulent Prandtl number, 217
turbulent Schmidt number, 217
turbulent spray combustion, 285
2D axisymmetric process, 107
two-fluid model, 81
typical homogeneous reactor systems, 290

uniform reactor model, 285
universal gas constant, 215, 287
unsolvable-scale fluctuations, 214
unsolvable sub-grid fluctuations, 223
upper-limit, 53

valve, 4
vapor, 16, 365
vapor component, 105
vapor concentration, 100
vaporization, 283
velocity field, 283
velocity fluctuation, 26
vibration mode, 297
vicinity, 346
viscosity, 40, 153, 308
viscous drag, 78
volume of fluid (VOF), 89

wake combustion, 109
wall losses, 268
wave-energy feedback, 117
wavelength, 38, 238
weak coupling case, 293
WENO, 88

Zeldovich conversion, 195
ZKS theory, 110